CONTACT MECHANICS — FRICTION:

MODELING AND EXPERIMENT

presented at

2003 STLE/ASME JOINT INTERNATIONAL TRIBOLOGY CONFERENCE
OCTOBER 26–29, 2003
PONTE VEDRA BEACH, FLORIDA

organized and sponsored by

THE TRIBOLOGY DIVISION, ASME

edited by

GEORGE G. ADAMS
NORTHEASTERN UNIVERSITY

IZHAK ETSION
TECHNION – ISRAEL INSTITUTE OF TECHNOLOGY

ITZHAK GREEN
GEORGIA INSTITUTE OF TECHNOLOGY

LIOR KOGUT
UNIVERSITY OF CALIFORNIA AT BERKELEY

ANDREAS A. POLYCARPOU
UNIVERSITY OF ILLINOIS AT URBANA-CHAMPAIGN

ISBN No. 0-7918-3706-8

• • MESSAGE FROM THE CHAIR • •

Welcome to the 2003 STLE/ASME International Joint Tribology Conference (IJTC), and thank you for choosing this meeting, amongst the many possible tribology meetings, to present your work. This year we offer talks covering relatively new areas of tribology, such as Nanotribology and Biotribology. This is your chance to learn about these emerging areas from the leading researchers in the country and the world.

We have over 150 talks scheduled for this October 27–29, 2003, conference. Many of them will cover the traditional areas such as wear, magnetic storage tribology, hydrodynamic lubrication, elastohydrodynamic lubrication, machine components, rolling element bearings, contact mechanics, and surface chemistry. Furthermore, we have a student poster program scheduled that reflects a strong interest in tribology coming from the tribologists of tomorrow. Please find some time to interact with these young people so they receive their first lesson in networking -- an important skill to learn. As you can see, the IJTC Planning Committee has developed a program that will satisfy all tribologists, whether they are students, or government, industry, or academic tribologists.

The Awards Luncheon is on Tuesday, October 28, 2003, and the Keynote Address will be given by Dr. Ken Ludema, who is a retired professor from the University of Michigan. He will talk on "The Many Faces of Tribology." Please be sure to come and listen to his pearls of wisdom.

I have always thought that a learning environment should be pleasant so that people will be encouraged to exchange ideas about various tribological problems. Toward that end, the IJTC Planning Committee has chosen the Sawgrass Marriott Resort and Beach Club in Ponte Vedra, Florida, as an ideal setting for this Conference. Whether you prefer a poolside conversation or a walk along the beach, discussing Integrated Engineered Surfaces, for example, with an expert is quite possible in this collegial atmosphere.

Many of you have brought your families, and I am pleased to say that this Conference location offers delights for them too. Tennis courts, golf courses, and pools await the exercise enthusiasts, and historic St. Augustine is only a short drive away. I am certain that your family members will find something that entices them.

Thank ou for your participation in this Conference. Without your fine efforts, this Conference would not offer so many learning opportunities for all of us.

Sincerely,

Elaine Yamaguchi
Chair 2003 STLE/ASME IJTC

CONTENTS

Proceedings of
2003 STLE/ASME Joint International Tribology Conference
Ponte Vedra Beach, Florida USA, October 26-29, 2003

2003-TRIB-0256

LESSONS LEARNED FROM 25 YEARS OF FRICTION MODELING

Andres Soom
Department of Mechanical and Aerospace Engineering
University at Buffalo, Buffalo, NY 14260

ABSTRACT

We have long known that tangential contact forces and the friction coefficient at a sliding contact arises not due to some "law" but as a result of contact geometric and material properties in combination with physical processes that are not always accessible to direct experimental verification. While tribologists have occasionally sought alternatives to the coefficient of friction as the primary means to quantify tangential forces at sliding contacts, its convenience, simplicity, long history and lack of viable alternatives assure that it will be remain in use for some time to come. Sometimes the coefficient can remain constant with normal load over orders or magnitude. One can be led to believe that the coefficient is a property of the interface or there is a "law" at work. At other times, the coefficient, usually measured, is said to vary with time, sliding distance or some other variable with no explanation as to why the variation occurs. In the present paper we discuss aspects of friction behavior and modeling, focusing on mechanics and the nature of the friction coefficient. The point of departure is a case where the adhesion theory of friction can be used to understand the mechanics of friction. One can then consider a number of situations and concepts that arise as variations of this case including average versus instantaneous friction; cases when Amontons-Coulomb relations do or don't apply; static versus dynamic friction; scale effects at rough surfaces; limits of applicability of the friction coefficient; friction measurements, thermal and velocity effects; and local versus global friction and constitutive relations.
Keywords: Friction, model, Coulomb, mechanics, friction coefficient

INTRODUCTION

Understanding and modeling of friction should go hand in hand. However, the coefficient of friction has been present in physics, engineering mechanics and machine design much longer than its physical understanding. Historically, e.g., in elementary mechanics, the coefficient of friction has been described and applied in the context of rigid body mechanics. These days, the friction coefficient has to bear a much heavier burden. Often it is more than it can bear.

During the 20th century Bowden and Tabor [1], followed by Archard [2] and Greenwood and Williamson [3] laid much of the groundwork for our understanding of dry friction between sliding surfaces through the so-called adhesion theory of friction. The basic notion is that friction forces at an interface arise when the real area of contact, which has some strength, is sheared or ruptured. The nature of the "material" at the interface or the bonds between the surfaces, that determine the strength of the interface are not explicitly considered, any more than in the case of the strength of engineering materials in general. From the point of view of the engineer, predicting the strength of the interface from first principles is a pointless and, probably, hopeless task. The value of the macroscopic interfacial strength or the friction coefficient under some base or standard geometry and operating condition remains a matter of testing. While not all resistance to sliding arises at the interface, we limit our discussion here to those situations where that is the case.

REAL AREA OF CONTACT AREA AND SHEAR STRENGTH

As a model problem we consider a smooth metallic sphere pressed into elastic contact with a smooth and flat countersurface. In this case, the real area of contact, A, can be calculated from Hertzian theory (4) as $A= KN^{2/3}$, where N is the normal load and K includes the elastic modulus and sphere radius. According to the adhesion theory the friction force is $F= \tau A = K\tau N^{2/3}$. The shear strength of the interface is τ. In an experiment with this geometry, the area and normal load will be known, and the shear strength can be can be calculated. A constant shear strength, independent of load, leads to a coefficient of friction, $\mu=F/N= K\tau/N^{1/3}$. The coefficient of friction decreases with load. As early as 1957 Archard [2] showed this to be the case, at loads above some value but below those required to cause plastic deformation. This result is also consistent with Greenwood and Tripp [5] who showed that, above some normal load, N_c the real area of contact between and sphere and a rough surface is the one obtained from that from Hertzian theory, i.e., the surface roughness is flattened. Obviously, the friction is load dependent and there is no "Coulomb's law" at work here.

However, Archard's measurements [2], shown schematically in Fig. 1, also indicated that, at low loads, the friction coefficient becomes nearly constant, independent of load. This is in contrast to the friction coefficient relation above which predicts that the friction coefficient approaches infinity as the normal load goes to zero. In the low load range, a type of Coulomb friction seems to apply. This can be explained by the contact theory of Greenwood and Williamson [3] who showed that, for randomly rough surfaces, the real area of contact is directly proportional to normal load, i.e., A=kN. In this relation, k includes elastic and surface roughness parameters. If, again, we consider the adhesion theory to apply, then $F= \tau A =\tau kN$. The coefficient of friction, $\mu=\tau k$ which is constant as long as the shear strength doesn't change. Interestingly, if the shear strength has the same value as at the higher loads, the coefficient, k, and the real area of contact can be determined. Of course the constancy of the friction coefficient with load for flat rough surfaces is largely explained by the Greenwood-williamson theory.

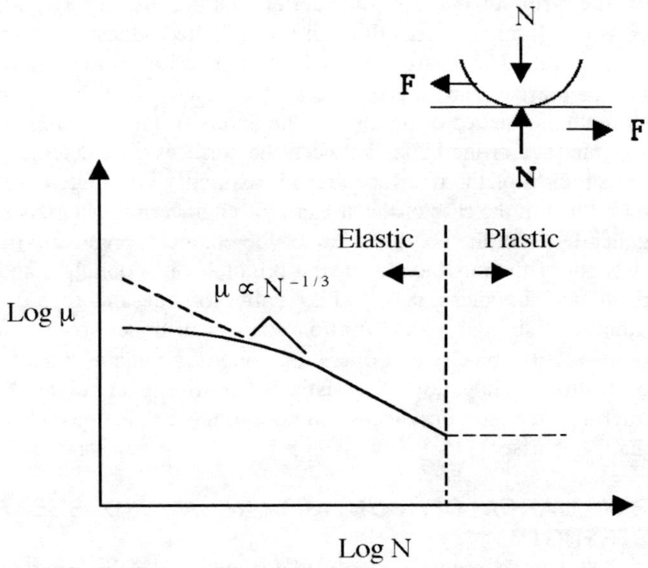

Fig. 1: Measured friction coefficient (solid line) with normal load for a metallic Hertzian ball on flat contact, based on Archard[2].

While it may seem a bit pedantic to be repeating these ideas with such emphasis in the year 2003, I would maintain that in modeling and measuring dry friction in general cases, it is incumbent upon the researcher or engineer to demonstrate that more complex principles need to be invoked. This could be the case if the materials are hysteretic, if significant plastic deformation occurs, if high intermolecular (true) adhesive forces are present or if capillary forces act at the asperity junctions, to name a few possibilitees. Otherwise, we have found these basic ideas of real area of contact and shear strength of the interface to be useful in modeling and understanding friction under a variety of circumstances. In this context it is also of value to measure friction at a number of loads, speeds and geometries to ascertain if the friction behavior is as expected.

Clearly it is the shear strength of the interface and not the friction coefficient that is the primary quantity of interest. However, since the real area of contact is seldom known, the above example being an exception, we are stuck with having to use the friction coefficient; but it must be used and interpreted with care and understanding.

MEASURING FRICTION COEFFICIENTS

Since friction characterization so often involves measurement, a few words must be said. First, one can ask, what is the normal load. In our work we have preferred to consider the normal load at the contact, rather than the more classic externally applied load. While they may be the same, they are not always. The adhesion theory of friction clearly refers to what is happening at the contact. When vibrations occur, inertial and mechanical damping forces are present in the system and must be properly accounted for if time-varying normal loads and friction forces are to be measured.

It is also not unusual to measure the friction force but not the normal load. On should somehow verify, not assume, that the normal load remains at its nominal value if that is the nature of the test to be run. Unknown or inadvertent changes in friction force due only to normal load fluctuations can be erroneously interpreted as fluctuations in the coefficient of friction.

It is also possible that changes in the alignment of the sliding components or the measurement system can occur and, hence, the directions of the contact forces can change over short or long times. This may be due to vibration, mechanical coordinate coupling, or wear. Also the equilibrium positions and alignments of the contacting components are usually different when stationary versus sliding.

INSTANTANEOUS AND AVERAGE FORCES

When normal and frictional forces fluctuate over time during tests, for whatever reason, one should either ignore the temporal fluctuations, only retaining average values, or one must measure them with sufficient bandwidth and fidelity to capture the quantities to be measured. Most tribometers and related instrumentation do not have sufficient bandwidth fidelity. Caution must be exercised. However, most test apparatuses do measure average friction and normal loads satisfactorily. We note parenthetically, that, according to the adhesion theory, and for rough surfaces, the average $<\mu>=<F(t)>/<N(t)>$, and instantaneous ($\mu(t)=F(t)/N(t)$ friction coefficients, under time-varying loads are identical.. The <> denotes time average. When the friction coefficient changes with normal load, as, for example, in the case of the ball on flat example, the value of the fluctuating friction coefficient is difficult to interpret and generally not meaningful. The average friction coefficient, as defined above, can change, usually decreasing somewhat, with increasing in normal vibrations. An occasional and subtle from of average friction reduction occurs when normal vibrations are sufficient to cause systematic or random loss of contact. Such loss of contact is not always obvious in test setting. A compact and excellent summary of dynamic effects in friction, (and vice versa), has been provided by Johnson [6].

LONG TERM CHANGES IN FRICTION

When the "average" coefficient of friction changes with time, over seconds or hours, it is most likely to be due to changes in the shear strength, in the context of this discussion. Surface films, e.g., oxides, are removed or deposited. Glazes or other layers form on the surface. Moisture increases or is removed. Chemical reactions may take place. Environmental may affect the interface properties. Another possibility is a change is the contact geometry, e.g., change in roughness, during operation. This would result in a change in the parameter k. Or both could occur.

LOCAL FRICTION

In continuum mechanics and computational mechanics analyses of frictional contacts, it is common to use a frictional boundary condition, $\mu=\tau/p$. The tangential stress is obtained by multiplying the normal pressure by a friction coefficient. In the context of the discussion above, this is only appropriate if the surfaces are rough. Strictly speaking, this type of boundary condition should be coupled with some type of contact compliance to account for the finite stiffness of the surface roughness. If the surfaces are smooth, a shear strength, independent of the normal pressure, would be more appropriate. However, since most engineering surfaces are rough at some level, and the tangential tractions are usually considerably lower than the normal pressures, the errors encountered may not be large.

SCALE EFFECTS

There has been interest in the role of small scale roughness, especially in the context of fractal representation of surfaces, suggesting, for example, that the real area of contact may decrease as progressively smaller length scales of roughness are considered. While such effects are certainly possible the behavior of the sphere on flat at discussed above suggest that this is not the case. The amplitude of the small scale short wavelength roughness seem to decrease fast enough with decreasing wavelength (increasing wavenumber) that wavelengths, shorter than some cut-off value can safely be ignored.

STATIC VERSUS DYNAMIC FRICTION

While dynamic friction under steady sliding condition can usually be interpreted in the straightforward manner discussed above, the same is not true of static friction. So-called static friction, the tangential resistance to macroscopic sliding motion starting from rest is normally presented as being larger than the dynamic by some tens of percent. In the context of friction arising due to the shear of the real area of contact, a larger static friction is explained by an increase in the shear strength or real contact area, due, for example to creep during rest. Some uncertainty is always present in static friction measurements due to the possibility of transient normal or angular motions. Also it should be recognized that some parts of a contact will slip before others.

VELOCITY AND TEMPERATURE DEPENDENCE OF FRICTION

The concepts discussed above apply quite widely when conditions are isothermal (no frictional heat generation), at low speeds or short duration sliding. As sliding speed increases, beyond, say one m/s for steel surfaces, asperity contact temperature can quickly reach a few hundred degrees. The strength of the interface can decrease as can the contact area due to thermal expansion), often at time intervals of the order of fractions of milliseconds. Both effects tend can significantly reduce the coefficient of friction with increasing sliding speed compared with isothermal sliding. An important issue in brake design is to achieve sufficient and consistent friction over a range of speeds and environmental conditions. The modeling and control of friction in high speed sliding has received little attention considering its broad significance.

CONSTITUTIVE RELATIONS

Attempts have been made over the years to develop constitutive relations that purport to describe the frictional interface and friction itself from a dynamic systems perspective. Indeed this is use in working with viscoelastic materials and geophysical rock mechanics but tends not to be appropriate for metallic or other similarly behaving contacts [6]. In the author's experience, the most useful parameter in addition to an accurate friction coefficient is the distributed (nonlinear) finite contact stiffness, which, for hard materials accounts for the compliance of the asperity contacts [7,8]. This allows one to relate the local normal pressure to the local contact deflection (compression), but does not attempt to ascribe dynamic behavior to the interface itself. Tangential compliance of asperities tends to play a role during stick-slip transitions but is usually not important during sliding.

CONCLUSIONS

This has been an incomplete and personal view of some principles and issues related to friction modeling as I have encountered them in engineering and research. It is definitely not a literature review. It is, however meant to suggest that in describing friction, sliding contact or interface behavior, one should keep in mind the mechanics and material behavior that is being modeled, how the system or test apparatus behaves and what the goal of the whole exercise is. Tribologists tend, naturally, to focus on the details and complexities of the friction process. Dynamicists and engineers seek simple friction models and parameters to insert into analyses or designs, with less regard for the mechanisms. Some do not even distinguish between dry and fluid lubricated conditions, (not a topic of the present paper). A certain amount of empiricism is inevitable. The average friction coefficient of a rough contact operating under some steady state condition, can, indeed, be used to characterize friction and can provide a point of reference. However the sensitivity of the interfacial shear strength to environmental factors, the presence or absence of contaminants, operating conditions or wear can lead to large variations in friction under nominally similar conditions. It would seem that understanding, modeling and, in engineering practice, controlling those variations needs be more of a focus than it has been up to now.

REFERENCES

[1] Bowden, F. P. and Tabor, D., 1939, "The Area of Contact Between Stationary and Between Moving Surfaces," *Proceedings of the Royal Society of London*, Vol. A169, pp. 391-413.

[2] Archard, J. F., 1957, "Elastic Deformation and the Laws of Friction," *Proceedings of the Royal Society of London*, Vol. A243, pp. 190-205.

[3] Greenwood, J. A. and Williamson, J. B. P., 1966, "Contact of Nominally Flat Surfaces," *Proceedings of the Royal Society of London*, Vol. A295, pp. 300-319.

[4] Hertz, H., 1882, "Über die Berührung Fester Elastischer Körper (On the Contact of Elastic Solids)," *Journal von Reine und Angewandte Matchematik*, Vol. 92, pp. 156-171 (For English translation see Miscellaneous Papers by H. Hertz, Eds: Jones and Schott, 1896, Macmillan, London).

[5] Greenwood, J. A. and Tripp, J. H., 1967, "The Elastic Contact of Spheres," *Journal of Applied Mechanics*, Vol. 34, pp. 153-159.

[6] Johnson, K. L., 2001, "Dynamic Friction" *in Tribology Research: from Model Experiment of Industrial Problem,"* (27th Leeds-Lyon symposium), Elsevier 2001, pp. 37-45.

[7] Hess, D.P. and Soom, A., 1991 "Normal Vibrations and Friction under Harmonic Loads: Part II- Rough Planar Contacts" J. Trib vol 113 no 1, pp. 87-92.

[8] Hess, D.P. and Soom, A., 1992, " Unsteady Friction in the Presence of Vibrations" *in Fundamentals of Friction: Macroscopic and Microscopic Processes*, (NATO Advanced Studies Institute) I. Singer and H. Pollock Eds. Kluwer Academic press, pp. 535-552.

Proceedings of
2003 STLE/ASME Joint International Tribology Conference
Ponte Vedra Beach, Florida USA, October 26-29, 2003

2003-TRIB-0257

MULTIPLE SOLUTIONS IN DYNAMIC CONTACT PROBLEMS WITH FRICTION

Yuri Kligerman
Faculty of Mechanical Engineering
Technion – Israel Institute of Technology
Haifa, 32000, Israel
Fax: + 972-4-829-5711
E-mail: mermdyk@tx.technion.ac.il

ABSTRACT

A frictional force resists the relative motion of the two surfaces. An understanding of the difference between static and dynamic contact with friction can lead to methods, which reduce friction and wear. The main goal of the present work is to evaluate the various solutions for the dynamic contact problem with friction and to compare theirs with the solution of the static contact problem. Another goal of the present work is to clarify whether the perturbation of some problem parameters during sliding with time can lead to the difference between solutions of the static and dynamic contact problems with friction. The essential parameter determining friction is the number of asperities in contact. The number of asperities in contact changes as a consequence of the wear. As follows from the classical wear models the perturbation frequency of the number of asperities in contact depends on the sliding velocity. The model of dynamic contact with friction based on the perturbation of the number of asperities in contact, along with the alternative models, is discussed in the present work.

KEYWORDS: frictional contact, dynamic friction, multiple solutions, adhesion friction.

INTRODUCTION

The sliding of two surfaces upon each other is such a common occurrence, yet this phenomenon remains one which is not well understood. The relative motion of the two surfaces is resisted by a tangential force, which is called the frictional force. The ratio of this tangential force to the normal force acting at the interface between contacting sliding surfaces is called the coefficient of kinetic (or dynamic) friction. Many attempts have been made to appreciate the difference between static and dynamic contact with friction. An understanding of the sliding friction nature can lead to methods, which reduce friction and wear.

The contact force and the adhesion force acting at the interface between sliding surfaces and also external force and inertial force acting in the direction perpendicular to the sliding surfaces define the normal force between these surfaces. The normal force and, friction force, and consequently friction coefficient are complicated nonlinear functions of the separation distance between sliding surfaces.

One of the possible sources for the difference between static and dynamic contact with friction is that in the last case the frictional characteristics of the system vary with time and average friction force and friction coefficient differ from those in the case of static friction.

There are several possible causes for the change of the forces under consideration with time during sliding. These may be plasticity index time variation, sliding speed dependent frequency of normal vibration and a time dependent friction originating from a stick-slip motion. Other possible causes are motion instability following from the coupling between system degrees of freedom (in the sliding direction and the normal one) or perturbation of some problem parameters during the sliding.

The main goals of the present paper are:

- to evaluate various solutions for the dynamic contact problem with friction and to compare them with the solution of the static contact problem; and
- to clarify whether the perturbation of some problem parameters during sliding with time can lead to the difference between solutions of the static and dynamic contact problems with friction.

One of the essential parameters determining the separation distance between two contacting surfaces and forces appearing between them is the number of asperities in contact. As a consequence of the contacting surfaces wear, for example, the number of asperities in contact can change. As follows from the classical wear models, discussed in the present work, the perturbation frequency of the number of asperities in contact depends on the sliding velocity. As a result, the separation distance between two contacting surfaces and forces appearing between them vary with the frequency defined by the sliding velocity. The time average value of the contact force, friction force and the friction coefficient can be other than the contact force, friction force and the friction coefficient for the static case, since they are nonlinear functions of the separation distance between contacting surfaces. The mathematical model realizing this approach and results of its analysis are presented in the present paper.

The present research incorporates analysis of dynamic contact with friction during sliding of rough surfaces for different dynamic conditions.

There have been a number of papers published recently that concern the modeling of friction, contact forces, adhesion forces acting between contacting surfaces, wear phenomena, self-excited oscillation and dynamic instabilities in sliding. The majority of studies, related to the tribologycal aspects of dynamic friction are experimental. Hence, they do not provide theoretical explanation to the difference between static and dynamic contact with friction.

Lyuksyutov [1] analyzed dynamic friction and instability of interface motion. The author considered the dynamic friction mechanism for a moving interface, and found the conditions for motion instability. The work of Oancea and Laursen [2] systematically examined the stability of constant velocity frictional sliding. The authors used their earlier proposed dynamic friction model featuring viscous response to highly transient slip events, and attempted to characterize the types of instabilities that can be observed. A non-linear dynamic stability study was presented for two systems: a single degree of freedom system governed by a velocity and state dependent friction model, and a two degree of freedom system in which the coupling between the normal and tangential degrees of freedom was explicitly included. In the problem treated analytically by Adams [3] two flats layered elastic half-spaces, of different material properties, were pressed together and slid against each other with a constant coefficient of friction. The results of this work demonstrated that unstable motion occurs for a wide range of material combinations. The physical mechanism responsible for this instability is that of a slip-wave destabilization, see, e.g. [4]. The influence of the properties of the layers on the destabilization of sliding motion was investigated.

A significant volume of works on contact and friction was done by Oden and Martins. They formulated continuum models of a large class of dynamic frictional phenomena and developed computation methods for analyzing these phenomena [5]. Of particular interest are the theories which can adequately predict stick-slip motion, frictional damping in structural dynamics, and sliding resistance. Friction modeling was based on a generalization of the Coulomb's friction law, allowing the friction coefficient dependence on the normal contact pressure. Martins et al. [6] pursued the same line of inquiry to investigate the role played by the various physical and geometric governing parameters on the qualitative behavior of simple sliding systems.

During the last decade Martins et al. published several works concerning dynamic and stability problems of frictional contact, e.g., [7, 8]. In all these works the authors used modified forms of Coulomb's law for the friction modeling, assuming power law dependency of the normal and tangential stresses at the interface on the approach of the contacting surfaces.

In 1998 Tworzydlo et al. [9] used a different approach to predict macroscopic, phenomenological behavior of the interface. This approach was an extension of previous asperity-based models of contact and friction [13, 14 and 17] considering realistic, complex shapes and mechanical properties of surface asperities. This approach differs from the Coulomb's friction models that are based on a phenomenological law, which couples local normal and tangential stresses in the contact.

The foregoing literature review demonstrates that dynamic contact with friction is a very actual interesting scientific and engineering problem. However, the above works do not settle the following questions from basic principles of mechanics:

- how and why does the dynamic contact with friction, corresponding to the sliding of two rough surfaces, differ from the static contact with friction?
- how do the dynamic contact and friction forces change with changes in the sliding velocity?

NOMENCLATURE

A_n	nominal contact area;
E	Hertz elastic modulus;
F_e	external force, pressing the slider against the table;
F_e^*	nondimensional external force;
$F_s(y)$	adhesion force;
$F_s^*(Y)$	nondimensional adhesion force;
k	equivalent viscous damping coefficient;
m	mass of the slider;
$P(y)$	contact load;
$P^*(Y)$	nondimensional contact force;
Q	tangential friction force;
y	mean separation between the two contacting rough surfaces;
$\dot{y}$	velocity of the slider in the direction normal to the contact;
$\ddot{y}$	acceleration of the slider in the direction normal to the contact;
Y	nondimensional coordinate, y/σ;
Y_{eq}	nondimensional mean separation, corresponding to static equilibrium state;
Y_o	initial separation;
Y'	nondimensional velocity of the slider in the direction normal to the contact;
Y''	nondimensional acceleration of the slider in the direction normal to the contact;
λ	nondimensional perturbation frequency of the number of contacting asperities;
ΔY	disturbance, $Y - Y_{eq}$;
ΔY_o	initial disturbance, $Y_o - Y_{eq}$;
ε	nondimensional perturbation amplitude of the number of contacting asperities;
ξ	is a nondimensional equivalent viscous damping coefficient;
σ	is the standard deviation of surface heights.

ANALYTICAL MODEL OF DYNAMIC FRICTION

The model of dynamic contact with friction, presented in this paper, describes the contact, adhesion and friction force changes in time during the relative frictional sliding of two surfaces. As shown by Greenwood and Tripp [10] the contact of two rough surfaces can be modeled by an equivalent single rough surface contacting a smooth plane. This result can be applied for the analysis of adhesion friction phenomenon.

The basic assumptions of asperity based model of Greenwood and Williamson [11] (GW model) are adopted:

1. The rough surface is isotropic.
2. Asperities are spherical near their summits.
3. All asperity summits have the same radius R but their heights vary randomly following normal distribution law.
4. Asperities are far apart and there is no interaction between them.
5. There is no bulk deformation. Only the asperities deform during contact.

We assume that a relatively light slider of finite mass slides over a surface of an infinite mass table as shown in Fig. 1. The slider is pressed against the table by an external force, F_e. We assume that, like in the classic problem of static friction, three forces act between the contacting surfaces during sliding. These are: the normal contact force P, adhesion force, F_s, and tangential friction force, Q. These three forces are affected by the separation, y, of the two surfaces. We assume that during frictional sliding the two contacting surfaces have constant relative sliding velocity. During the sliding these two surfaces can also move relative to one another in the direction normal to the plane of contact. As a result, the tangential friction force acting in the contacting plane changes, along with the contact force and adhesion force acting in the direction normal to contacting plane.

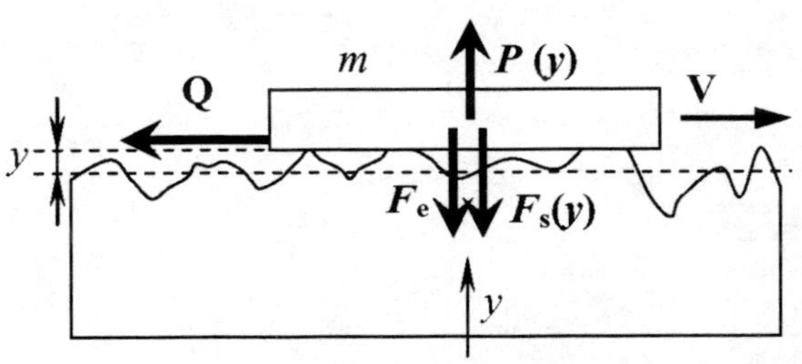

Figure 1. Layout of Contact During Frictional Sliding

Several potential directions were investigated for the formulation of the time dependent contact problem of rough surfaces in relative sliding. These include the possibility of plasticity index time variation, sliding speed dependent frequency of normal vibration and a time dependent friction originating from a stick-slip motion. All these attempts failed to relate the friction force and sliding speed and thus a different approach was taken.

Since we assume that the rough surface is isotropic, the contact problem should not be affected by a change of position of the slider on the surface of the table. However, one can assume that some parameters of the problem (e.g., the number of asperities in contact) are perturbed during the sliding. The frequency of these perturbations depends on the sliding velocity. Perturbation of the number of contacting asperities can cause relative vibration of the two surfaces in the direction normal to contacting plan. This phenomenon is known as parametrically excited oscillation (cf. Nayfeh and Balachandran, [12]).

The motion of the slider in the normal direction follows the equation:

$$m\ddot{y} + k\dot{y} + F_S(y) - P(y) = F_e \qquad (1)$$

Eq. (1) can be rewritten in nondimensional form as

$$Y'' + \xi Y' + (1 + \varepsilon\cos\lambda\tau)\frac{\left(F_s^*(Y) - P^*(Y)\right)}{F_e^*} = -1 \qquad (2)$$

where:

nondimensional time τ is given as $\tau = \omega t$ and $\omega = \sqrt{F_e/(m\sigma)}$;

$F_s^*(Y)$, $P^*(Y)$, and F_e^* are nondimensional forces obtained by normalizing the relevant dimensional forces with respect to $A_n E$. Models for the elastic-plastic contact, $P^*(Y)$, and adhesion, $F_s^*(Y)$, of rough surfaces were developed by Chang et al. [13, 14] and modified by Kogut and Etsion [15, 16];

E is the Hertz elastic modulus (cf., Chang et al., [13]);

Since the number of contacting asperities does not increase and does not decrease monotonically, as a first approximation, we assume that the perturbation of the number of asperities in contact can be described as a simple periodic trigonometric function. In the subsequent investigation we will consider other types of perturbation function as well.

Eq. (2) is a nonlinear ordinary differential equation with nonconstant coefficients. Solving this equation by the fourth order Runge-Kutta method we obtain the change of the mean separation between the two contacting surfaces. This allows determining changes of the tangential friction force during the sliding. The model of the tangential friction force acting between rough surfaces as a function of the mean separation was developed by Chang et al., [17] and modified by Kogut and Etsion [18].

Analysis of the average tangential friction force during sliding as a function of the perturbation amplitude and frequency of the number of contacting asperities is used for evaluation of the friction force dependency on the sliding velocity.

NUMERICAL RESULTS

The numerical analysis of the foregoing model was carried out for the following parameters:

- plasticity index of the friction pair, defined by Greenwood and Williamson, [11], 0.39;
- radius of curvature of asperity summits 46.1 μm;
- standard deviation of surface heights 2.95 nm;
- areal density of asperities 0.35 μm^{-2};
- energy of adhesion of an interface 0.045 $J \cdot m^{-2}$;
- Young moduli of contacting surfaces 100 GPa and 450 GPa;
- Poisson's ratios of contacting surfaces 0.2 and 0.21;
- interatomic spacing 0.5 nm;
- hardness of softer material 2.0 Gpa.

All these data are taken from Polycarpou and Etsion, [19]. The dependence of the dimensionless contact load and adhesion force on the dimensionless mean separation Y is presented in Fig. 2. The nondimensional external normal load was assumed constant and equal to $F_e^* = 6.25 \cdot 10^{-6}$. The state of static equilibrium with this external normal load was evaluated from Eq. (2) for $Y' = 0$; $Y'' = 0$; $\varepsilon = 0$; $\lambda = 0$. The dimensionless static separation was found as $Y_{eq} = 2.217$.

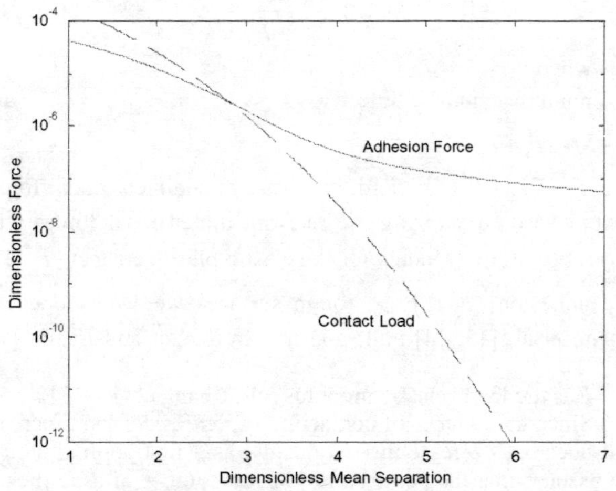

Figure 2. Dimensionless Forces vs. Dimensionless Mean Separation

Fig. 3 presents the dimensionless force $P^*(Y) - F_s^*(Y$ appearing in Eq. (2), vs. the dimensionless mean separation where $P^*(Y) - F_s^*(Y)$ represents the external force in any give static equilibrium position *Y*.

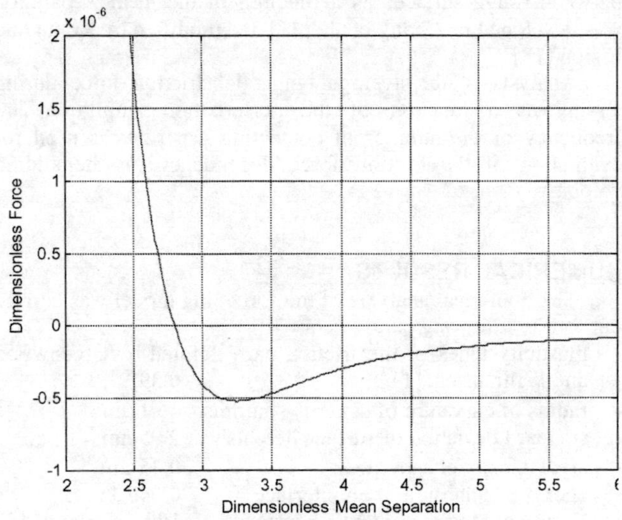

Figure 3. Static External Force vs. Dimensionless Mean Separation

From Figs. 2 and 3 it can be seen that the system under consideration is essentially nonlinear with the mean separation. However during vibration with small amplitude around the state of static equilibrium the forces $P^*(Y)$ and $F_s^*(Y)$ may be linearized relative to the *Y* change, and there can be little doubt that the behavior of the system will be as for linear system.

In solving the dynamic equation (Eq. 2) it is required to evaluate the forces $P^*(Y)$ and $F_s^*(Y)$ at each time step according to the corresponding *Y* value (Chang et al., [13, 14]). This procedure is very tedious and consumes a great deal of computing time. When moving perpendicular to the contact surface, the vibrating slider passes the same *Y* states each period of vibration. Hence, contact force and adhesion force can be evaluated for a wide range of possible values of mean separation, and approximated using a seventh order polynomial splines. Now, evaluation of the forces during numerical integration in time of Eq. (2) can be carried out by the substitution of the necessary time in the simple seventh order polynomials. This accelerated substantially the calculation procedure.

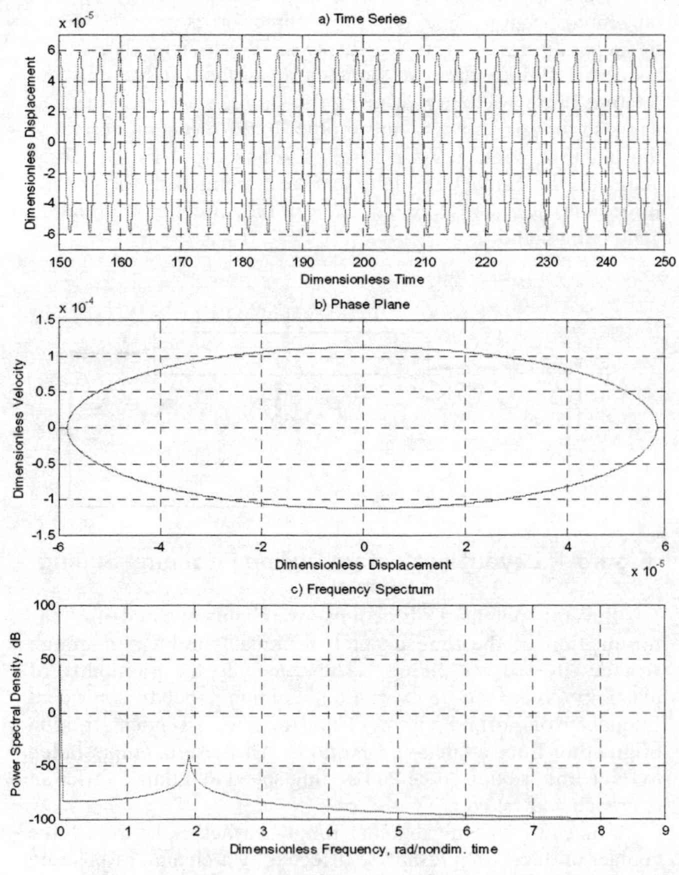

Figure 4. Free Vibration due to Initial Conditions: $\Delta Y_0 = 5.85 \cdot 10^{-5}$; $Y_0' = 0.0$

Before embarking on the complete solution of Eq. 2 it is worthwhile to investigate the free vibration response to an initial disturbance in the absence of viscous damping. This is a common practice in the solution of nonlinear dynamic systems (cf. Nayfeh and Balachandran, [12]). It allows construction of a "skeleton curve" relating the amplitude of free vibration to their frequency and providing valuable information on the possible system's behavior.

Figs. 4 and 5 give examples of the free vibration response for the time-series, phase plane, and frequency spectrum of response corresponding to two sets of initial conditions. In order to demonstrate the response of linearized system the initial position of the system was chosen to be very close to its equilibrium state $\left(\Delta Y_0 = 5.85 \cdot 10^{-5}\right)$ with zero initial velocity ($Y_0' = 0.0$). Fig. 4a presents the dimensionless displacement measured from the static equilibrium position. In this case the behavior of the system appears linear (the time series is sinusoidal, the trajectory in the phase plane (Fig. 4b) is elliptical, and the spectrum (Fig. 4c) includes only one frequency close to the eigenfrequency of linearized system).

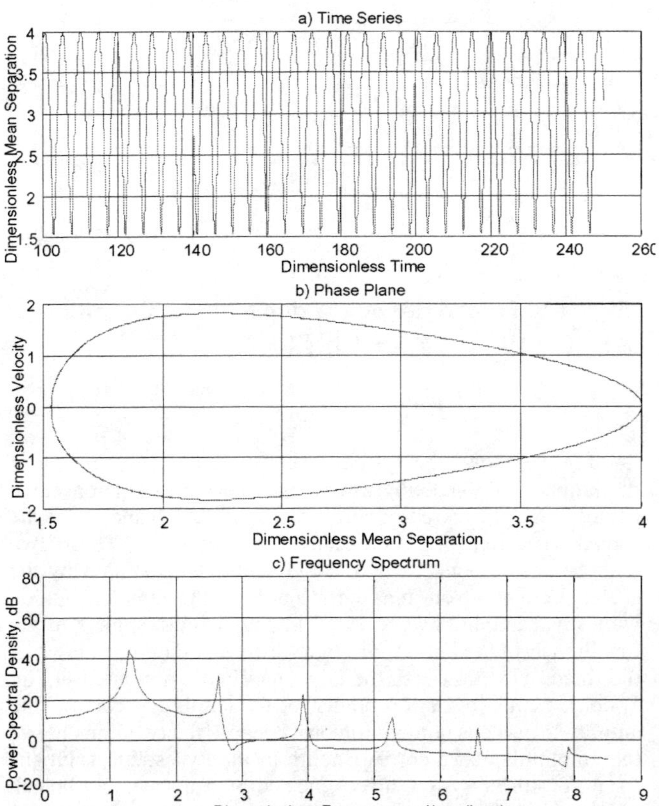

Figure 5. Free Vibration due to Initial Conditions: $Y_0 = 4.0;\ Y_0' = 0.0$

Fig. 5 presents free response of the system due to much larger initial disturbance $\Delta Y_0 = 1.783$ with initial position $Y_0 = 4.0;\ Y_0' = 0.0$. The response to these initial conditions is periodic with greater amplitude and a smaller frequency than in the previous case. Now, in the range of the mean separation variation the force $P^*(Y) - F_s^*(Y)$ is nonlinear with the mean separation (see Figs. 3 and 5a), and it is reasonable that the behavior of the system is typical for nonlinear systems. The trajectory in the phase plane, Fig. 5b, is noncircular and nonelliptic (because of the variable stiffness of the system with the mean separation); the content of the spectrum, Fig. 5c, includes the frequency of the nonlinear system free vibration, corresponding to specific amplitude, and multiple frequencies.

Fig. 6 shows the amplitude curve vs. frequency of free vibrations (skeleton curve). The shape of the skeleton curve testifies a coexistence of at least two stable solutions (above and below the skeleton curve) corresponding to forced vibrations for perturbed system with the same perturbation frequency of the number of contacting asperities. These two stable solutions are separated by an unstable solution and occur in the bounded range of low frequencies below the eigenfrequency of linearized system. The type of these nonlinear systems is known as a softening-type (cf. Nayfeh and Balachandran, [12]).

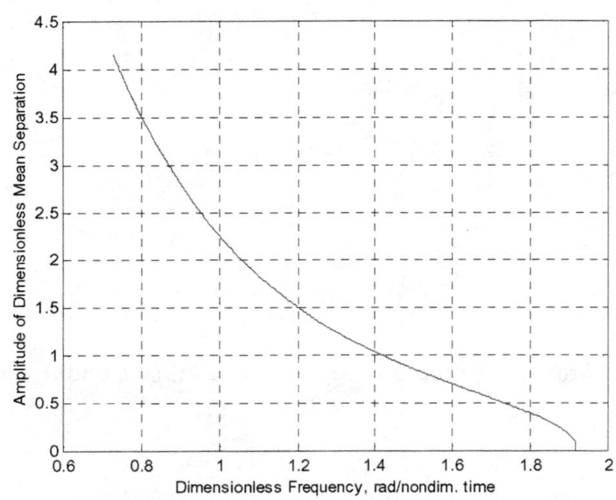

Figure 6. Skeleton Curve vs. Dimensionless Frequency of Free Vibration

In order to evaluate the influence of the perturbation of the number of contacting asperities on the time average friction force, forced vibration of the system under parametric excitation (nonzero ε and λ in Eq. (2)) was tested numerically. The parameter ε was assumed to be equal to 0.1 corresponding to a deviation of 10% from the nominal number of contacting asperities. Numerical evaluation was carried out for nondimentional viscous damping coefficient 0.02. Two examples of the response due to parametric excitation are shown in Figs. 7 and 8. These solutions were obtained for the same value of nondimensional frequency λ=1.5 but different initial conditions, showing the coexistence of two stable solutions at this frequency.

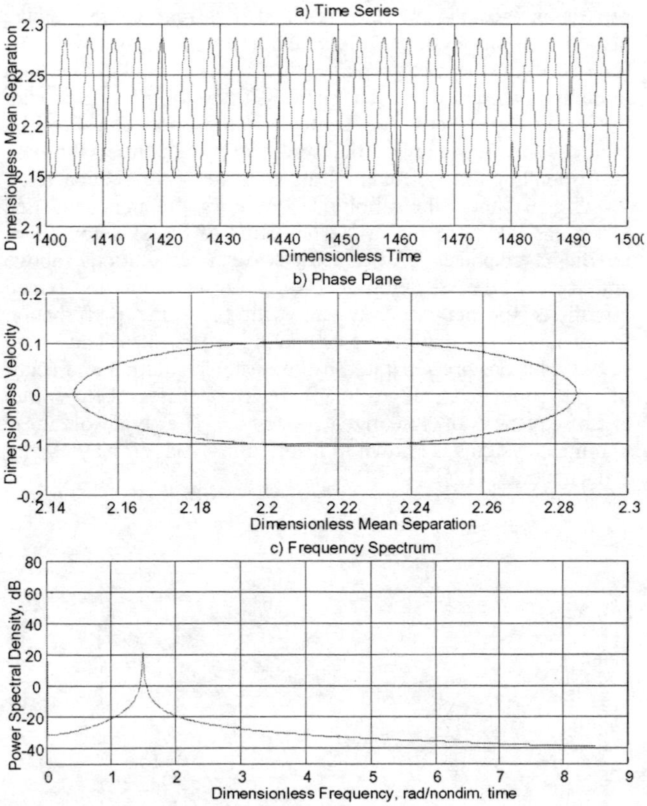

Figure 7. Forced Response due to Initial Conditions:
$Y_0 = 2.2;\ Y_0' = 0.1$

Fig. 7 presents a system response corresponding to a stable solution with relatively smaller amplitude (below the skeleton curve), where the behavior of the system appears linear (the time series is sinusoidal, Fig. 7a, the trajectory in the phase plane is elliptical, Fig.7b). A spectrum of this response includes only one frequency equal to the excitation frequency λ=1.5, Fig. 7c.

Fig. 8 presents a system response corresponding to a stable solution with relatively larger amplitude (above the skeleton curve). The large amplitudes of vibration correspond to a nonlinear variation of the force $P^*(Y)-F_s^*(Y)$ and, as a result, nonlinear behavior of the system. The topology of the trajectory on the phase plane, Fig. 8b, corresponds to a nonlinear system. The content of the spectrum, Fig. 8c, includes multiple frequencies in addition to the perturbation frequency of the number of the asperities in contact (in this particular case 1.5).

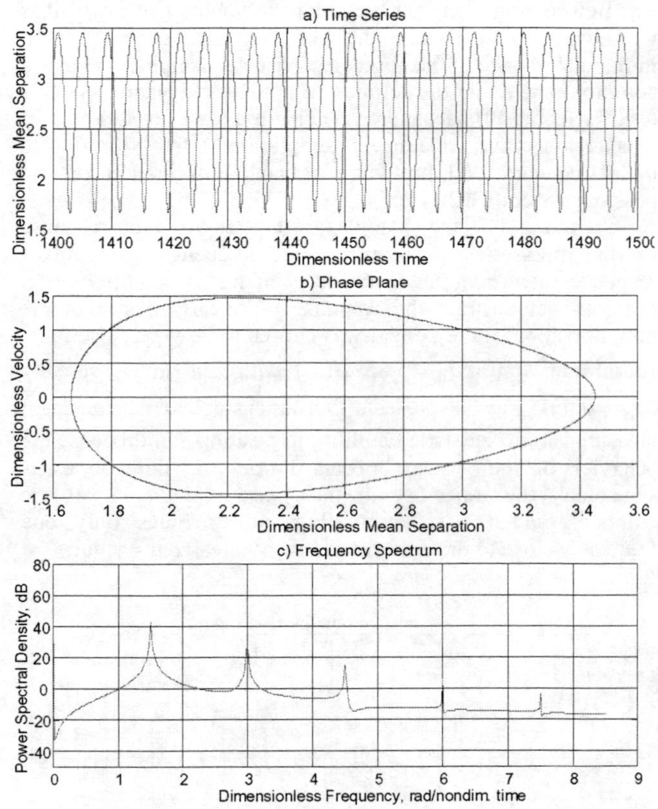

Figure 8. Forced Response due to Initial Conditions:
$Y_0 = 2.1;\ Y_0' = 1.1$

Fig. 9a shows the system response due to perturbation of the number of contacting asperities, where the skeleton curve (dashed line) is presented along with the two branches of the forced vibration amplitude curve (solid lines). These two branches correspond to two stable solutions separated by the skeleton curve. Note that the example of the forced vibration with small amplitude, presented in Fig. 7, corresponds to the low branch of the amplitude curve, Fig. 9a, and the example of the forced vibration with the large amplitude, presented in Fig. 8, corresponds to the top branch of the amplitude curve. The amplitude curve is non-continuous, because the two branches of the amplitude curve corresponding to the two stable solutions for a non-linear system are connected by a non-visible branch of amplitude curve corresponding to unstable solution (cf. Nayfeh and Balachandran, [12]). If experiments were conducted to construct the amplitude curve, the classical "jump" phenomenon would be observed at the ends of each of the two branches (jump from the end of the low branch to large-amplitude vibration branch at the same frequency and vice versa).

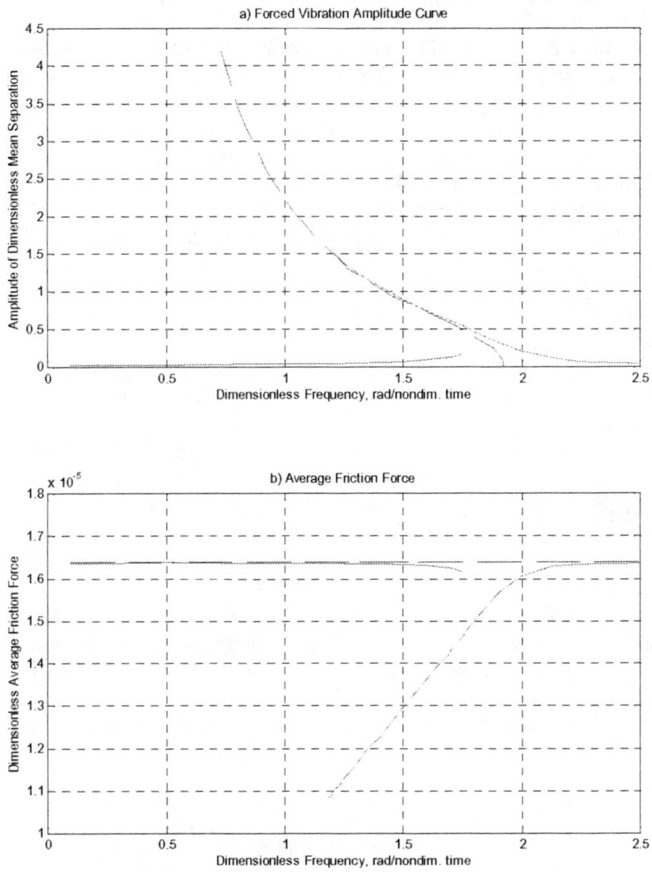

Figure 9. Forced Vibration Amplitude Curve (a) and Average Friction Force (b) vs. Dimensionless Frequency of Vibration

The variation of the mean separation with time defines the variation of the tangential force (friction force) between the sliding surfaces (see Chang et al., [17]). Results of the friction force averaged in time for the various values of the perturbation frequency are presented in Fig. 9b. The dashed line presents the value of the nondimensional static friction force corresponding to the constant value of the nondimensional external force $F_e^* = 6.25 \cdot 10^{-6}$ that is also used in the dynamic problem. The solid line, close to the dashed one, corresponds to the average dynamic friction force associated with the low branch of the amplitude curve in Fig. 9a. As can be seen the dynamic friction force is negligibly different from the static one in this case. The second solid line in Fig. 9b corresponds to the top branch of the amplitude curve in Fig. 9a. For this case the average dynamic friction force differs substantially from the static friction for frequencies below the eigenfrequency of the linearized system. The dynamic friction is always smaller than the static one.

DISCUSSION AND CONCLUSION

Presented model can explain the difference between static and dynamic contact with friction, based on the assumption that the perturbation frequency of the number of asperities in contact depends on the sliding velocity. This dependence can be explained, for example, by the wear model, proposed by Archard, [20]. According to the Archard theory each junction may be assumed to remain in existence during a sliding distance equal to the junction diameter, after which it is broken and a new junction takes up its load-carrying capacity. Applying the Archard theory to frictional sliding it can be concluded that, since the time to pass the distance equal to the junction diameter depends on the sliding velocity, then the perturbation frequency of the number of the asperities in contact increases with increasing sliding velocity.

Our approach explains the essential fall of the dynamic friction force relative to the static friction in the range of high sliding velocities. Low values of friction force correspond to the top branch of the amplitude curve (see Fig. 9). As the sliding velocity is reduced, the amplitudes of vibration corresponding to the top branch of the amplitude curve increase beyond all bounds and the model does not give realistic values of friction force. The dependence of the dynamic friction force on sliding velocity corresponding to the low branch of the amplitude curve is insignificant and can not explain the difference between static and dynamic friction in this range. The examination of our approach for other values of the external force led to similar results. It means, that small perturbations of the number of contacting asperities do not influence the average friction force at low values of the sliding velocity.

Modification of this model can clarify the picture of dynamic contact with friction at low values of the sliding velocity. For example, an addition of a rotational degree of freedom (angular tilt of the slider) coupled with a normal displacement could result in self-excitation of the system. In this case relative large amplitudes of vibration at low sliding velocities can be obtained. These vibrations, which increase the average mean separation between the sliding surfaces also, reduce the average dynamic friction in comparison to the static one.

ACKNOWLEDGMENTS

This research was supported in parts by the Israel Science Foundation and by the German-Israeli Foundation for Scientific Research and Development. This support is gratefully acknowledged. The author wishes to express his gratitude to Prof. I. Etsion of Technion – Israel Institute of Technology for his invaluable advice.

REFERENCES

[1] Lyuksyutov, I. F., 1995, "Dynamical Friction and Instability of Interface Motion," J. of Physics-Condensed Matter, **7**, pp. 7153-7161.

[2] Oancea, V. G., and Laursen, 1997, T. A., "Stability Analysis of State Dependent Dynamic Frictional Sliding," Int. J. of Non-Linear Mech., **32**, pp. 837-853.

[3] Adams, G. G., 1998, "Dynamic Instabilities in the Sliding of Two Layered Elastic Half-Spaces," ASME J. of Tribology, **120**, pp. 289-295.

[4] Adams, G. G., 1995, "Self-Excited Oscillations of Two Elastic Half-Spaces Sliding with a Constant Friction Coefficient," ASME J. Appl. Mech. **62**, pp. 867-872.

[5] Oden, J.T., and Martins, J.A.C., 1985, "Models and Computational Methods for Dynamic Friction Phenomena," Computer Methods in Applied Mechanics and Engineering, **52**, pp. 527-634.

[6] Martins, J.A.C., Oden, J.T., and Simões, F.M.F., 1990, "A Study of Static and Kinetic Friction," Int. J. of Engineering Science, **28**, pp. 29-92.

[7] Martins, J.A.C., Guimarães, J., and Faria, L.O., 1995, "Dynamic Surface Solutions in Linear Elasticity and Viscoelasticity with Frictional Boundary Conditions," ASME J. of Vibration and Acoustics, **117**, pp. 445-451.

[8] Martins, J.A.C., Barbarin, S., Raous, M., and Pinto da Costa, A., 1999, "Dynamic Stability of Finite Dimensional Linearly Elastic Systems with Unilateral Contact and Coulomb Friction," Computer Methods in Applied Mechanics and Engineering, **177**, pp. 289-328.

[9] Tworzydlo, W.W., Cecot, W., Oden, J.T., and Yew, C.H., 1998, "Computational Micro- and Macroscopic Models of Contact and Friction: Formulation, Approach and Applications," Wear, **220**, pp. 113-140.

[10] Greenwood, J. A., and Tripp, J. H., 1971, "The Contact of Two Nominally Flat Rough Surfaces," *Proceedings of the Institution of Mechanical Engineers*, **185**, pp. 625-633.

[11] Greenwood, J. A., and Williamson, J. B. P., 1966, "Contact of Nominally Flat Surfaces," Proc. Roy. Soc. (London), **A295**, pp. 300-319.

[12] Nayfeh, A. H., and Balachandran, B., 1995, *Applied Nonlinear Dynamics. Analytical, Computational, and Experimental Methods*, John Wiley & Sons, Inc., New York.

[13] Chang, W. R., Etsion, I., and Bogy, D. B., 1987, "Elastic-Plastic Model for the Contact of Rough Surfaces," ASME J. of Tribology, **109**, pp. 257-263.

[14] Chang, W. R., Etsion, I., and Bogy, D. B., 1988, "Adhesion Model for Metallic Rough Surfaces," ASME J. of Tribology, **110**, pp. 50-56.

[15] Kogut, L. and Etsion, I., 2002, "Elastic-Plastic Contact Analysis of a Sphere and a Rigid Flat," ASME J. of Applied Mechanics, **69**, pp. 657-662.

[16] Kogut, L. and Etsion, I., 2003, "Adhesion in Elastic–Plastic Spherical Microcontact," J. of Colloid and Interface Science, **261**, pp. 372–378.

[17] Chang, W. R., Etsion, I., and Bogy, D. B., 1988, "Static Friction Coefficient Model for Metallic Rough Surfaces," ASME J. of Tribology, **110**, pp. 57-63.

[18] Kogut, L. and Etsion, I., 2003, "A Semi-Analytical Solution for the Sliding Inception of a Spherical Contact," ASME J. of Tribology, **125**, pp. 499-506.

[19] Polycarpou, A. A., Etsion, I., 1998, "Comparison of the Static Friction Subboundary Lubrication Model with Experimental Measurements on Thin-Film Disks," STLE Tribology Transaction, **41**, pp. 217-224.

[20] Archard, J. F., 1953, "Contact and Rubbing of Flat Surfaces," J. Appl. Phys. **24**, pp. 981-988.

Proceedings of
2003 STLE/ASME Joint International Tribology Conference
Ponte Vedra Beach, Florida USA, October 26-29, 2003

2003-TRIB-0258

ASYMMETRIC ASPERITY HEIGHT DISTRIBUTIONS IN A SCALE-DEPENDENT MODEL FOR CONTACT AND FRICTION

George G. Adams
Department of Mechanical, Industrial and Manufacturing Engineering
Northeastern University, Boston, MA 02115, USA
Email: adams@neu.edu

Sinan Müftü
Department of Mechanical, Industrial and Manufacturing Engineering
Northeastern University, Boston, MA 02115, USA

ABSTRACT

The effect of an asymmetric distribution of asperity heights is accounted for in a recently developed scale-dependent multi-asperity model of contact and friction. A Weibull distribution of asperity heights is used which allows the skew and kurtosis to be varied, but not independently of each other. The contact and friction model used includes the effects of adhesion and of scale-dependent friction. The results obtained demonstrate that positive/negative skew decreases/increases both the friction coefficient and its dependence on the magnitude of the normal load.

INTRODUCTION

Contact and friction affect the operation of many machines and tools that we use every day, as well as some of the most basic activities in nature. Examples range from belt drives, brakes, tires, and clutches in automobiles and in other machines; gears, bearings and seals in a variety of mechanical systems; electrical contacts in motors; slider-disk interactions in a computer disk drive; various MEMS devices; a robotic manipulator joint; the motion of a human knee-joint (natural or artificial); and walking/running.

The friction force F is the tangential force resisting the relative motion of two surfaces which are pressed against each other with a normal force P. Amontons, in 1699, and Coulomb in 1785, developed our phenomenological understanding of *dry* friction between two contacting bodies. Amontons-Coulomb friction states that the ratio of the friction force (during sliding) to the normal force is a constant called the coefficient of *kinetic* friction. Similarly the coefficient of *static* friction is the ratio of the maximum friction force F that the surfaces can sustain, without relative motion, to the normal force. These friction laws can be summarized by defining the coefficient of friction μ as

$$\mu = \frac{F}{P} \tag{1}$$

without distinguishing between static and low-speed sliding friction. Although Eqn. (1) provides an extraordinarily simple *phenomenological* friction law, the *nature* of the friction force is not well-understood.

Contact modeling is an essential part of any friction model. It consists of two related steps. First, the equations representing the contact of a single pair of asperities are determined. For nanometer scale contacts the effect of adhesion on the contact area is important. Second, the cumulative effect of individual asperity contacts is determined. Such contact models are uncoupled and represent surface roughness as a set of asperities, often with statistically distributed parameters. The effect of each individual asperity contact is local and considered separately from the other asperities; the cumulative effect is the sum of the actions of individual asperities (e.g. the well-known Greenwood Williamson model, [1]).

For sufficiently small size contacts, the adhesion forces between the surfaces affect the contact conditions. Various adhesion models, typically between an elastic sphere and a flat, have been introduced. The model by Johnson, Kendall and Roberts (JKR) assumes that the attractive intermolecular surface forces cause elastic deformation beyond that predicted by the Hertz theory, thereby producing a subsequent increase of the contact area [2]. The model by Derjaguin, Muller and Toporov (DMT), on the other hand, accounts for the adhesive stress outside of the contact area, but assumes that the contact stress profile remains the same as in the Hertz theory [3]. Due to the assumptions involved, the JKR/DMT models are most suitable when the range of surface forces is small/large compared to the elastic deformations, as pointed out by Tabor [4]. Another model, introduced by Maugis [5], describes a continuous transition between the JKR and DMT models.

Contact and friction models which deal with adhesion in multi-asperity contacts have also been developed. In the first of a series of papers Chang, Etsion and Bogy (CEB) [6] developed an elastic-plastic multi-asperity contact model for normal loading based on volume conservation of a plastically deformed asperity control volume. In [7], the effect of adhesion was included by using the DMT model for contacting asperities and the Lennard-Jones potential between non-contacting asperities. Finally a model for calculating the coefficient of friction was given in [8]. It assumed that once plastic yielding is initiated in a pair of contacting asperities, no further tangential force can be sustained. Fuller and Tabor [9] investigated the effect of roughness on the adhesion between elastic bodies. Experiments were conducted between rubber spheres and a hard flat surface with controlled roughness. A theoretical model which used the JKR model of adhesion along with a Gaussian distribution of asperity heights was developed.

Stanley, Etsion and Bogy (SEB) [10] developed a model for the adhesion of two rough surfaces, affected by sub-boundary layer lubrication, in an elastic-plastic multi-asperity contact. Polycarpou and Etsion [11] used the SEB model to predict the static friction coefficient. The tangential load was found using the same procedure as in the CEB model [8] for solid-solid contact. Kogut and Etsion developed multi-asperity contact [12] and friction models [13] which included the effects of plastic deformation. The maximum shear load that an asperity can sustain is limited by the combined normal and shear load which causes the plastic deformation zone to reach the surface. Thus the friction analysis [13] predicts higher friction than the related work in [8].

The scale dependence of the friction stress for single asperity contacts has recently been investigated by Hurtado and Kim (HK), [14,15]. They presented a micromechanical dislocation model of frictional slip between two asperities for a wide range of contact radii. According to the HK model, if the contact radius *"a"* is smaller than a critical value, the asperities slide past each other in a concurrent slip process where the adhesive forces are responsible for the shear stress; hence the shear stress remains at a high constant value. On the other hand, if the

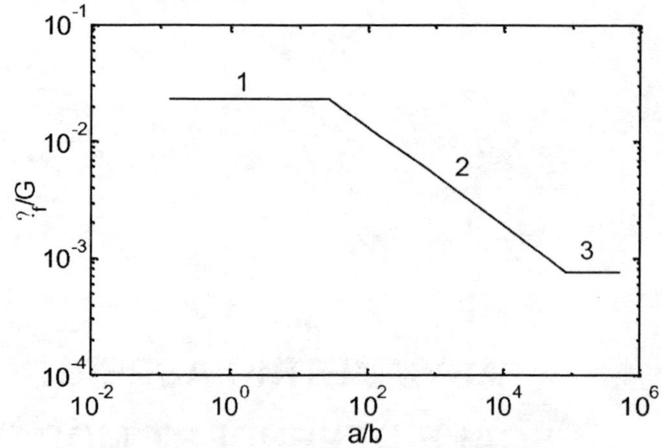

Figure 1. Relationship between friction stress and contact radius according to HK model.

contact radius is greater than that critical value, the shear stress decreases for increasing values of contact radius until it reaches a second constant, but lower value. The relationship between the non-dimensional friction stress ($\overline{\tau}_f = \tau_f / G^*$) and the non-dimensional contact radius ($\overline{a}=a/b$) is approximated in Fig. 1. The contact radius *"a"* is normalized by the Burgers vector b and the friction stress τ_f is normalized by the effective shear modulus $G^*=2G_1G_2/(G_1+G_2)$ where G_1 and G_2 are the shear moduli of the contacting bodies. Adams, Müftü, and Mohd Azhar (AMM), [16] incorporated the HK model and the adhesion contact model of Maugis, into a statistical multi-asperity model for contact and friction. The relationship between the friction force and the normal load between two rough surfaces during a slip process was determined. *Three key dimensionless parameters* representing the surface roughness, the friction regime of the contacts, and the surface energy of adhesion, were seen to influence the value of the friction coefficient.

The previously cited works used a Gaussian distribution of asperity heights. The effect of asymmetry in the height distribution has been accounted for by a few authors. Kotwal and Bhushan [17] modeled this effect and found that there are load-dependent optimal values of skew and kurtosis which minimize the real area of contact. McCool [18] extended the GW model by using a Weibull distribution of asperity heights and by including the effect of a surface coating. The multi-asperity elastic-plastic CEB contact model was generalized to include a Weibull distribution by Yu and Polycarpou [19]. Skew was found to greatly affect the relation between the force, contact area, and the number of contacting asperities

In this paper the effect of an asymmetric distribution of asperity heights is accounted for in the recently developed AMM scale-dependent multi-asperity model of contact and friction. A Weibull distribution of asperity heights is used which allows the skew and kurtosis to be varied, but not independently of each other. The contact and friction model used includes the effects of adhesion and of scale-dependent friction. The results demonstrate that positive/negative skew decreases/increases both the friction coefficient and its dependence on the magnitude of the normal load.

DEVELOPMENT OF THE ASYMMETRIC MODEL
Contact Model

The scale-dependent multi-asperity contact and friction model developed by Adams, Müftü, and Mohd Azhar [16] will be extended to include asymmetric asperity height distributions. For two real surfaces separated by a distance d (defined from the mean of asperity heights) the number of contacting asperities n is

$$n = N\int_{\bar{d}}^{\infty}\bar{\phi}(\bar{z})d\bar{z} \tag{2}$$

where N is the total number of asperities, σ is the standard deviation of asperity peak heights, $\bar{z} = z/\sigma$ is the dimensionless height coordinate measured from the mean of asperity heights, $\bar{\phi}(\bar{z})$ is the probability density of asperity peaks, and $\bar{d} = d/\sigma$ is the non-dimensional separation between the two surfaces (Fig. 2).

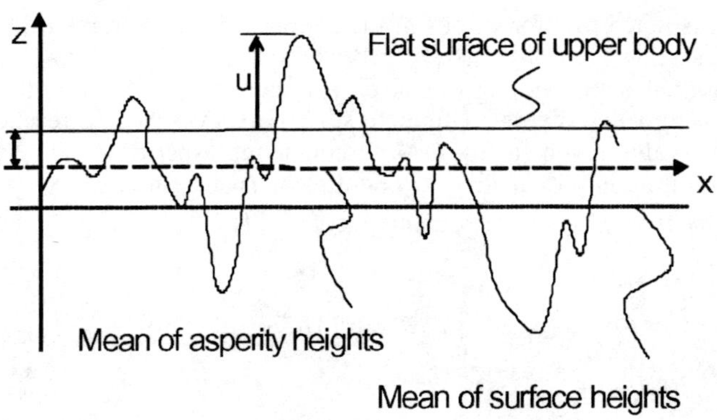

Figure 2. Contact of a rough surface (lower body) with a flat surface (upper body).

The relation between the normal load P and deformation $u = z - d$ of two contacting spherical asperities with adhesion is given by the Maugis model [5]. In that model, a uniform tensile stress σ_0 exists between the contacting asperities just outside the contact zone, $a \le r \le c$, where c is the radial extent of the adhesion zone. The separation between the two surfaces at $r = c$ is equal to the prescribed maximum adhesion distance h. Thus the work of adhesion is given by $w = \sigma_0 h$. In [5] the following non-dimensional relations among the asperity contact radius (A_M), the asperity contact force (P_M), and the asperity defomation (Δ_M) were obtained

$$\frac{?A_M^2}{2}\left[\sqrt{m^2-1}+\left(m^2-2\right)\tan^{-1}\sqrt{m^2-1}\right] + \frac{4?^2A_M}{3}\left[\sqrt{m^2-1}\tan^{-1}\sqrt{m^2-1}-m+1\right]=1 \tag{3}$$

$$P_M = A_M^3 - \lambda A_M^2\left[\sqrt{m^2-1}+m^2\tan^{-1}\sqrt{m^2-1}\right] \tag{4}$$

$$\Delta_M = A_M^2 - \frac{4}{3}A_M\lambda\sqrt{m^2-1} \tag{5}$$

where λ is the non-dimensional Maugis adhesion parameter, and m is the non-dimensional adhesion radius. The various non-dimensional quantities used in the Maugis model are defined as

$$A_M = a\left(\frac{K}{\pi w R^2}\right)^{1/3},\quad \lambda = 2\sigma_0\left(\frac{R}{\pi w K^2}\right)^{1/3},$$
$$m = \frac{c}{a},\quad P_M = \frac{P}{\pi w R},\quad \Delta_M = u\left(\frac{K^2}{\pi^2 w^2 R}\right)^{1/3} \tag{6}$$

where $K = (4/3)\,E^*$.

The simultaneous solution of Eqs. (3)–(5) gives the relations between m, P_M, A_M and Δ_M for a given value of λ. In practice it is convenient to vary m and solve the resulting quadratic equation for the only positive root A_M in (3). Then (4) and (5) can be solved explicitly for P_M and Δ_M respectively. The non-dimensional Maugis adhesion parameter λ is expressed in terms of the surface roughness parameter α, the friction regime parameter β, the surface energy of adhesion parameter γ, the maximum adhesion distance h, and the Burgers vector b, as

$$\lambda = \left(\frac{b}{h}\right)\left(\frac{9\beta\gamma^2}{2\pi\alpha}\right)^{1/3}, \tag{7}$$

where the three key parameters α, β and γ are

$$\alpha = \left(\frac{\sigma}{R}\right)^{1/2},\quad \beta = \frac{(R\sigma)^{1/2}}{b},\quad \gamma = \frac{w}{E^* b}. \tag{8}$$

A physical interpretation of the surface parameters α and β is provided by noting that in a simple vertical scaling of the surface by a factor k, the standard deviation of asperity heights σ is scaled by k but the asperity radius of curvature R is scaled by $1/k$. Thus, α is scaled by k, but β remains constant. Hence α is a representation of the surface roughness, and is referred to as the *surface roughness parameter*. The parameter β describes the ratio of the contact radius (due to an asperity penetration equal to σ) to the Burgers vector length. Thus small β are expected to be indicative of nano-scale asperity contacts and progressively larger values of β correspond to transition and larger values of the contact radius (Fig. 2). Therefore β is referred to as the *friction regime parameter*. The *surface energy parameter* γ represents the ratio of the adhesive stress to the product of the composite Young's modulus and the Burgers vector.

It is further noted that for the case considered here, in which one of the surfaces is assumed to be rigid and flat, $G^*=2G$ and the composite Young's modulus is given by $E^* = E/(1-\nu^2)$. Furthermore the relation $G=E/2(1+\nu)$ has been used in obtaining (7).

The non-dimensional contact radius, normal load, and asperity penetration used in the Maugis model (A_M, P_M and Δ_M, respectively) can be expressed in terms of the corresponding non-dimensional parameters defined in this work ($\bar{a}$, $\bar{P}$, $\bar{u}=u/\sigma$, respectively) according to

$$\bar{u} = \Delta_M \hat{\delta}, \quad \bar{P} = P_M / \hat{P}, \quad \bar{a} = A_M / \hat{A} \tag{9}$$

where

$$\hat{\delta} = \left(\frac{3\pi\gamma}{4\alpha^2\beta}\right)^{2/3}, \quad \hat{P} = \frac{(1-\nu)\alpha^2}{2\pi\beta\gamma}, \quad \hat{A} = \left[\frac{4}{3\pi\gamma}\left(\frac{\alpha}{\beta}\right)^2\right]^{1/3}.$$

An expression for the non-dimensional normal force $\bar{P}$ acting on the nominal contact area is obtained by integrating the normal force on individual asperities from the Maugis model (P_M), i.e.

$$\bar{P} = \frac{1}{\hat{P}} \int_{\bar{d}}^{\infty} P_M \phi(\bar{z}) d\bar{z} \tag{10}$$

It is noted that due to adhesion, asperities may come into contact even if the asperity overlap u is negative. This effect has been included in the evaluation of the integrals in Eqs. (2) and (10) by varying the adhesion radius ratio m in Eqs. (3)–(5) when evaluating the force and contact area. Thus the lower limits of the integrals in Eqs. (2) and (10) are effectively slightly less than $\bar{d}$.

Friction Model

Although adhesion affects the relationship between the normal force and the contact radius, it does not affect the relation between the friction force and contact radius. From Fig. 1, the dimensionless shear stress is a function of the dimensionless contact radius and is approximated by

$$\log(\bar{\tau}_f) = \begin{cases} \log\bar{\tau}_{f_1}, & \bar{a} < \bar{a}_1 \\ M\log\bar{a} + B, & \bar{a}_1 < \bar{a} < \bar{a}_2 \\ \log\bar{\tau}_{f_2}, & \bar{a} > \bar{a}_2 \end{cases} \tag{11}$$

where the left and right limits of region-2 are $(\bar{a}_1, \bar{\tau}_{f_1})$ and $(\bar{a}_2, \bar{\tau}_{f_2})$ respectively. The constants of Eq. (11) are given by

$$M = -\left(\log(\bar{\tau}_{f_1}/\bar{\tau}_{f_2})\right)/\left(\log(\bar{a}_2/\bar{a}_1)\right)$$
$$B = \left(\log(\bar{\tau}_{f_1})\log(\bar{a}_2) - \log(\bar{\tau}_{f_2})\log(\bar{a}_1)\right)/\left(\log(\bar{a}_2/\bar{a}_1)\right) \tag{12}$$

where M and B are, respectively, the slope and y-intercept of the line in region-2 of the log-log plot of Fig. 1. The friction force F_f acting on a single asperity can be determined from Eq. (11) by using $F_f = \pi a^2 \tau_f$ resulting in

$$\frac{F_f}{G^* b^2} = \begin{cases} \bar{\tau}_{f_1}\bar{a}^2, & \bar{a} < \bar{a}_1 \\ 10^B \bar{a}^{M+2}, & \bar{a}_1 < \bar{a} < \bar{a}_2 \\ \bar{\tau}_{f_2}\bar{a}^2, & \bar{a} > \bar{a}_2 \end{cases} \tag{13}$$

The total shear force F acting on the nominal contact area can be calculated by integrating the shear forces acting on each asperity against the probability density function, i.e.

$$F = N\int_{\bar{d}}^{\infty} F_f \phi(\bar{z}) d\bar{z}. \tag{14}$$

It is noted that for values of the applied tangential force F less than that given by Eq. (14), the distribution of tangential and normal forces among the asperities may cause some asperities to slip while others continue to stick. However when F reaches the value given in Eq. (14) all contacting asperities will slide resulting in global slip. In non-dimensional form, the force per asperity $(\bar{F})$ becomes, using Eqs. (13-14)

$$\bar{F} = \frac{F}{NGb^2} = 2\pi\int_{\bar{d}}^{\bar{z}_1} \bar{\tau}_{f_1}\bar{a}^2\phi(\bar{z})d\bar{z} + 2\pi\int_{\bar{z}_1}^{\bar{z}_2} 10^B \bar{a}^{M+2}\phi(\bar{z})d\bar{z} + 2\pi\int_{\bar{z}_2}^{\infty} \bar{\tau}_{f_2}\bar{a}^2\phi(\bar{z})d\bar{z} \tag{15}$$

Thus, the coefficient of friction μ for two real surfaces separated by a distance $\bar{d}$, can be obtained from Eqs. (1), (10), and (15).

Weibull Distribution

In [16] a Gaussian distribution of asperities heights was assumed. In this paper we investigate the effect of asymmetry by using a Weibull function, i.e.

$$\phi(\hat{z}) = \frac{\omega \hat{z}^{\omega-1}}{\eta^\omega} e^{-(\hat{z}/\eta)^\omega}, \quad \hat{z} > 0 \tag{16}$$

to represent the asperity height distribution. For this two-parameter Weibull function the mean (z_M), standard deviation of asperity heights (σ), skew (S), and kurtosis (K) are given by in terms of the two Weibull parameters (ω,η), [20].

The Weibull distribution variable ($\hat{z}$) is then shifted by

$$z=\hat{z}-z_M,\quad z_M=\eta B_1,\quad B_n=\Gamma(1+\frac{n}{\omega}), \tag{17}$$

where Γ is the gamma function, so that $z = 0$ corresponds to the mean of asperity heights.

Furthermore using the non-dimensional variable $(\bar{z}=z/\sigma)$

$$\bar{\phi}(\bar{z})=\omega C^{\omega}\bar{z}^{\omega-1}e^{-(C\bar{z})^{\omega}},\quad C=(B_2-B_1^2)^{1/2} \tag{18}$$

is obtained. Note that this form of the Weibull distribution depends on only one parameter (ω) and that $\bar{z}=0$ corresponds to the mean of asperity heights. The skew, standard deviation, and kurtosis are given by

$$\begin{aligned} S&=(B_3-3B_2B_1+2B_1^3)/C^3,\\ \sigma&=\eta C \\ K&=(B_4-4B_3B_1+6B_2B_1^2-3B_1^4)/C^4 \end{aligned} \tag{19}$$

respectively. It follows that these non-dimensional quantities also depend only on the parameter ω. As ω is varied both the skew and kurtosis change. However it is not possible to change the skew without also changing the kurtosis as indicated by Eqn. (19).

RESULTS AND DISCUSSION

Results have been obtained for the friction coefficient as a function of the normal load, for various values of the skew $(\bar{S})$ and the three key parameters (α,β,γ) defined in the AMM model. Figure 3 shows the effect of skew for the default values $(\alpha = 0.01,\ \beta = 1000,\ \gamma = 0.001)$ used in [16]. Positive/negative skew is seen to decrease/increase both the friction coefficient as well as its variation with normal load. This result is due to negative skew causing more asperities to be above the mean of asperity heights than below. Correspondingly there is less of a variation in height distribution for the z > 0 portion than for z < 0. The more lightly loaded the contact, the smaller is the fraction of the Weibull distribution which is in contact. Thus the asymmetry in the height distribution causes the surface to appear as if its roughness is reduced for lightly loaded contacts.

Figures 4, 5, and 6 show similar results for a lower value of the surface roughness parameter $(\alpha = 0.006)$, for a lower value of the friction regime parameter $(\beta = 500)$, and for a higher value of the adhesion parameter $(\gamma = 0.002)$ respectively. The smoother surface shows a greater effect of skew than the rougher one. This result may be expected based on the previous discussion. The effect of a reduction of the friction regime parameter is to increase friction for the Gaussian distribution. However as the skew decreases to the value of -1 the effect of β on friction decreases. The effect of an increase in the surface energy parameter is to increase friction for both positive and negative skew values.

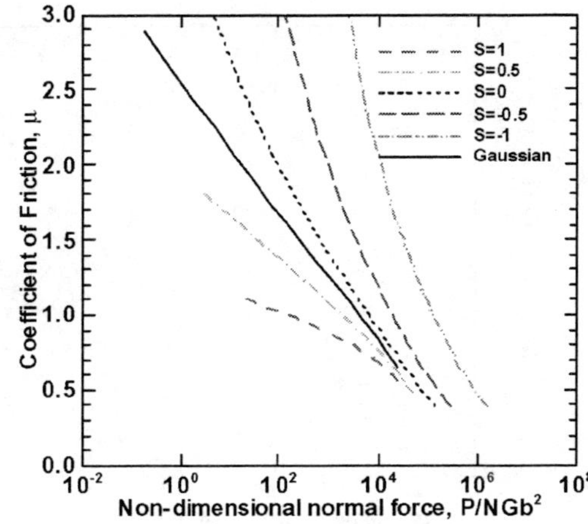

Figure 3. Coefficient of friction vs. normal load for various skew and $\alpha = 0.01,\ \beta = 1000,\ \gamma = 0.001$

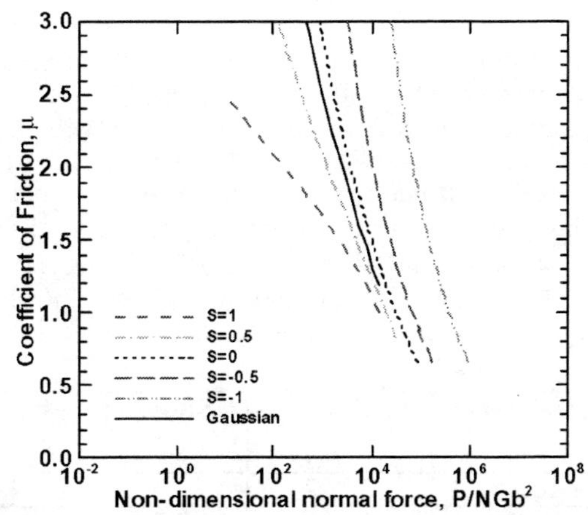

Figure 4. Coefficient of friction vs. normal load for various skew and $\alpha = 0.006,\ \beta = 1000,\ \gamma = 0.001$

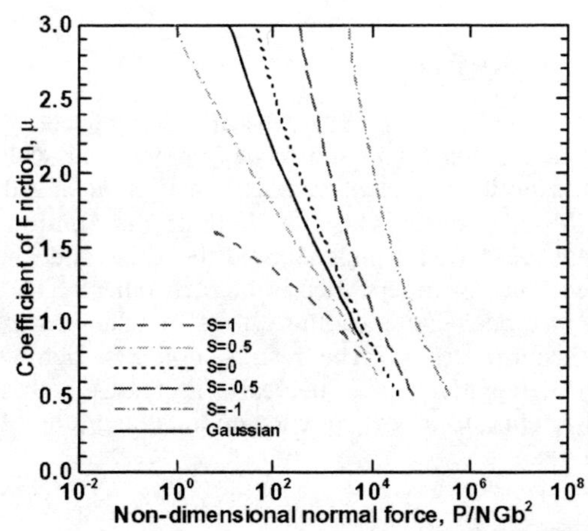

Figure 5. Coefficient of friction vs. normal load for various skew and $\alpha = 0.01,\ \beta = 500,\ \gamma = 0.001$

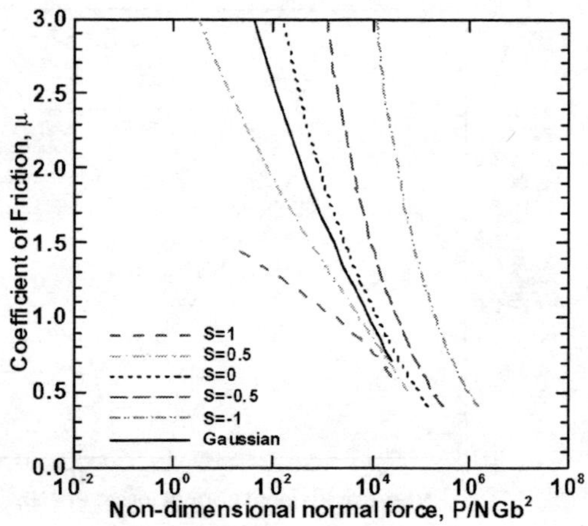

Figure 6. Coefficient of friction vs. normal load for various skew and $\alpha = 0.01$, $\beta = 1000$, $\gamma = 0.002$

It is noted that the friction results for the Gaussian distribution do not correspond exactly to the zero skew case of the Weibull distribution. This difference is because the Weibull distribution only approximates a Gaussian distribution. For lightly loaded contacts the difference between these two distributions in their "leading edge" is sufficient to produce these discrepancies. The values of the Weibull shape parameter (ω), the mean ($\bar{z}_M$) the skew ($\bar{S}$), and kurtosis ($\bar{K}$) are shown in Table 1 below.

Parameter ω	Mean $\bar{z}_M$	Skew $\bar{S}$	Kurtosis $\bar{K}$
40.74	32.32	-1	4.773
7.493	6.340	-0.5	3.258
3.602	3.243	0.0	2.716
2.216	2.098	0.5	3.028
1.564	1.530	1.0	4.159

CONCLUSIONS

The effect of an asymmetric distribution of asperity heights has been accounted for in the recently developed scale-dependent multi-asperity model of contact and friction of Adams, Müftü and Mohd Azhar [16]. A Weibull distribution of asperity heights was used which allowed the skew and kurtosis to be varied, but not independently of each other. The contact and friction model included the effects of adhesion and of scale-dependent friction. The results obtained demonstrated that positive/negative skew decreases/increases both the friction coefficient and its dependence on the magnitude of the normal load.

REFERENCES

[1] Greenwood, J.A., and Williamson, J.B.P., 1966, "Contact of Nominally Flat Surfaces," *Proceedings of the Royal Society of London,* **A295**, pp. 300-319.

[2] Johnson, K.L., Kendall, K., and Roberts, A.D., 1971, "Surface Energy and the Contact of Elastic Solids," *Proceedings of the Royal Society of London,* **A324**, pp. 301-313.

[3] Derjaguin, B.V., Muller, V.M., and Toporov, Y.P., 1975, "Effect of Contact Deformations on the Adhesion of Particles," *Journal of Colloid and Interface Science,* **53**, pp. 314-326.

[4] Tabor, D., 1976, "Surface Forces and Surface Interactions," *Journal of Colloid and Interface Science,* **58**, pp. 2-13.

[5] Maugis, D., 1992, "Adhesion of Spheres: The JKR-DMT Transition Using a Dugdale Model," *Journal of Colloid and Interface Science,* **150**, pp. 243-269.

[6] Chang, R.W., Etsion, I., Bogy, D.B., 1987, "An Elastic-Plastic Model for the Contact of Rough Surfaces," *ASME Journal of Tribology,* **109**, pp. 257-263.

[7] Chang, R.W., Etsion, I., Bogy, D.B., 1988, "Adhesion Model for Metallic Rough Surfaces," *ASME Journal of Tribology,* **110**, pp. 50-56.

[8] Chang, R.W., Etsion, I., Bogy, D.B., 1988, "Static Friction Coefficient Model for Metallic Rough Surfaces" *ASME Journal of Tribology,* **110**, pp. 57-63.

[9] Fuller, K.N.G and Tabor, D., 1975, "The Effect of Surface Roughness on the Adhesion of Elastic Solids," *Proceedings of the Royal Society of London,* **A345**, pp. 327-342.

[10] Stanley, H.M., Etsion, I. and Bogy, D.B., 1990, "Adhesion of Contacting Rough Surfaces in the Presence of Sub-Boundary Lubrication," *ASME Journal of Tribology,* **112**, pp. 98-104.

[11] Polycarpou, A.A., and Etsion, I., 1998, "Static Friction of Contacting Real Surfaces in the Presence of Sub-Boundary Lubrication," *ASME Journal of Tribology,* **120**, pp. 296-303.

[12] Kogut, L., and Etsion, I., 2003, "A Finite Element Based Elastic-Plastic Model for the Contact of Rough Surfaces," *Tribology Transactions,* **46**, pp. 383-390.

[13] Kogut, L., and Etsion, I., 2003, "A Static Friction Model for Elastic-Plastic Contacting Rough Surfaces," *ASME Journal of Tribology,* in press.

[14] Hurtado, J.A,. and Kim, K.-S., 1999, "Scale Effects in Friction of Single Asperity Contacts: Part I; From Concurrent Slip to Single-Dislocation-Assisted Slip," *Proceedings of the Royal Society of London,* **A455**, pp. 3363-3384.

[15] Hurtado, J.A., and Kim, K.-S.,1999, "Scale Effects in Friction in Single Asperity Contacts: Part II; Multiple-Dislocation-Cooperated Slip," *Proceedings of the Royal Society of London,* **A455**, pp. 3385-3400.

[16] Adams, G.G., Müftü, S., and Mohd Azhar, N., 2003, "A Scale-Dependent Model for Multi-Asperity Model for Contact and Friction," *ASME Journal of Tribology,* **125**, in press.

[17] Kotwal, C.A., and Bhushan, B., 1996, "Contact Analysis of Non-Gaussian Surfaces for Minimum Static and Kinetic Friction and Wear," *Tribology Transactions,* **39**, pp. 890-898.

[18] McCool, J.I., 2000, "Extending the Capability of the Greenwood Williamson Microcontact Model," *ASME Journal of Tribology,* **122**, pp. 496-502.

[19] Yu, N., and Polycarpou, A.A., 2002, "Contact of Rough Surfaces With Asymmetric Distribution of Asperity Heights," *ASME Journal of Tribology,* **124**, pp. 367-376.

[20] Johnson, N., and Kotz, S., 1970, *Continuous Univariate Distributions-I,* John Wiley and Sons, Inc., New York.

Proceedings of
2003 STLE/ASME Joint International Tribology Conference
Ponte Vedra Beach, Florida USA, October 26-29, 2003

2003-TRIB-0259

AN EXPERIMENTAL INVESTIGATION OF ELASTIC PLASTIC CONTACT AND FRICTION OF A SPHERE ON FLAT

Ofer Levinson Izhak Etsion* and Grigori Halperin
Department of Mechanical Engineering
Technion ,Haifa, 32000, Israel
***E-mail: etsion@tx.technion.ac.il**

ABSTRACT

An experimental investigation is presented to evaluate recently published models for the contact and sliding inception of a deformable sphere loaded against a smooth rigid flat. The effects of the normal load on the contact area, junction growth, and the static friction force in the elastic-plastic contact regime are presented. Very good correlation is found between the predicted and measured contact area. A dramatic decrease of the static friction coefficient with increasing normal loading is observed, similar to the trend predicted by the model. The quantitative agreement is, however, less satisfying. Some possible reasons for the poor agreement are pointed out.

KEYWORDS: spherical contact, elastic-plastic contact, junction growth, friction laws, static friction.

INTRODUCTION

It is well known that relative sliding of two surfaces under a compressive force is opposed by friction. The classical Coulomb's laws of friction state that the friction force is proportional to the normal load and is independent of the apparent contact area of the surfaces. Indeed, from many theoretical works on the real contact area of rough surface (see review by Liu et al. [1]) it became clear that the real contact area, which is the main contributor to the friction force, is independent of the apparent contact area. Many of these works are based on modeling the contact behavior of a single spherical asperity, which is then incorporated in a statistical model of multiple asperity contact e.g., the pioneering work of Greenwood and Williamson [2] (GW model) for mainly elastic contact or the work of Pullen and Williamson [3] (PW model) for pure plastic contact. Chang et al. [4] extended the GW model by developing an elastic plastic contact model (CEB model) that was then used to predict static friction of contacting rough surfaces (Chang et al. [5]). Sliding inception was treated as a failure mechanism rather then resorting to a coulomb friction law with a certain friction coefficient. The results in [5] revealed a behavior of the friction coefficient that differs from the teaching of the classical laws. It was shown (as was already known from some experiments) that the friction coefficient depends on the normal load. According to the CEB model [4] the sphere remains in elastic Hertzian contact until a critical interference is reached above which volume conservation of the sphere tip is imposed. Also, according to the assumptions in [5] an asperity is incapable of supporting additional tangential load as soon as this critical interference is reached. The above conservative assumptions resulted in unrealistic underestimated friction coefficient values. More recently, finite element based elastic-plastic solutions by Kogut and Etsion [6, 7] provided models for the contact parameters and sliding inception of a sphere and a rigid flat. This approach enabled to predict the static friction coefficient of a spherical contact more accurately than in [5].

Several recent experimental investigations of static friction, that cover a large variety of test conditions e.g. [8]–[10], also showed behavior of the friction force and friction coefficient that contradicts the classical laws of friction. Hence, the main purpose of the present work is to further explore the peculiarities of static friction behavior and to investigate the validity of the contact and sliding inception models presented in Refs. [6] and [7].

EXPERIMENTAL PROCEDURE

Friction force and relative displacement of a sphere specimen loaded on a rigid flat specimen were measured simultaneously as shown schematically in Fig. 2. The flat specimen was firmly held on a plate mounted on a linear bearing. The plate was drawn by a weight, which was released by electrical motor at a controlled constant velocity of 0.5 mm/Sec. A PC controlling the weight release was also used for data acquisition of the output from the force and displacement transducers and for further data analysis.

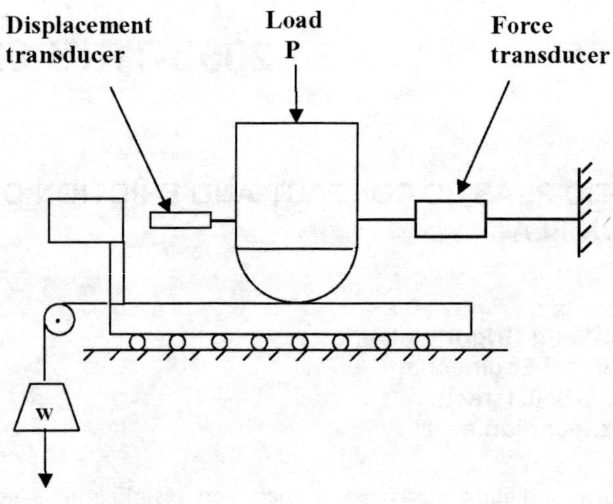

Figure 1. Schematic of the experimental system

Specimen Preparations

A cold drawn, 10 mm diameter, copper rod with negligible strain hardening was used to produce the spherical specimens with diameters, D, of 3, 5, 10, and 15 mm. The hardness, H, of the copper was 1GPa and its modulus of elasticity, E, was 106GPa. This particular copper was selected to comply with the assumption of an elastic-perfectly-plastic behavior of the sphere material that was made in Refs. [6,7]. All the spherical specimens were polished to a high surface finish of roughness average Ra=0.02μm to comply with another assumption of perfectly smooth surfaces.

Both hardened steel (H=6.2GPa, E=205GPa) and sapphire (H=190GPa, E=435GPa) were used for the rigid flat specimens. The hardened steel was also polished to a high surface finish (Ra=0.01μm) and the sapphire flat had extremely fine surface with very low roughness average Ra=0.003μm.

Experimental Details

Prior to any test, the spherical and flat specimens were ultrasonically cleaned with acetone and than dried in air. Friction tests were performed at different normal load values, P, ranging from 1 to 120N. After normal loading of the sphere, the weight release mechanism was activated and the increase in the friction force and relative displacement with time were recorded at a very high sampling rate of 2kHz. This procedure enables the detection of sliding inception when a sudden increase in the relative displacement rate occurs. Fig. 2 presents a typical behavior of the friction force and relative displacement as a function of time. As can be seen the full sliding begins about 0.5 Sec. after the beginning of the test when the friction force starts to increase. At this instant the friction force reaches a maximum (the static friction) and then falls to a slightly lower value (the dynamic friction). The test shown in Fig. 2 was performed with a sphere having a diameter D=15mm and a normal load P that is 30 percent higher than the critical load P_c, which corresponds to the critical interference at which first yielding occurs in the copper sphere [6]. At this load ratio, P/P_c=1.3, the contact is just beginning to become elastic plastic.

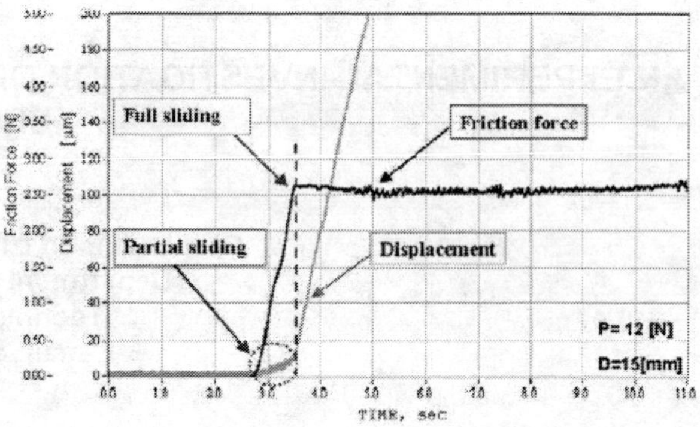

Figure 2. Friction force and relative displacement at P/P_c= 1.3

A completely different behavior was observed with a smaller diameter D=3mm at much higher load ratios P/P_c, when the contact is substantially elastic plastic [6]. In this case, which is typically represented in Fig. 3 for a load ratio P/P_c = 154, the friction force at the instant of sliding inception (static friction) was lower than the subsequent dynamic friction. A similar behavior was also observed by Xie et al. [8] who related it to plowing of a soft aluminum flat by a hard spherical tip in their experiment. Such plowing could not occur in the present tests since the copper sphere is softer than the steel or sapphire flat.

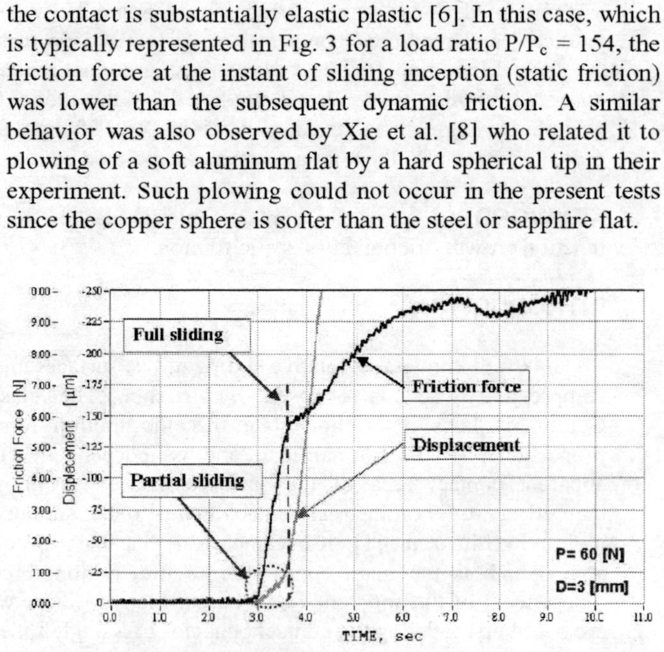

Figure 3. Static friction and relative displacement at P/Pc=154

Each friction test was repeated at least five times, depending on the scatter in the data to ensure consistency of the results. Each repetition was performed with fresh contacting surfaces.

The contact area was measured indirectly by first loading the sphere against the flat with a given load P, allowing a rest time of 30 Sec., and then unloading and measuring the residual plastic trace diameter on the copper sphere with an optical microscope. In some cases the sphere was slid on the flat under the normal load P before unloading. This was done to explore any potential junction growth.

EXPERIMENTAL RESULTS

The experimental results of the contact area measurements for three different sphere diameters, D, along with the theoretical predictions of Ref. [6] are shown in Fig. 4. The contact area, A, is normalized by A_c, the contact area corresponding to P_c. As can be seen the correlation is very good. The experimental results are somewhat lower than the theoretical ones as would be expected from a post loading measurement where the elastic portion of the deformation is completely recovered. Indeed, the error made by this type of indirect measurement decreases as the dimensionless load P/P_c increases and the relative contribution of elastic deformation to the contact area diminishes.

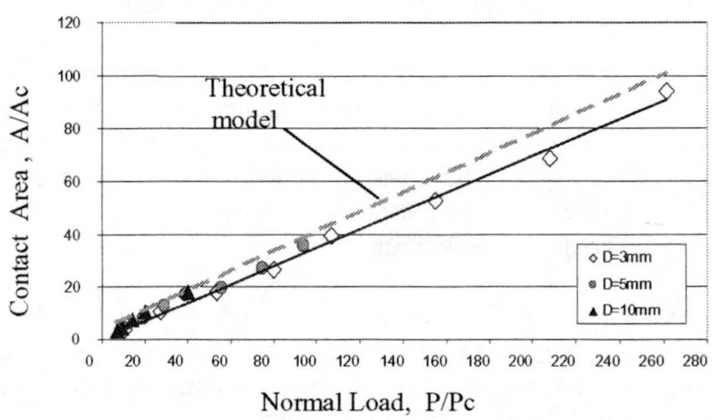

Figure 4. Dimensionless contact area vs. dimensionless load

A typical trace of the residual plastic deformation is shown in Fig. 5. The dimensionless normal load in this case was P/P_c=220, deep into the elastic plastic regime [6]. As can be seen, the diameter of the residual plastic trace in the pure normal loading case (a) is very much the same as that of the trace resulting from a combined normal and tangential loading case (b). The only observed difference between the two cases is in their surface finish. In the "no sliding" case the original surface finish of the sphere is maintained after the unloading as can be seen from the lapping marks inside and outside the contact area footprint. In the "after sliding" case these lapping marks are not visible any more inside the contact area footprint. Instead, very weak markings in the direction of sliding can be seen but the surface inside the contact footprint is much smoother indicating plastic flow during sliding. It seems therefore, that the present measurements do not support the Tabor [11] idea of junction growth following Courtney-Pratt and Eisner [12] experiments. The contact area in [12] was derived from electrical conductance measurements and is less reliable than the present measurements.

A certain increase of the real contact area does occur inside the boundaries of the contact perimeter as the contact surface becomes smoother due to plastic flow resulting from the sliding. However, the overall diameter of the contact area is not affected by the addition of a tangential load.

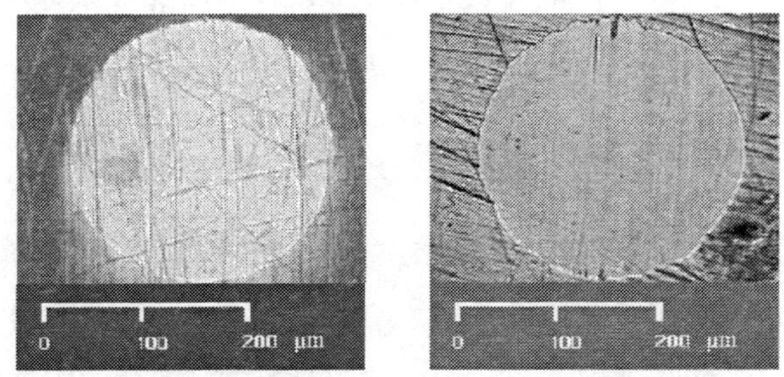

(a) No Sliding **(b) After Sliding**

Figure 5. Photographs of the contact area trace following normal load of P/P_c=220

Fig. 6 presents the results of the static friction coefficient μ_s for the copper on steel tests at three different sphere diameters, D. Clearly the behavior does not obey the classical laws of friction. At a given load, P, the static friction coefficient increases with increasing sphere diameter. For the two larger diameters the friction coefficient decreases with increasing normal load. Very similar behavior was obtained with the copper on sapphire tests.

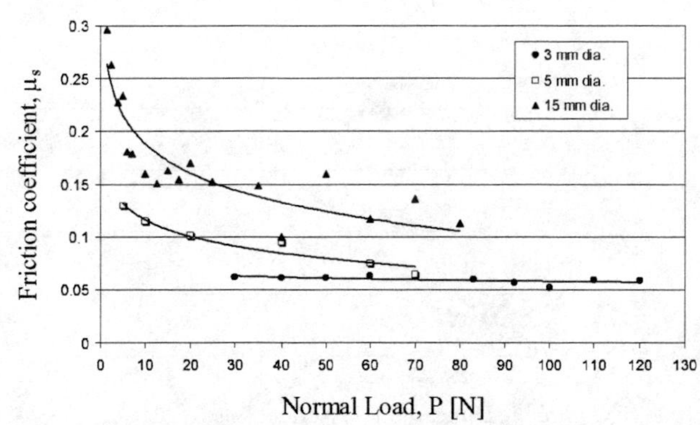

Figure 6. Static friction coefficient vs. normal load, copper against hardened steel

The sliding inception model of Kogut and Etsion [7] predicts a unique dependent of μ_s on the dimensionless load P/P_c. Hence, the static friction coefficient results from all the tests for all the sphere diameters in both the copper on steel and copper on sapphire tests were plotted vs. P/P_c in Fig. 7. Indeed, the individual effects of sphere diameter or material combination on the static friction coefficient merged into a unique relation between μ_s and P/P_c.

The effect of P/P_c on the static friction coefficient is remarkable. As can be seen from Fig. 7, the static friction coefficient is $\mu_s=0.3$ at the lower end of P/P_c but it decreases sharply as P/P_c increases up to about $P/P_c=150$. Thereafter the reduction of the static friction coefficient with increasing P/P_c becomes much more moderate reaching a value of $\mu_s=0.06$ at $P/P_c=425$. This represents a dramatic reduction of 80% in the static friction coefficient over the range of tested dimensionless normal loads.

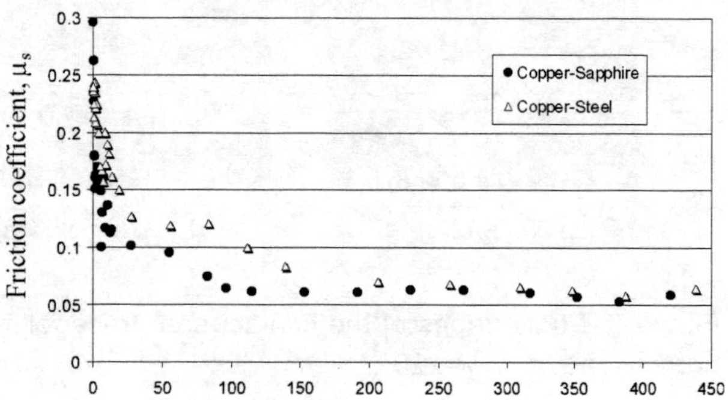

Figure 7. Static friction coefficient Vs. dimensionless normal load

A comparison of the static friction force, Q, measured in the present experiments and that predicted by the model of Ref. [7] is shown in Fig. 8. The results, in a dimensionless form Q/P_c are plotted against the dimensionless normal load P/P_c. As can be seen the theoretical model (dashed line) fairly predicts the experimental results at P/P_c values between 5 and 10. At lower values the model overestimates the friction force while at higher P/P_c values it severely underestimates the experimental results. The reason for the lower experimental friction force at $P/P_c < 5$ is probably due to the fact that the real area of contact is less than the apparent one due to the surface roughness, whereas the model assumes perfectly smooth surfaces. The different trends of the experimental and theoretical results at $P/P_c > 10$ could result from an inaccurate assumption made in [7] regarding the vanishing capability of supporting tangential force above a certain normal loading.

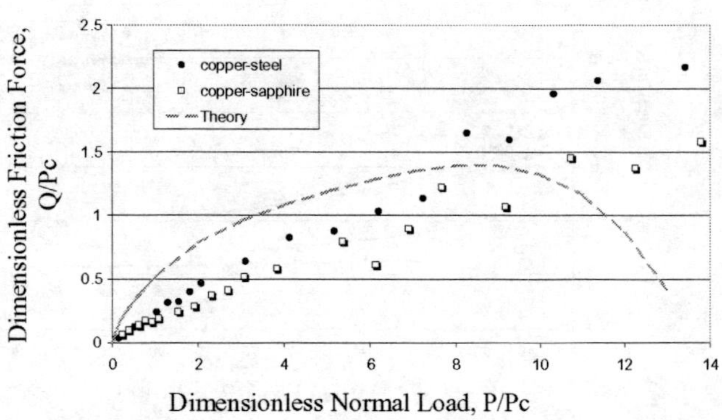

Figure 8. Dimensionless Friction Force vs. Dimensionless Normal Load.

CONCLUSION

An experimental investigation was carried out to explore the effect of normal loading on the contact area and static friction in a spherical elastic plastic contact. The experimental results were compared with two recently published models in order to evaluate the validity of these models. The main findings of the present experiments are the following:

- The theoretical model of Ref. [6] predicts very well the elastic plastic contact area between a deformable sphere and a rigid flat under pure normal loading.
- The present experiments did not confirm the Tabor perception of an overall junction growth under combined normal and tangential loading. Instead, an increase in the real contact area within the contact interface was observed as a result of plastic flow during sliding that makes the contact interface smoother.
- The static friction of a substantially elastic-plastic contact in the present experiments was found to be lower than the dynamic friction, contrary to the common practice of static and dynamic friction relation.
- The static friction coefficient can be presented as a unique function of the dimensionless normal load P/P_c in agreement with the model prediction in Ref. [7].
- The normal load has a dramatic effect on the static friction coefficient, contrary to the teaching of the classical laws of friction. A drastic reduction of about 80% in the static friction coefficient was observed in the present experiments as the dimensionless normal load increases to a value of 150. Above this load the static friction coefficient varies very little and remains very low at about 0.06.
- The model of Ref. [7] does not predict correctly the quantitative friction force. It could be improved by relaxing some of its assumptions.

ACKNOWLEDGMENT

The research reported here was partially supported by the Bernstein Research Fund.

REFERENCES

[1] Liu, G., Wang, Q. J., and lin, C., 1999," A Survey of Current Models for Simulating the Contact between Rough Surfaces," Tribol. Trans., **42**, pp 581-591.

[2] Greenwood, J.A., and Williamson, J.B.P., 1966, "Contact of Nominally Flat Surfaces," Proc. R. Soc. London, **A295**, pp. 300-319.

[3] Pullen, J., and Williamson, J.B.P., 1972, "On the Plastic Contact of Rough Surfaces," Proc. R. Soc. London, **A327**, pp. 159-173.

[4] Chang, W.R., Etsion, I., and Bogy, D.B., 1987, "Elastic Plastic Model for the Contact of Rough Surfaces," ASME Journal of Tribology, **109**, pp. 257-262.

[5] Chang, W.R., Etsion, I., and Bogy, D.B., 1988, "Static Friction Coefficient Model Metallic Rough Surfaces", ASME Journal of Tribology, **110**, pp. 57-63.

[6] Kogut, L., and Etsion, I., 2002, "Elastic-Plastic Contact Analysis of a Sphere and a Rigid Flat," ASME Journal of Applied Mechanics, **69**, pp. 657-662.

[7] Kogut, L., and Etsion, I., 2003, "A Semi-Analytical Solution for the Sliding Inception of a Spherical Contact," ASME Journal of Tribology,

[8] Xie, W., De Meter, E.C., and Trethewey, M.W., 2000, "An Experimental Evaluation of Coefficients of Static Friction of Common Workpiece-Fixture Element Pairs", International Journal of Machine Tools & Manufacture, **40**, pp. 467-488.

[9] Liu, Z., Neville, A., and Reuben, R.L., 2002, "The Effect of Film Thickness on Initial Friction of Elastic-Plastically Rough Surface With a Soft Thin Metallic Film", ASME Journal of Tribology, **124**, pp. 627-636.

[10] Johnson, K.L, 2000," The Contribution of Micro/Nano-Tribology to The Interpretation of Dry Friction", Proc. Instn. Mech. Engers., **214**, Part C, pp. 11–22.

[11] Tabor, D., 1959, "Junction Growth in Metallic Friction: The roll of Combined Stresses and Surface Contamination," Proc. R. Soc. London, **A251**, pp. 378-393.

[12] Courtney-Pratt, J.S., and Eisner, E., 1957, "The Effect of Tangential Force on the Contact of Metallic Bodies," Proc. Roy. Soc. A, **238**, pp. 529-550.

Proceedings of
2003 STLE/ASME Joint International Tribology Conference
Ponte Vedra Beach, Florida USA, October 26-29, 2003

2003-TRIB-0260

FINITE ELEMENT SIMULATION OF DYNAMIC INSTABILITIES IN FRICTIONAL SLIDING CONTACT

L. BAILLET
INSA Lyon, LaMCoS
69621 Villeurbanne Cedex FRANCE
Email: Laurent.baillet@insa-lyon.fr

V. LINCK
INSA Lyon, LaMCoS
69621 Villeurbanne Cedex
FRANCE

S. D'ERRICO
INSA Lyon, LaMCoS
69621 Villeurbanne Cedex
FRANCE

B. LAULAGNET
INSA Lyon, LVA
69621 Villeurbanne Cedex
France

Y. BERTHIER
INSA Lyon, LaMCoS
69621 Villeurbanne Cedex
FRANCE

Keywords: friction, instabilities, contact, dynamic, finite element.

ABSTRACT

This paper describes tools for numerical modelisation which enable the understanding of the appearance of the vibration of structure generated by the frictional contact between two bodies (the excitation source being friction). The dynamic finite element code PLASTD is used to reproduce transitory phenomena generated at the contact interface. This code includes contact and friction algorithms based upon the formulation of Lagrange multipliers.

A numerical study of the dynamic response of a 2D mechanical model composed of a deformable plane in relative translation and unilateral contact with Coulomb friction with a rigid surface is presented. The steady sliding solution is generically unstable and leads to a dynamic response which leads to the generation of instabilities characterized by the appearance of sliding-sticking or sliding-sticking-separation waves. It is important to notice that those instabilities appear even with a constant friction coefficient.

These simulations have permitted to obtain the local contact conditions (kinematics, tribological state, contact stresses, etc). The kinematics shows the existence of local impacts and sliding at high frequencies. Furthermore, local normal pressure is much higher than that expected for a smooth surface.

Finally, a 3D simulation of braking is carried out. We focused on the vibration of the disc and the brake pad which caused noises due to the generation of interface instabilities.

INTRODUCTION

Whatever the language or country, squealing, hammering, hooting, juddering… these are the many names describing noises induced by the friction of bodies in contact. In the case of braking, these noises are in fact produced at the interface between the disc and the brake pad.

A review of the work carried out in the field of vibrations due to friction between two bodies in relative movement was proposed by [1]. The presence of instabilities characterized by the propagation of waves in the interface was studied analytically by [2-6], and using the finite element method by [7,8]. The unstable state generated in the interface of the bodies in friction contact appears at the level of the mechanism caused by the vibration of the body surfaces and therefore by the appearance of noise.

More often it is these consequences, the noises, which are studied using the vibrations of the parts around the contact zone. The friction coefficient is then used as an adjusting parameter. This way of treating the problem leads regretfully to treating them at the level of their consequences (vibrations, noises) and not at the level of the tribological causes (interface between two bodies), leading to expensive and difficult to control solutions.

To reverse this tendency this paper describes tools for numerical modelisation, these enable the understanding of braking noises generation at the tribological level.

FORMULATION PROBLEM OF PLASTD

The explicit dynamic finite elements code PLASTD in 2 and 3 dimensions, see Baillet et al. [9], is used to simulate the frictional contact between two bodies. This software uses a forward incremental Lagrange multiplier method, see Carpenter et al. [10], which enables the evaluation of the normal and tangential contact stresses in the contact region as well as the determination of whether the contact surfaces locally stick, slide or separate.

The forward increment Lagrange multiplier method equations set is constructed using the equations of motion developed via the principle of virtual work equation at time step t_n ($t_n = n\Delta t$) augmented by displacement constraints acting on contacting surfaces at time t_{n+1} ($t_{n+1} = t_n + \Delta t$):

$$\begin{aligned} & M\ddot{u}_n + C\dot{u}_n + F_n^{int} + G_{n+1}^T \lambda_n = F_n^{ext} \\ & G_{n+1}\{X_n + u_{n+1} - u_n\} \leq 0 \end{aligned} \quad (1)$$

where λ_n is the contact force vector acting on the nodes at the contact surface, G_{n+1} is the global assembled matrix of the constraint elementary matrices, $X_{n+1} = X_n + u_{n+1} - u_n$ is the coordinate vector at time t_{n+1}, M and C are respectively the mass and the damping matrices, F^{int} and F^{ext} are the nodal vectors of internal and external forces, $\dot{u}$ and $\ddot{u}$ the velocity and the acceleration vectors. The contact algorithm uses slaves nodes and target surfaces. In 3 dimensions the target surfaces are described by four-node quadrilateral elements, as the 3D element used is an isoparametric hexahedra. An elementary target surface defined by four nodes is decomposed into a bi-cubic Ferguson patch with a C^1 continuity across the adjacent boundary. The friction law used is the standard Coulomb friction model without regularisation of the tangential stress against the tangential velocity component [9].

In order to eliminate the high frequencies during the simulation, a numerical damping can be used thanks to the β_2 ($\in [0.5;1[$) time integration scheme. With that scheme, the velocity and acceleration of the surface at time t_n is expressed as:

$$\begin{aligned} & \dot{u}_n = \left[\dot{u}_{n-1} + \Delta t(1-\beta_2)\ddot{u}_{n-1} + \frac{2\beta_2}{\Delta t}(u_{n+1} - u_n)\right]/(1+2\beta_2) \\ & \ddot{u}_n = 2(u_{n+1} - u_n - \Delta t\,\dot{u}_n)/\Delta t^2 \end{aligned} \quad (2)$$

The value β_2 equal to 0.5 corresponds to the central difference method (no numerical damping) and the value 1 corresponds to the greatest numerical damping.

SIMULATION OF A 2D MECHANICAL MODEL

Description of the model

The simulation presented considered the contact with friction of a deformable pad on a rigid disc. The model of the pad (Fig.1) is composed of 400 elements (40 elements along the length and 10 elements along the height). Both surfaces are considered flat as the initial roughness is erased in the first moment of contact between the two first bodies.

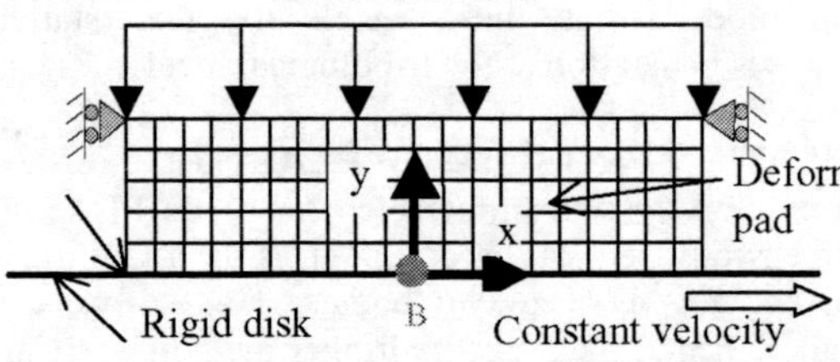

Figure 1. Model and boundary condition of the 2D mechanical model

At the initial time an increasing force F is applied in the y direction until it reaches a pre-selected value in order to correspond to a uniformly distributed normal stress on the top of the pad. The application of this force puts the pad in friction contact with the disc whose linear speed is kept constant throughout the simulation time. The results are studied at the center node (node B). The material characteristics and dimensions of the pad are summarised in Table 1. The third bodies, the thermal and physico-chemical effects are not taken into account in these simulations.

Young Modulus E (MPa)	10000
Poisson coefficient ν	0.3
Density ρ (kg/m^3)	2000
Length of the pad (mm)	100
Height of the pad (mm)	20

Table 1. Material and geometric characteristics of the pad

Sliding-sticking-separation waves

The simulation using the parameters described in Table 2 has shown instabilities on the contacting surface of the pad. Those instabilities are characterised by the appearance of sliding-sticking-separation waves, see Baillet et al [11], Oueslati et al [8] and Adams et al [4-6,12]. It is important to notice that those instabilities appear even with a constant friction coefficient on the whole contact surface. A stable state can be characterized by a stationary sliding contact of the surface of the pad on the moving disc, see Adams [2,3]. On the other hand, an unstable state is characterised by the formation of contact zones (sliding or sticking) on the pad surface and of zones separated from the disc surface (Fig.2).

Those instability waves propagate from the rear to the front of the pad at a speed close to the sound celerity in the material (≈ 2000 m/s). Due to these separation waves, the normal contact stress can locally reach 10 MPa, instead of the 1 MPa applied on the top surface of the pad as shown on Figure 2. This increase is due as much to the reduction of the contact area as to the kinematics ("impact") of the surfaces.

Figure 3 shows the trajectory at the center of the contacting surface of the pad. It can be shown [13] that at whatever point on the surface of the pad there exists a limit displacement cycle. This cycle decomposes into a displacement with contact (sliding or sticking) 1 in the direction of the disc velocity and a negative displacement corresponding to the elastic return during which the node is no longer in contact with the disc 2. In this simulation, the maximum amplitude of the displacement along the x-axis is about 5 μm whenever situated on the surface of the pad. The maximum separation between the pad and the disc is around 2.5 μm.

Figure 4 represents the normal and sliding velocities at the center of the pad surface in contact. When the surface is just arriving in contact with the disc after the separation, the local impact velocity is around 1.7 m/s. During the contact phase, the relative sliding velocity between the pad and the disc reaches a value of 2.5 m/s (for a disc velocity of 2 m/s). The frequency of impact is around 68 kHz.

An investigation on the influence of different parameters in 2 [13] and 3 [14] dimensions has shown the importance of friction, speed, applied load and material behavior in the local contact conditions (stress, deformation...). During the braking simulation between two deformable bodies (disc and pad), one finds the same instabilities on the contact level. Thereafter, one will thus be interested in the consequences of these instabilities on the vibration of the bodies in contact, therefore on the generation of the noise.

Friction coefficient μ	0.4
Distributed normal load F [N]	100
Equivalent normal stress on the top [MPa]	1
Velocity of the disc [m/s]	2
Simulation time step Δt [s]	1.10^{-7}
Numerical damping β_2	0.8

Table 2. Simulation parameters

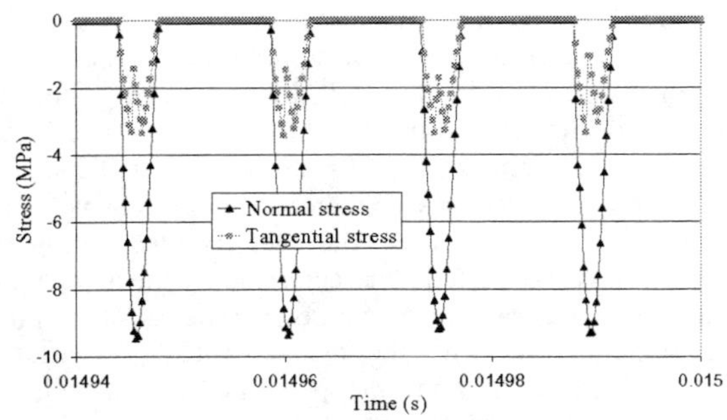

Figure 2. Contact stresses evolution as a function of time at node B

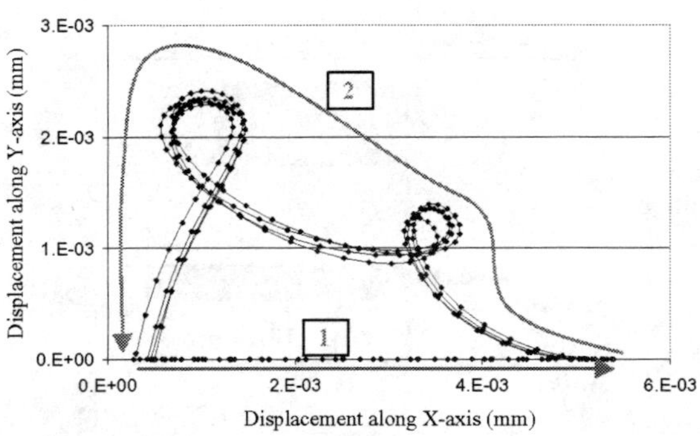

Figure 3. Limit cycle of the node B

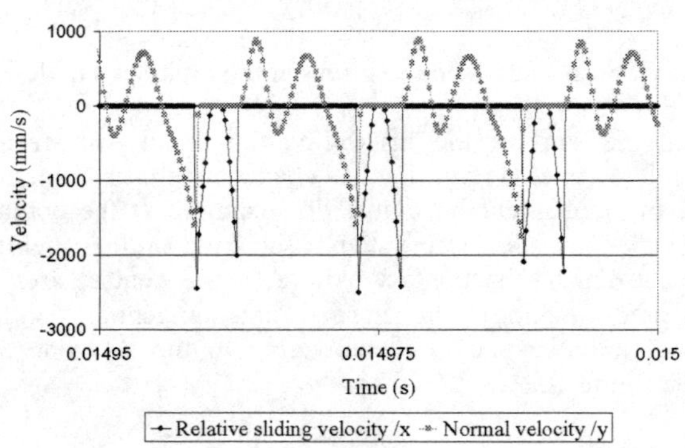

Figure 4. Normal and Relative velocities at the node B

BRAKING SIMULATION

Model of simulation

The PLASTD software was used in 3D for the braking simulation. The aim is to analyze the instability phenomenon generated within the contact of the two bodies (brake pad and disc) and also the level of the surface acceleration generated. In this paper, results of the first bodies acceleration will be presented, assuming that the temporal and frequencial behavior in acceleration is a faithful representation of the noise radiation. For more detailed developments concerning vibro-acoustic transfer mechanisms, one can refer to B. Laulagnet [15]. The finite element model that has been used is represented in figure 5. The two deformable bodies are mesh with cubic elements of eight nodes. Surfaces are smooth (roughness, superficial imperfections and third bodies are neglected) and the heating and physico-chemical effects are not taken into account.

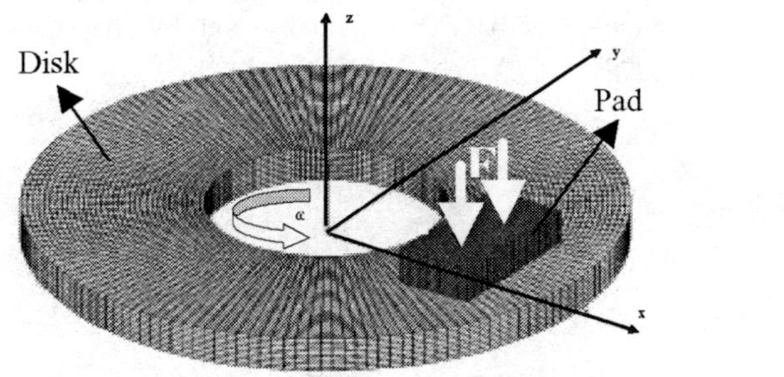

Figure 5. Finite elements mesh of a brake pad in contact on a rotating disc

The code being in an explicit dynamic formulation, the chosen time step Δt is inferior to the critical one, which depends on the materials (Young's modulus, Poisson's coefficient, volumic mass and damping) and on the length of the element of the body.

The boundary conditions of the model are as follows:

- the basic force F is applied to all the nodes of the upper area of the brake pad; at the beginning of the simulation, this force is applied in a linear way, then it stays constant;
- the upper area nodes of the brake pad are constrained depending on x and y.
- the nodes belonging to the inner disc radius are constrained depending on the z-axis and have an imposed rotative speed ω.

Disc :	
Young's modulus E [MPa]	210000
Poisson's coefficient ν	0.3
Volumic mass ρ [Kg/m^3]	7800
Interior radius r_i [mm]	60
Exterior radius r_e [mm]	150
Thickness e [mm]	20
Pad :	
Young's modulus E [MPa]	10000
Poisson's coefficient ν	0.3
Volumic mass ρ [Kg/m^3]	2500
Width (x), length (y), height (z) [mm]	40 x 80 x 20

Table 3. Characteristics of the brake pad and the disc

The main parameters of a simulation are listed in table 4.

Parameter	Value
Friction coefficient of Coulomb type	0.5
Normal force imposed on the pad F [N]	25000
Equivalent pressure [MPa]	7.8
Disc speed ? [rad/s]	47.6
Time step of the simulation ? t [s]	$0.1*10^{-6}$
Numerical damping $ß_2$	0.6
Viscous damping $ß_v$	$0.25*10^{-6}$

Table 4. Simulation parameters

Three different mesh, type Mi, (Fig.6) (with different element numbers) have been used. For each one, a convergence study has been carried out for both the mechanical aspect (stresses, displacements, speed) and for the vibration aspect (surface acceleration, frequency calculation and natural modes). The figure 7 shows the percentage error of the natural frequencies for the first 50 natural modes (mode n°50, f = 20897 Hz) and for the different mesh under consideration. These calculations have been carried out by the ABAQUS finite element code.

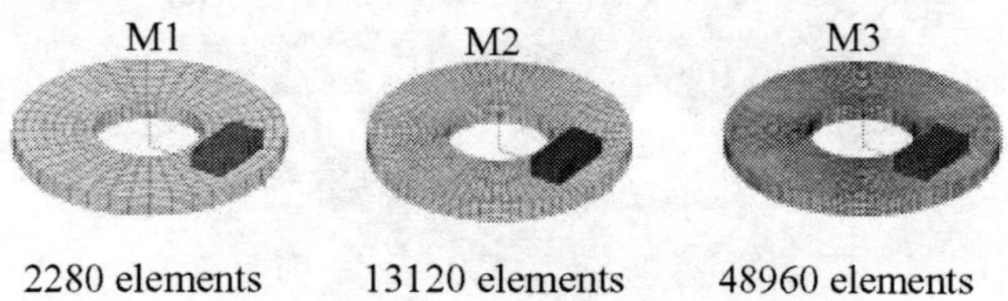

Figure 6. The three different mesh used

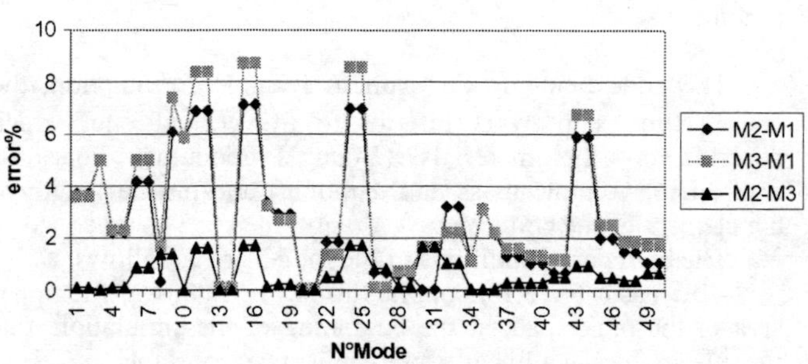

Figure 7. Percentage error of the natural mode frequencies for the three grids, type M1, M2, M3

The error in the frequencies obtained with the M3 and M2 grids remains inferior to 2%, the results obtained with these 2 grids can be considered to be almost identical. The final model M3 is then chosen, it is formed of 40960 elements and a brake pad of 8000 elements.

The characteristics of the two bodies (brake pad/disc) in contact are given in table 3. At the initial instant, a F force is applied to all the nodes of the braking pad upper face, putting the braking pad in contact with the disc, this rotates at a constant applied speed of ω. The analysis of what is happening within the contact and the study of the mechanical coupling between the mechanism and the first body have been carried out [16]. In the following part of this paper, an attempt is made to understand how the instabilities that appear on the contact area affect the contact area cinematic, and also affect the surface acceleration of the bodies. The analysis of this last value will enable us to find at what frequencies these vibrations occur and to know if there are audible frequentiel components (below 20 kHz). The acoustic radiation being proportional to the normal nodal acceleration, so the normal nodal acceleration will be presented in temporal and frequentiel fields.

Cinematic of the real contact area

During the simulations, the friction between the two bodies generates instabilities on the contact area of the braking pad and the disc that can be characterized by the appearance of stick-slip-separation waves (see Baillet, Oueslati [8]). These instabilities in the contact area are the originating causes of the vibrations in the mechanism. They are the noise source. The analysis of the contact cinematic enables the realisation that the phenomenon that exists in the contact area reverberates into all the mechanism. Figure 8 shows the sum of the contact forces, depending on x and y, of the nodes of the lower area of the pad and show clearly the change from a stable state to an unstable state after an elapsed time of t=0.005 seconds. A stable state is characterized by a contact area where the nodes do not stick and stay in a sliding contact on the moving disc during the simulation. An unstable state is characterised by the appearance on the braking pads areas of contact zones (sliding or sticking) and of separated zones with the disc area [2]. The friction coefficient and the disc rotation speed being constant, the sliding or the sticking of an area depends among other things on the normal constraint variation of the contact due to the propagation of the wave, called the dynamic effect [8].

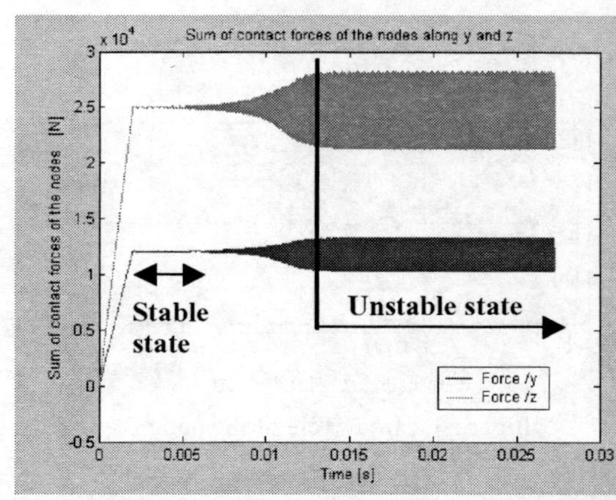

Figure 8. Sum of the contact forces at the brake pad nodes

From the start of the instability, the brake pad area is divided into 4 zones (Fig.9). Each is characterised, as a function of time, by a different behavior. For example at the contact entry (Z1), the surface has a system associating sliding, sticking and separating. At the contact exit (Z3), the contact area is slipping and sticking. The normal and tangential contact stresses of a brake pad node belonging to the Z4 zone are shown on figure 10.

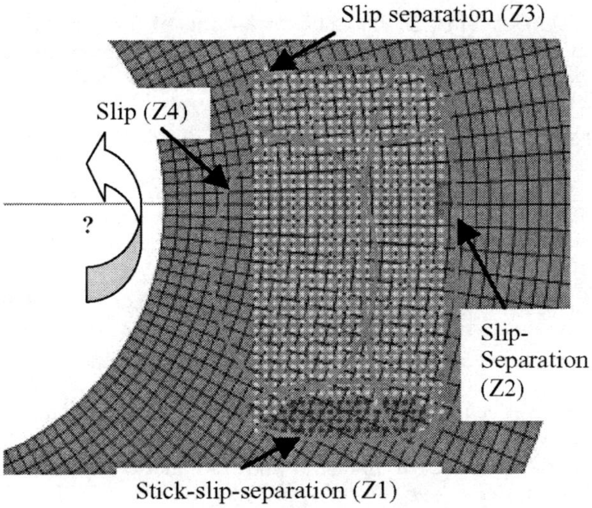

Figure 9. The different contact zones of the brake pad and their status

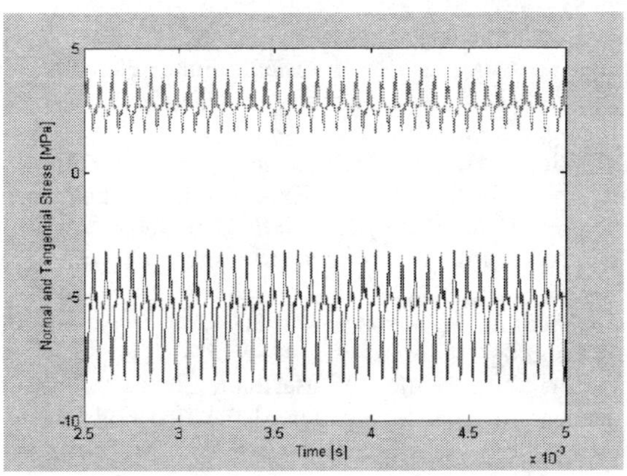

Figure 10. Normal and tangential contact stresses of a brake pad node on the contact area (zone Z4)

The normal contact pressures of brake pad zones can therefore reach superior values to those calculated theoretically. This increase is due to both the contact area decrease and the surface impact that occurs in some contact area zones. The instability that appears begins at 15 kHz and continues with a periodic phenomenon whose basic periodicity is an unstable frequency mode but which accumulates all the frequency harmonics which caused this instability (15, 30, 45 etc.). In what follows, the study of the first body vibration will show that the normal acceleration of the surfaces happens at the same frequencies.

Vibration study

The PLASTD code allows us, as a function of time, to obtain the acceleration, the speed and other parameters at different nodes. The analysis of the normal nodal acceleration, informs us about the vibration of the 2 bodies (disc/pad) and informs us about the acoustic radiation during braking. Figure 11 shows the accelerations at one surface point of the disc and one of the pad. Figure 12 shows the Fourier Transform of the accelerations and gives periodic spectrum with the main peak placed at 15 kHz and whole number multiples.

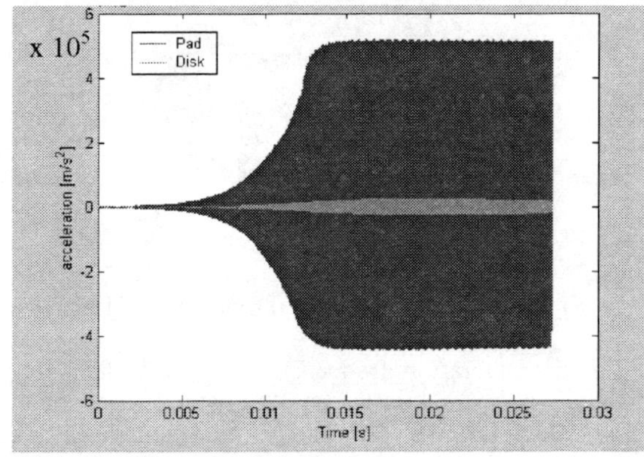

Figure 11. Acceleration of the disc and brake pad facet as a function of time

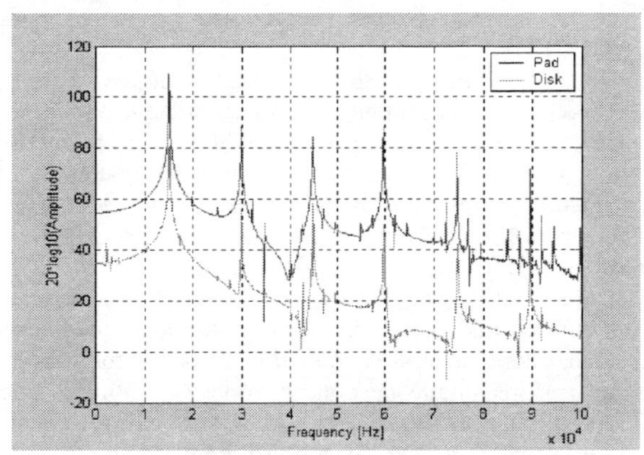

Figure 12. Fourier transforms of the accelerations

The two spectrum are identical but have different amplitudes, the brake pad has higher amplitude levels than the disc. In the case studied, the main vibration source is localised on the brake pad face at the entry. The normal speed representation of the two bodies (depending on z) enables the disc vibration to be visualised (Fig.13).

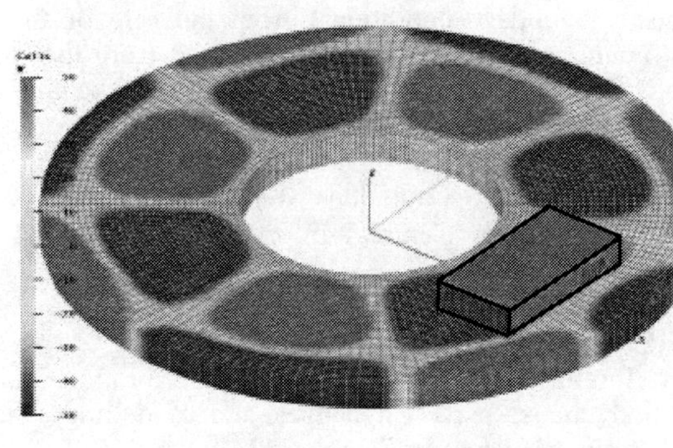

Figure 13. Speed depending on z

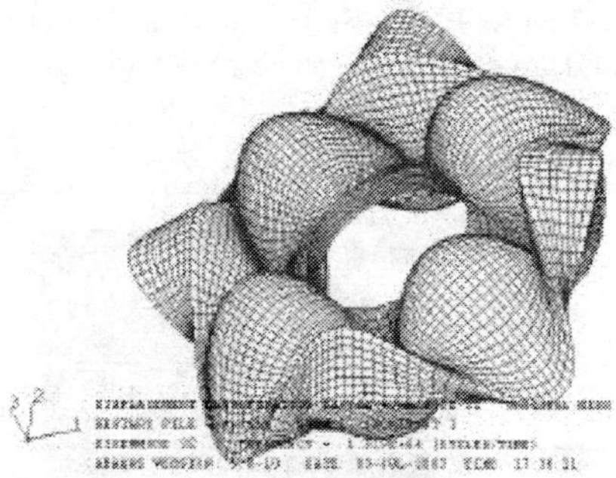

Figure 14. 32nd disc natural mode, f = 15195 Hz

The disc vibration frequency is 15 kHz. This frequency compounds more or less to the 32nd natural mode of the disc (Fig.14) which is 15195 Hz. The natural frequencies calculated with ABAQUS are not damping frequencies. It can clearly seen that the disc has just started vibrating with a 15kHz mode which is the same frequency as the contact area phenomenon of the brake pad and the disc.

CONCLUSION

The simulation of sliding (braking) of the two first bodies (pad, disk) with a constant friction coefficient and a smooth surface have shown the appearance of first body vibrations. Friction, originally used as an adjusting parameter to obtain experimentally observed behavior, has now become a true parameter like the behaviour and geometry of the first bodies, the test conditions, the pressure, speed, etc. .

In the contact, these vibrations define the local cinematic of the surface and therefore 1- the distribution of the contact pressures, thus the stresses and deformation, 2- the tribological state of the instantaneous contact zones, i.e. sliding, sticking and separation, thus the zones where it will be necessary to take into account the action of a third body [17], 3- how the instantaneous zones, which relay one another, thereby ensuring continuous macroscopic sliding. The acceleration calculation of the first body surfaces makes it possible to incorporate the acoustic elements.

The scale of this mechanical model is thus significant (some micrometers, a few microseconds) for studying at the same time what occurs in the contact (tribology) and exchanges with the outside (acoustic). This "reliable" model is also a tool which can be parameterized to separate the role of the mechanism (boundary conditions) from the role of the first bodies (Young modulus, Poisson's ratio) and from that of the third body (rheology) [18] in the behavior of a sliding contact.

ACKNOWLEDGMENTS

The 3D braking simulations were mostly done on the computer of the Centre Informatique National de l'Enseignement Supérieur.

REFERENCES

[1] R.A. Ibrahim, 1994, "Friction-induced vibration, chatter, squeal, and chaos: part II-dynamics and modelling", *ASME Journal Applied Mechanics Reviews,* **47 (7)**, pp. 227-253.

[2] G.G.Adams, 1995, "Self-Excited Oscillations of Two Elastic Half-Spaces Sliding With a Constant Coefficient of Friction", *ASME Journal Applied Mechanics Reviews,* **62**, pp. 867-872.

[3] G.G.Adams, 1998, "Steady Sliding of Two Elastic Half-Spaces Friction Reduction due to Interface Stick-Slip", *ASME Journal Applied Mechanics Reviews,* **65**, pp. 470-475.

[4] F.M.F. Simões, J.A.C. Martins, 1998, "Instability and ill-posedness in some friction problems", *International Journal of Engineering Science,* **49**, pp. 1265-1293.

[5] A. Cochard, J.R. Rice, 2000, "Fault rupture between dissimilar materials: Ill-posedness, regularization, and slip-pulse reponse", *Journal of Geophysical Research,* **105**, pp. 891-907.

[6] K. Ranjith, J.R. Rice, 2001, "Slip dynamics at the interface between dissimilar materials", *Journal of the Mechanics and Physics of Solids*, **49**, pp. 341-361.

[7] Q.S. Nguyen, 2003, "Instability and friction", *Comptes Rendus Mecanique*, **331 (1)**, pp. 99-112.

[8] A. Oueslati, Q.S. Nguyen, L. Baillet, 2003, "Stick-slip-separation waves in unilateral and frictional contact", *Comptes Rendus Mecanique,* **331 (2)**, pp. 133-140.

[9] L.Baillet, T.Sassi, 2002, "Finite element method with Lagrange multipliers for contact problems with friction", *Comptes Rendus de l'Académie des Sciences Paris, Series I*, **334**, pp. 917-922.

[10] N.J. Carpenter, R.L. Taylor, M.G. Katona, 1991, "Lagrange constraints for transient finite element surface contact", *International Journal for Numerical Methods in Engineering*, **32**, pp. 130-128.

[11] L.Baillet, Y.Berthier, S.Descartes, 2002, "Modelling of the vibrations induced by friction. Experimental visualisation and identification of the relays between the first bodies and the third body", *Proceedings of European conference on braking*, Lille, pp. 181-188.

[12] G.G.Adams, M.Nosonovsky, 2000, "Contact modelling-forces", *Tribology International,* **33**, pp. 431-442.

[13] V. Linck, L. Baillet, Y. Berthier, 2003, "Modeling the consequences of local kinematics of the first body on friction and on third body sources in wear", *Wear,* **255 (1-6)**, pp. 299-308.

[14] L. Baillet, B. Laulagnet, S. d'Errico, *Journal of Sound and Vibration*, in preparation.

[15] B.Laulagnet, 1998, "Sound radiation by a simply supported unbaffled plate", *Journal of the Acoustic Society of America,* **103 (5)**.

[16] Y. Berthier, Background on friction and wear, Lemaître *Handbook of Materials Behavior Models*, Academic Press, 2001, Chap. 8.2, pp. 676-699.

[17] I. Iordanoff, B. Sève, Y. Berthier, *ASME Journal of Tribology*, 2002, **124**, pp. 530-538.

[18] S. Descartes and Y. Berthier, "Rheology and flows of solid 3rd bodies: background and application to a $MoS_{1.6}$ coating", *Wear,* 2002, **252**, pp. 546-556

Proceedings of
2003 STLE/ASME Joint International Tribology Conference
Ponte Vedra Beach, Florida USA, October 26-29, 2003

2003-TRIB-0261

UNDERSTANDING THE EFFECTS OF FRICTION AND SURFACE ROUGHNESS ON NOISE GENERATION FROM ELASTIC COMPONENTS IN SLIDING CONTACT

Andres Soom
Department of Mechanical and Aerospace Engineering
University at Buffalo
Buffalo, NY 14260

Gary F. Dargush
Department of Civil, Structural and Environmental Engineering
University at Buffalo
Buffalo, NY 14260

Catalin I Serpe
Seagate
Bloomington, MN 55435

ABSTRACT

One type of troublesome friction-induced noise, common in brakes, clutches and mechanical seals, is high frequency chirp or squeal. The frequencies at which these noises and underlying vibrations occur typically range from around 1 kHz to more than 10 kHz. We have found that the essential physical ingredients needed to model this problem are two finite elastic systems coupled by friction and a distributed interfacial contact stiffness, transverse to the direction of sliding. The contact stiffness is associated with the roughness of the sliding surfaces and, sometimes, with the presence of wear particles within the contact. Our approach is to perform an eigenvalue analysis, using finite elements, of pairs of coupled sliding elastic rings. Due to the presence of friction, the stiffness matrix is asymmetric and mode coupling or mode splitting can occur. Typically ten per cent of the first forty or so vibratory modes are potentially unstable. Generally one or two of these appear as instabilities in the actual physical system being modeled. No stick-slip action needs to be invoked and these instabilities can occur with a single constant coefficient of friction.

Keywords: Friction, instability, roughness, vibration, noise

INTRODUCTION

High frequency noise and vibration mechanical systems with sliding contacts often involve the elastic (as opposed to rigid body) deflections of the contacting bodies. In this presentation we are concerned with such systems with the additional stipulation that the contact is of an extended conformal nature, rather than line or point. In addition, the contact compliance due to surface texture, and, possibly, wear debris, is modeled by linearized distributed stiffness per unit area. Typically, the linearized contact stiffness is inversely proportional to the surface roughness and directly proportional to the contact pressure. In this context, a smooth and flat surface would have an infinite contact stiffness. Johnson [1] provides a concise summary of the various categories of interactions that can occur between friction and overall dynamic system behavior.

Problems of this type have been studied theoretically by Nichiwaki [2] with both theory and experiments by Nakai and Yokoi [3]. The use of finite element codes has also allowed investigations of detailed system geometry and various other parameters (Liles 1987). We have found this modeling approach to agree quite well with some practical problems which will be shown in the presentation.

Coupled system analysis for dynamic instabilities

The dynamic analysis for the sliding ring on ring system is performed as a complex eigenvalue analysis using the NASTRAN finite element package. Complex eigenvalue analysis can predict the existence of unstable or undamped roots, which will be the cause of noise. The theory underlying the frequency analysis, the perturbation analysis, is presented in this section.

At any given time during sliding process the rings move with some relative angular velocity ω. At that instant, there is a contact stiffness k_n and friction coefficient μ that can be determined at each point on the nominal contact area.

Frequency instability analysis is performed as a perturbation from that configuration. The perturbation analysis is posed as

$$\left(M\lambda^2 + (C_s + C_\mu)\lambda + (K + K_\mu)\right)\Phi = 0 \tag{1}$$

a matrix eigenproblem governed by Eq. (1). M is the mass matrix, C_s represents the system damping matrix and K is the standard stiffness matrix. Meanwhile, C_μ and K_μ are the contributions from friction between the rings proportional to velocity and displacement perturbations, respectively. The presence of these frictional effects determine the eigenvalues λ and the mode shapes Φ of Eq. (1) to be, in general, complex quantities.

For frictionless problems, the eigenvalues of Eq. (1) become purely imaginary when the damping is zero and are usually rewritten as $\lambda = i\omega$, where ω is the natural frequency of the system. If the damping C_s is proportional to M and K (i.e., Rayleigh damping), the eigenvalues of the frictionless system become complex but always have a negative real part. On the other hand, in the presence of friction, the eigenvalues could have positive real parts. These modes are unstable and therefore capable of self-excitation.

The contribution of the C_μ and K_μ terms can be illustrated by using the simple example from Fig. 1. Let the elements E1 and E2 with nodes 1 and 2 represent a moving contact. Consider N to be the total normal contact force, k_n the local contact stiffness and μ the local friction coefficient. Let x_i and y_i represent the tangential and normal displacement perturbations at node i of the interface, while N_i and T_i are the perturbed normal and tangential forces, as indicated in the diagram.

The normal force N_i at node i is related to the normal displacement through the normal contact stiffness:

$$N_1 = k_n(y_1 - y_2)$$

(2)

$$N_2 = k_n(y_2 - y_1)$$

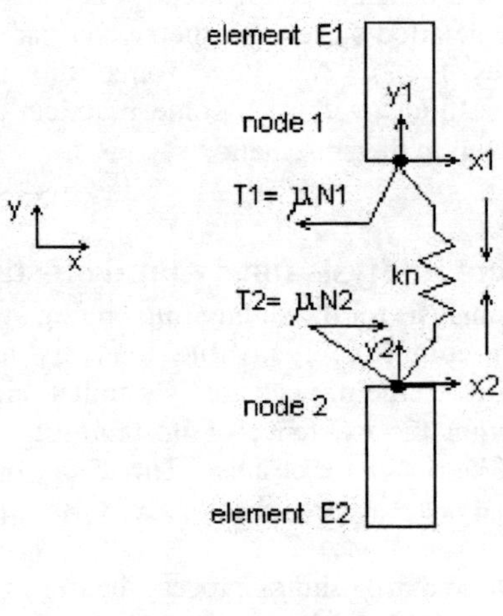

Fig.1 Sliding contact interface model

This provides a symmetric contribution to the overall stiffness matrix which can easily be included in K as defined above. On the other hand, with frictional sliding, the tangential force at node i becomes:

$$T_i = \mu N_i \tag{3}$$

Therefore, the frictional contribution to the stiffness matrix becomes asymmetric.

$$\begin{Bmatrix} N_1 \\ T_1 \\ N_2 \\ T_2 \end{Bmatrix} = \begin{bmatrix} 0 & 0 & 0 & 0 \\ \mu k_n & 0 & -\mu k_n & 0 \\ 0 & 0 & 0 & 0 \\ -\mu k_n & 0 & \mu k_n & 0 \end{bmatrix} \begin{Bmatrix} y_1 \\ x_1 \\ y_2 \\ x_2 \end{Bmatrix} \tag{4}$$

In the finite element eigenfrequency analysis, these contributions are included in the K_μ matrix of Eq. (1). As a result of the asymmetry, the eigenproblem defined in Eq. (1) can have complex eigenvalues with positive real parts. Thus, friction can lead to instabilities in the overall system.

In the finite element eigenfrequency analysis, these contributions are included in the K_μ matrix of Eq. (1). As a result of the asymmetry, the eigenproblem defined in Eq. (1) can have complex eigenvalues with positive real parts. Thus, friction can lead to (flutter) instabilities in the overall system.

In order to facilitate this analysis a FORTRAN program was developed to automatically generate the necessary MSC/NASTRAN input file based upon information on geometry and parameters such as friction coefficient, contact stiffness and damping. The contact stiffness at the interface is defined by the use of linear spring elements. Typical stiffness values (per unit area) for a one micrometer roughness amplitude at an applied normal pressure of 1 MPa is 10^{12} N/m^3. The possible negative damping contribution of friction-speed characteristics is ignored. The friction coefficient specified as a constant. The terms relating the forces at the interface elements may be input to the problem for each point directly as DMIG (direct input matrix related to grid) data.

The complex Lanczos method is used because it is robust and suitable for moderate to large models. The method takes advantage of the bandedness and sparsity of the input matrices. It is implemented as a block shifted method. Trial eigenvectors are computed by inverse iteration after a trial eigenvalue is assumed. These vectors are used to reduce the problem to tridiagonal form using the Lanczos reduction method. The roots are computed and the eigenvectors are calculated by inverse iteration. The eigenvectors are then back-transformed to the original basis. The trial eigenvalues are referred to as shift points. The method is called block method because it extracts several eigenvectors within a frequency block close to the trial eigenvalue.

Results and Conclusions

Relative stability is established by the eigenvalue analysis. We show that as the contact stiffness is increases, i.e., as the surfaces become smoother, instabilities increase. For very high contact stiffness, essentially smooth surfaces, the majority of modes are unstable. At low contact stiffness, i.e., rougher surfaces or surfaces with wear particles, very few, if any, unstable modes are present. Similarly, an increase in friction, leads to greater instability (larger values of the negative modal damping) of the unstable modes, but within a range of normal friction values, the number of unstable modes does not change significantly. At zero friction coefficient the instabilities disappear. It is important use reasonable contact stiffness values to obtain good agreement with actual systems. We conclude that, as friction coefficients and surface roughness changes with time or operating conditions these types of instabilities can come and go in sporadic manner. Many if not all such sliding systems are marginally stable at best. More extensive results can be found in Soom et al [5].

REFERENCES

[1] Johnson, K. L., 2001, "Dynamic Friction" *in Tribology Research: from Model Experiment of Industrial Problem,"* (27th Leeds-Lyon symposium), Elsevier, pp. 37-45.

[2] Nishiwaki, N., 1993, "Generalized Theory of Brake Noise," *Proc. Instn. Mech. Engrs.*, Vol. 207, pp. 195-202.

[3] Nakai, M., and Yokoi, M., "Band Brake Squeal," *Journal of Vibration and Acoustics*, Vol. 118, pp. 190-197.

[4] Liles, G.D., 1989 "Analysis of Disc Brake Squeal Using Finite Element Methods," S.A.E. Paper 89-1150.

[5] Soom, A., C. I. Serpe, and G. Dargush, 2003, 'High Frequency Noise Generation from Elastic Components in Sliding Contact: Flutter Instabilities Including the Role of Surface Roughness and Friction," *Proceeding of the 30th Leeds-Lyon Symposium on Tribology*, Elsevier (in press).

Proceedings of
2003 STLE/ASME International Joint Tribology Conference
Ponte Vedra Beach, Florida USA October 26-29, 2003

2003-TRIB-0262

FRICTION INDUCED TRANSVERSE VIBRATIONS OF AN AXIALLY ACCELERATING STRING

Giampaolo Zen and Sinan Müftü
Northeastern University
Department of Mechanical Engineering
Boston, MA 02115

ABSTRACT

The purpose of this study is to investigate the dynamic response of axially translating continua undergoing both the effect of friction and axial acceleration. The axially moving continuum is initially modeled as a string, neglecting its flexural stiffness; the response, with particular interest given to transverse vibrations and dynamic stability, is studied through numerical methods. A finite element method is employed to discretize the space domain and an implicit α–method is employed to integrate the resulting matrix equation in the time domain. Results are given through time history diagrams and stability considerations.

INTRODUCTION

Translating continua are encountered in numerous machinery such as power transmission belts, band saw blades, and processing systems such as magnetic tapes, thread lines, paper and photographic film.

Many researchers have addressed the axially moving continuum problems, but most of them have considered conservative systems without friction [1,2]. However, in many applications stationary frictional guides, which are potential sources of undesired vibrations and dynamic instability, exist. The response of a frictional, non-conservative system has been studied by Chen [3], and Cheng and Perkins [4], among others, who considered a string moving with constant axial velocity under the effect of a stationary dry friction load.

Accelerations and decelerations, as well as the undesirable fluctuations of the translational speed could have significant effects on the vibrational behavior of the translating systems. Miranker derived the governing equations of an accelerating string using energy methods [5]. Solutions of this and similar systems have been studied by Pakdemirli et al. [6], among others. To the best of our knowledge the combined effect of frictioned guides and acceleration has not been addressed.

In systems requiring frequent start-stop operations, such as high capacity data tape-recorders, simultaneous consideration of friction and acceleration is necessary. Here a generic model and a numerical method are presented to address such a system. The primary goal of this paper is to show the validity of numerical model by comparing the numerical results with that of reference [6].

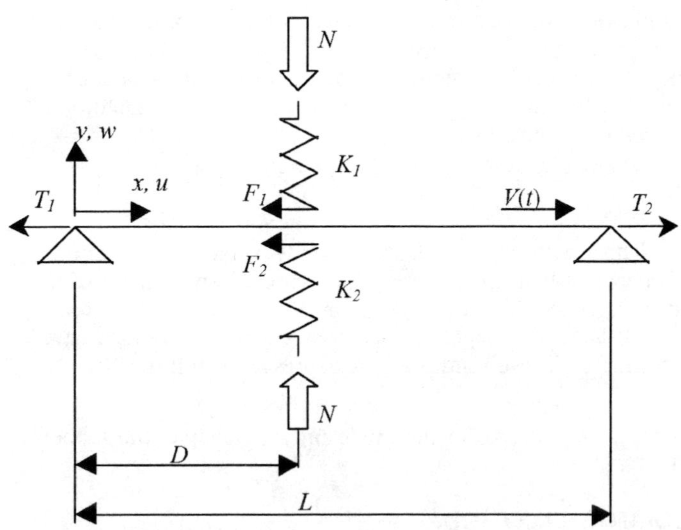

Figure 1. System modeled as an accelarating string subjected to stationary frictional loads F_1 and F_2.

EQUATION OF MOTION

A tensioned flexible string is translating with time dependent velocity $V(t)$ between two fixed supports at $x = 0$ and $x = L$. At a distance D from the left support there is a frictional guide resting on two springs. The surfaces of the guides are assumed to remain in contact with the string because of a static force N and the stiffnesses K_1 and K_2 of the guide supports. F_1 and F_2 are the top and the bottom friction forces due to the guide. The longitudinal and the transverse displacements are u and w, respectively.

In this paper the equation of motion is derived using the extended Hamilton's principle, assuming a time dependent translation velocity, a Lagrangian strain and a concentrated elastically supported frictioned guide. Equations have been derived both for the longitudinal and for the transverse vibration, ending up with a system of two coupled non-linear partial differential equations with time dependent coefficients:

$$\rho\left[u_{tt}+V_t(t)(1+u_x)+2V(t)u_{xt}+V^2u_{xx}\right]-EAu_{xx}=-T(t)_x -(F_1+F_2)\delta(x-D) \quad (1a)$$

$$\rho\left[w_{tt}+2V(t)w_{xt}+V^2(t)w_{xx}+V_t(t)w_x\right]-EA(u_xw_x)_x+\left[T(t)w_x\right]_x +kw\delta(x-D)=-q \quad (1b)$$

where the subscripts x and t indicate partial derivatives with respect to space and time respectively, w is the vertical displacement, ρ is the linear density, k the global stiffness of the guide supports, δ the Dirac function, q the eventual distributed load, H the Heaviside function, T is the tension and E is the elastic modulus of the string. In general, the friction forces F_1 and F_2 depend on the vertical displacement w as follows,

$$F_1=\mu\left(N+k_1w(D,t)\right) \text{ and } F_2=\mu\left(N-k_2w(D,t)\right) \quad (2)$$

In the time scale of the transverse motion, the longitudinal inertia term is neglected considering that the speed of propagation of longitudinal waves $(EA/\rho)^{1/2}$ greatly exceeds that of the transverse waves $(T/\rho)^{1/2}$. Thus the study reduces to the transverse displacement equation, that is now uncoupled from the longitudinal displacements. The resulting equation of motion is given below,

$$\rho\left[w_{tt}+2Vw_{xt}+\left\{V^2-T-(F_1+F_2)H(x-D)\right\}w_{xx}+V_tw_x\right] +kw\delta(x-D)=-q \quad (3)$$

This equation is solved subject to two boundary conditions. On the right-hand side support the vertical displacement of the string is constrained to zero. A harmonic motion is applied to the left-hand side support in order to excite its various natural frequencies. These conditions are expressed as follows,

$$w(0,t)=A_o\sin(\omega_o t), \quad w(L,t)=0 \quad (4)$$

where A_o and ω_o are the amplitude and frequency of the support motion.

SOLUTION METHOD

Equations (2) - (4) are solved numerically. The total energy of the system is minimized as follows,

$$\delta\left(\int_{t_1}^{t_2}(E_K-E_P)\,dt\right)=0 \quad (5)$$

where E_K and E_P are the kinetic and potential energies of the system, given by,

$$E_K=\frac{1}{2}\rho\int_0^L\left[\left(u_t+V(t)\left(1+u_x^2\right)\right)^2+\left(w_t+V(t)w_x\right)^2\right]dx \quad (6)$$

$$E_P=\frac{1}{2}E\int_0^L\left[A\left(u_x^2+\frac{1}{4}w_x^4+u_xw_x^2\right)\right]dx-\int_0^L q(t)w\,dx-\int_0^L T(t)\left(u_x+\frac{1}{2}w_x^2\right)dx \quad (7)$$

Note that in this formulation the translation speed V, the tension T and distributed load q acting on the string are allowed to depend on time. Upon energy minimization, linear finite elements are used to discretize equation (5) in space, resulting in,

$$[M]\{a\}+[G]\{v\}+[K]\{w\}=\{q\} \quad (8)$$

where $[M]$, $[G]$ and $[K]$ are the mass, gyroscopic inertia and stiffness matrices, and $\{a\}$, $\{v\}$, $\{w\}$ and $\{q\}$ are the nodal normal acceleration, normal velocity, normal displacement and load vectors, respectively. Resulting set of ordinary differential equations were integrated in the time domain using the alpha-method (Hilbert-Hughes-Taylor method [7]) with an implicit integration algorithm as described in [8].

A Matlab program has been developed to solve the final equation. In order to check its correctness, the results are compared to Cheng and Perkins (CP), where an analytical solution to a simplified problem has been presented [4]. In this case the translation velocity and the friction forces F_1 and F_2 are assumed to be constant. The following non-dimensional variables are defined by CP,

$$X=x/L,\ d=D/L,\ c^2=T_2/\rho,\ \gamma^2=T_1/T_2,\ \eta=V/c,$$
$$T^*=tc/L,\ \Omega=\omega L/c,\ K=2KL/\rho c^2,\ W=w/L \quad (9)$$

For the comparison the following values are used d = 0.2 and 0.5, η = 0.3, γ= 0.5, K = 20. The excitation frequency Ω_o varies in the range 0 - 10 and A_o = 1. The external pressure q is zero. The time step Δt and the spatial node spacing Δx were chosen such that $c\Delta t/\Delta x = 1/\sqrt{3}$ for consistent mass matrices [7]. Relatively small amount of numerical damping was chosen by setting α = -0.05. The non-dimensional nodal spacing Δx was 1/100.

RESULTS

The transient response of the string is obtained numerically. The response envelope of the system is obtained by superposition of the string shape at different time steps. The excitation frequency corresponding to the maximum response envelope is determined.

Figure 2 shows comparison of the response envelopes for the case d = 0.2 and η = 0.3 predicted by CP and the numerical methods. There is a good agreement between the analytical and numerical results. The numerical method is capable of capturing the response envelopes, especially for the second, fourth and fifth mode shapes. In all cases the location of the bigger response envelope is predicted properly, however for the first and the third mode shapes, the relative amplitudes of the response envelope to the left and to the right of the frictional support (d = 0.2) are off, as compared to CP results.

d = 0.2		*d* = 0.5	
Cheng-Perkins	**Numerical**	**Cheng-Perkins**	**Numerical**
3.37	3.24	1.98	1.94
4.85	4.74	3.95	3.90
6.80	6.70	5.23	5.01
9.54	9.25	5.99	6.01
10.44	10.30	7.93	7.94

Table 1 Comparison of non-dimensional natural frequencies obtained by Cheng and Perkins [4] and the numerical method.

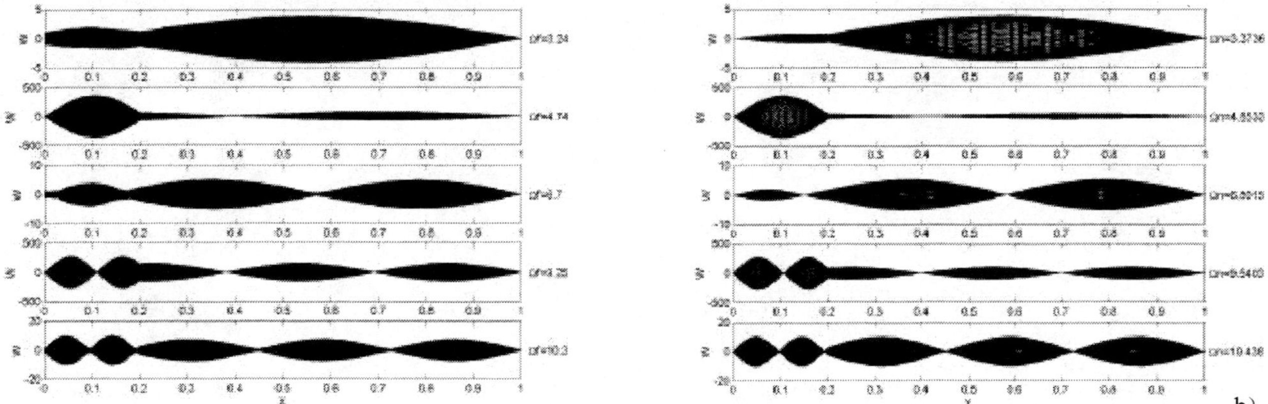

a) b)

Figure 2. The first five natural frequencies and corresponding moc Perkins' results, [4].

The maximum displacements W_{max} from Figure 2 are plotted as a function of the non-dimensional excitation frequency Ω_o in Figure 3. The first five natural frequencies of the system are reported in Table 1, which shows that the numerical method slightly underestimates the magnitudes of the natural frequencies.

SUMMARY AND CONCLUSION

A generic numerical method for analyzing the dynamics of a translating, tensioned string subjected to frictional point loads from top and the bottom is developed. The method is also capable of including the string acceleration. The effect of friction is tested with a known, relatively simpler analytical solution and good correlation is obtained. Future work will include the stability analysis of the numerical time integration method as applied to this problem, analysis of the dynamic stability of the accelerating system, as well as a system with acceleration and friction.

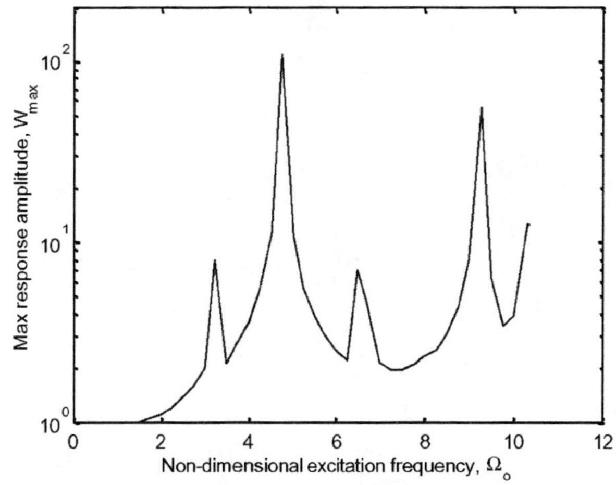

Figure 3. Numerically obtained non-dimensional frequency-amplitude diagram for $d = 0.2$ and $\gamma = 0.5$

REFERENCES

1. Ulsoy, A.G. and Mote, C.D., JR., (1982), "Vibration of Band Saw Blades," *ASME, Journal of Engineering for Industry*, **104**, pp. 1-8.
2. Wickert, J.A. and Mote, C.D., JR., (1988), "Current Research on the Vibration and Stability of Axially Moving Materials," *Shock and Vibration Digest*, **20**(5), pp. 3-13.
3. Chen, J.S., (1997) "Natural frequencies and stability of an axially – travelling string in contact with a stationary load system," *Journal of Vibrations and Acoustics*, **119**, pp.152-157.
4. Cheng, S.P., Perkins, N.C., (1991) "The vibration and stability of a friction-guided, translating string," *Journal of Sound and Vibration*, **144**, pp. 281-292.
5. Miranker, W.L. (1960) "The wave equation in a medium in motion," *IBM Journal of Research and Development*, **4**(1), pp. 36-42.
6. Pakdemirli, M., Ulsoy, A.G., Ceranoglu, A., (1994) "Transverse vibration of an axially accelerating string,"
7. Hughes, T.J.R, (1987), *The Finite Element Method*, Prentice-Hall, Englewodd Cliffs, New Jersey, U.S.A.
8. Müftü, S., (1994) *The Transient Foil Bearing Problem in Magnetic Recording*, Ph.D. thesis, University of Rochester, Rochester, NY, U.S.A..

Proceedings of
2003 STLE/ASME Joint International Tribology Conference
Ponte Vedra Beach, Florida USA, October 26-29, 2003

2003-TRIB-0263

STICK-SLIP VIBRATION BETWEEN TWO LARGE CONCENTRIC CIRCULAR DISCS IN ROTATIONAL CONTACT WITH MULTIPLE POINT LOADS

M. Baleri
Crossbow Technology, Inc,
41 Daggett Drive
San Jose, CA 95134

F. Sassani
Department of Mechanical Engineering
The University of British Columbia
Vancouver, BC, Canada, V6T 1Z4

P.L. Ko
Research Officer
National Research Council
3250 East Mall
Vancouver, BC, Canada, V6T 1W5

ABSTRACT

This paper presents a study of the stick-slip frictional phenomenon when large contact areas subjected to uneven contact loads are involved. The objective of the investigation is to gain better understanding of the phenomenon from experimental observations and to develop a mathematical representation that can be used for modeling, simulation and design purposes. A dynamic integral-model has been proposed and simulations have been carried out. The effects of various system parameters on the behavior of the system have been studied experimentally and analytically. The simulation results using the proposed integral-model are in good agreement with the experimental results. The latter also show that stick-slip vibrations can be influenced by the loading conditions.

INTRODUCTION

Stick-slip vibration, characterised by a saw-tooth type of periodic cycles of motion and arrest, occurs often between slowly moving bodies in dry or boundary lubricated sliding contact. It is believed to be caused by the friction characteristic of the contacting surfaces and it can be significantly influenced by the restraining parameters of the system.

The phenomenon has been studied and discussed in a broad range of publications spanning several decades [1-6]. Stick-slip is often attributed to the existence of a static coefficient of friction that is higher than the kinetic coefficient of friction in the low velocity region. As the sliding velocity is increased towards a critical value the stick-slip vibration would diminish and eventually disappear [7,8]. Many attempts have been made to analytically model the stick-slip phenomenon and, in the process, to study the influences of various system parameters [8-10]. More recently, with advances in control and robotic technologies, attaining precision positioning of these systems and devices has been of paramount importance. However, the unpredictable behaviour of friction at the contact surfaces and consequently the unstable movement can significantly affect the performance of the control and robotic devices. In the early '90s a series of papers dealing with friction and friction-induced vibration in control systems including a survey of models and compensation methods for the control of machines with friction was published by Armstrong-Hélouvry et al. [11,12], in which they also emphasized the importance of loading conditions. This is particularly true for large contact surfaces, such as excavators and large robotic manipulators.

The majority of the published work on stick-slip vibration deals with small contact areas where loading is considered to be evenly distributed and the whole contact area slides with the same velocity. It is only recently that Tseng and Wickert [13] reported some results on the stability analysis of large discs in rotational contact. The present paper presents an analytical approach for modeling the stick-slip phenomenon of large concentric discs subjected to multi-point loading in rotational contact and some experimental results.

MODELLING OF STICK-SLIP VIBRATION IN LARGE CONTACT AREAS

Figure 1 is a schematic diagram of the test device, which shows that the lower disc rotates continuously at a constant speed and the upper disc is loaded at a number of points on the surface with its movement restrained by an elastic system (a beam-spring). When the lower disc starts to rotate, the upper disc moves with it due to the interface friction and causes the elastic beam to deflect. As the spring force of the beam exceeds the friction force at the interface, the upper disc would either remain at that position allowing the lower disc to rotate freely or slip back thus reducing the spring force until the two forces are again equal. When the latter situation occurs, the upper disc will oscillate between the two friction forces exhibiting a stick-slip waveform.

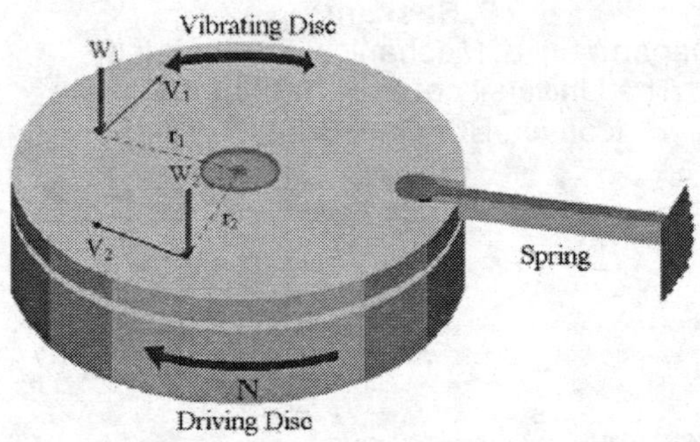

Figure 1. Stick-slip system for large contact areas.

The stick-slip phenomenon in large contact areas (especially for rotary contacts) is a complex one due to the fact that the tangential velocity at a point on the surface (viz. V_1 and V_2) varies with the radial distance (viz. r_1 and r_2). Also, the distribution of normal load over the surface would vary due to the multi-point loading pattern (viz. W_1 and W_2). Thus each point on the surface would have an individual combination of normal load and linear velocity. If these points were to act independently, some points may exhibit stick-slip while others may not, depending on their individual characteristics. But, because these points are kinematically constrained within the same disk, they collectively determine the overall behavior of the disk. Thus modeling and analysis of this system require new approach.

Contact Pressure Distribution

A finite element model in ANSYS® using non-linear contact elements was developed to obtain the distribution of the contact pressure at the interface between the plates due to the applied loads. The schematic representation of the plates used in ANSYS model is similar to that shown in Figure 1. It consists of two concentric plates of the same diameter positioned one on top of the other. The four loads can be applied individually at a specified radial distance from the centre and spaced 90°, 180° or 270° from one another. Deadweights are applied through a pivoted bracket to a roller bearing at the load point.

The plate segments are modeled by placing the top plate over the bottom one and applying contact elements at the interface. As both plates are elastic in nature, the contact problem is considered to be a flexible-to-flexible surface contact. The applied load is distributed over a 5 mm line of contact at a specified radial distance. The contact pressure distributions for the radial and tangential directions are shown in Figure 2.

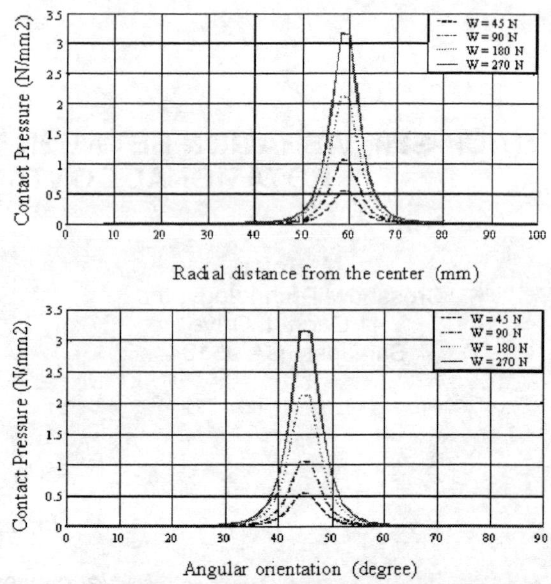

Figure 2. Radial and circumferencial contact pressure distributions (load at 59mm and 45^0).

It can be seen that the maximum contact pressure is at the load application point (at 59 mm and 45^0). The flattened peak is due to the "5 mm line of contact" from roller bearing arrangement. The contact pressure distribution is approximately normal around the load application point and decreases rapidly away from this point where the applied load would have negligible contribution. The mass of the plate contributes a uniform contact pressure covering the entire disc surface. Hence the contact pressure due to the combined point-load and the distributed weight of the disc would vary from a maximum value at the point of load application to a minimum value farther away from this point.

Based on the results of the finite element model, an empirical expression, Eq. (1), for the contact pressure distribution across the entire interface of the plate was developed. The contact pressure varies as a function of the radial and angular distances. The maximum value of the contact pressure was calculated in terms of the applied load and the weight of the plate. Figure 3 shows the pressure distribution over one quadrant of the plate. They are in good agreement with the finite-element results.

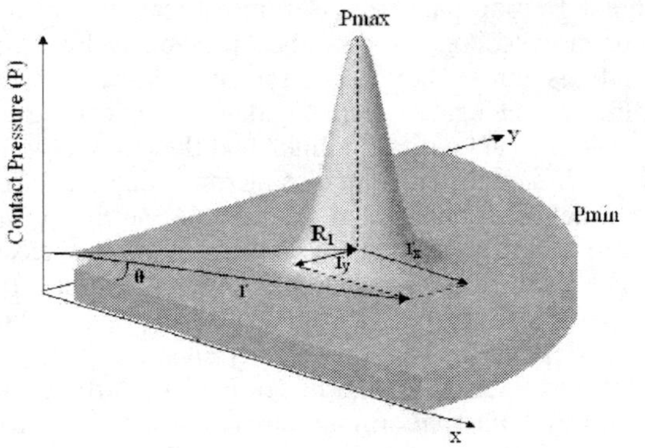

Figure 3. Model for the contact pressure distribution at the interface.

$$P(r,\theta) = P_{\min} + (P_{\max} - P_{\min})\exp\left[-\left(\left(\frac{r_x}{\sigma}\right)^2 + \left(\frac{r_y}{\sigma}\right)^2\right)\right] \tag{1}$$

where,

P_{min} minimum contact pressure due to self-weight of the plate
= Self-weight of the plate/Surface area of the plate;

P_{max} maximum contact pressure due to the applied load
= Factor X (Applied load + Self-weight of the plate)/Surface area of the plate;

r_x, r_y x & y co-ordinate distances from the load application point respectively;

σ variance of the normal distribution.

The values for "Load factor" and "Variance" of the normal distribution were computed using non-linear curve fitting techniques to make the total normal load resulting from this contact pressure distribution to be equal to the sum of the applied load and the weight of the plate. The load factor used in calculating P_{max} was found to be 365, and the variance was computed to be equal to 5.33. The variation of these parameters with respect to range of applied loads used in this work remained within 0.25 %.

The extended form of Eq. (1) that incorporates multiple point loads, n, can be expressed as,

$$P_n(r,\theta) = P_{\min} + \sum_i \left(\frac{P_{\max}}{n} - P_{\min}\right)\exp\left[-\left(\left(\frac{r_{x,i}}{\sigma}\right)^2 + \left(\frac{r_{y,i}}{\sigma}\right)^2\right)\right] \tag{2}$$

with the coordinate system shown as in Figure 3 and the loads applied along the angular bisector of each quadrant:

$$r_{x,i} = r\cos\theta - R_l \cos\left(\frac{i\pi}{2} - \frac{\pi}{4}\right)$$

$$r_{y,i} = r\sin\theta - R_l \sin\left(\frac{i\pi}{2} - \frac{\pi}{4}\right)$$

where i represents the quadrant over which the loads are applied (i.e. 1, 2, 3 and/or 4).

Modeling and Simulation of Interface Friction

A dynamic friction model for large contact areas was developed following the LuGre dynamic model described by Canudas de Wit et al. [14]. The LuGre dynamic model is simple yet it takes into consideration most of the frictional phenomenon including the static friction, viscous friction, Coulomb friction and the negative friction-velocity relationship. The schematic diagram in Figure 4 shows an elemental area on the friction plate described earlier. The plate is restrained by a spring and is subjected to stick-slip oscillation.

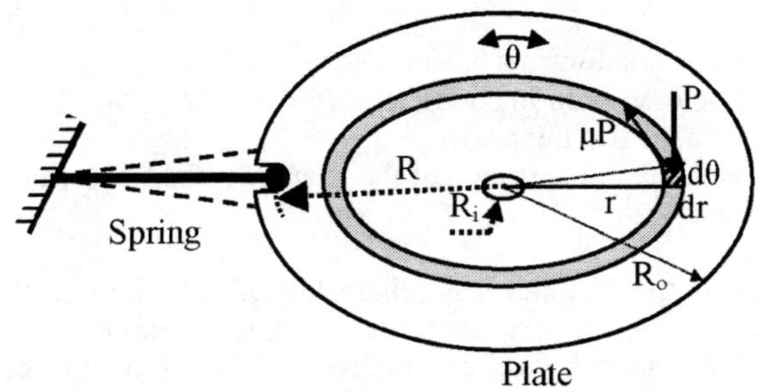

Figure 4. Schematic representation of the plate subjected to stick-slip motion.

The equation of motion for the oscillating plate can be written as

$$J\ddot{\theta}(t) + c\dot{\theta}(t) + kR^2\theta(t) = T_f(\omega_r) \tag{3}$$

where,

J mass moment of inertia of the disc;

c damping coefficient other than inter-surface friction;

k stiffness of the constraining spring;

R radial distance between the centre and the tip of spring;

ω_0 angular velocity of the driving disc;

ω_r relative angular velocity = ω_0 - $\dot{\theta}(t)$.

The term on the right-hand side of Eq. (3) represents the total frictional torque. It is assumed that contact occurs over the entire interface as revealed from the ANSYS results. Therefore, the total friction torque can be obtained by integrating an elemental area of radial length dr and angular width $d\theta$ at a distance r from the centre over the whole area of the disc. Equations (4) to (7), which are based on the LuGre model for localized contacts and proposed by Canudas de Wit et al. [14], are used to calculate the total friction torque.

$$T_f(\omega_r) = \sigma_0 z + \sigma_1 \frac{dz}{dt} + T_v \tag{4}$$

$$\frac{dz}{dt} = \omega_r - \frac{\sigma_0}{g} z \, |\omega_r| \tag{5}$$

$$g = \int_0^{2\pi} \int_{R_i}^{R_o} \left[\mu_c + (\mu_0 - \mu_c) \exp\left\{ -\left(\frac{v_r}{v_s} \right)^2 \right\} \right] P_n(r,\theta) r^2 \, dr \, d\theta \tag{6}$$

$$T_v = \int_0^{2\pi} \int_{R_i}^{R_o} \mu_v P_n(r,\theta) r^2 dr \, d\theta \tag{7}$$

Equation (5) represents the dynamics of the surface contact, z, which describes the average relative deflection of bristle-like surface asperities in contact during the stick phase [14]. The function g describes the part of the steady state characteristics of the model for constant velocity motions.

μ_0 Static friction coefficient;
μ_c Coulomb friction coefficient;
μ_v Viscous friction coefficient;
v_s Velocity at the trough of the friction-velocity curve.

Thus, the friction model is characterized by four static parameters: μ_0, μ_c, μ_v and v_s, and two dynamic parameters: σ_0 and σ_1. The parameter σ_0 can be thought of as a stiffness coefficient of some microscopic deformation, z, prior to gross sliding, while σ_1 is a damping coefficient associated with dz/dt.

Parameters	Description	Value	Units
σ_0	Bristle stiffness [14]	10000	Nm/rad
σ_1	Damping coefficient of bristle	100	Nm sec/rad
μ_o	Friction coefficient at $\omega_r = 0$	0.1919	
μ_c	Coulomb friction coefficient	0.1019	
μ_v	Viscous friction coefficient	0.0408	
v_s	Velocity where friction is minimum	0.01	m/sec
J	Mass moment of inertia	0.01826	Kg-m^2
c	Damping coefficient	0.0	N m sec/rad
W_s	Self-weight of the disc	14.331	N
R	Radius at contact with the spring	0.121	m

Table 1. Model parameter values used in all simulations

The other friction parameters, were estimated from non-linear identification techniques described previously by Canudas de Wit and Lischinsky [15]. The system parameters were either measured or calculated. The simulation was carried out using Equations (2) - (7) with Simulink® and the parametric values of Table 1. It started with the disc at rest. The displacement of the cantilevered beam and hence the spring torque was increased linearly. When the spring torque exceeded the frictional (sometimes called breakaway) torque, the plate started to slip and the friction decreased rapidly due to the negative slope of the friction-velocity characteristic. (Note: this is sometimes referred to as Stribeck Effect in articles published in control and automation journals, perhaps for the reason that the friction-velocity curve resembles that of the Stribeck diagram for lubricated bearings.) In the event the spring torque would decrease and the plate would slow down. As the velocity was decreased further the friction torque increased again due to the negative slope on the low-velocity part of the friction-velocity curve. When the friction torque equaled the spring torque, the motion stopped and the disc was once again at rest.

Simulated results of parametric studies

Different combinations of driving speed, applied load, loading radius and spring stiffness were applied to the model. The simulated results were plotted in Figures (5) and (6). The results show that the amplitude of the stick-slip motion increases and the stick period decreases when either the applied load W is increased or the spring stiffness of the system k is decreased or the speed of the driving disc ω_0 is decreased. The radial distance of the load-application points R_l also has some effect on the stick-slip amplitude, particularly for low disc velocities. The results show that moving the load-application point farther away from the centre increases the amplitude of stick-slip.

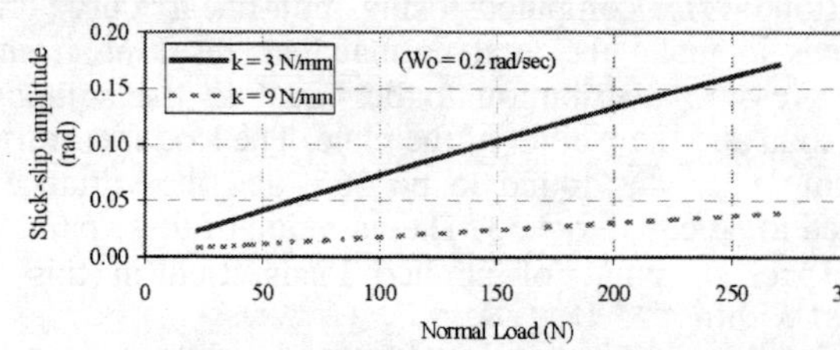

Figure 5. Simulation results for the variation of load and spring stiffness

The effects of three loading configurations, viz. single, two-point and four-point loading for the same total load, on the stick-slip vibration were studied. In this series of simulations, a given load was applied at a specified radial distance from the centre either as a single load or as two or four equally divided load points 90 degrees apart. The simulated results indicate that the stick-slip vibration remains the same for the three loading configurations.

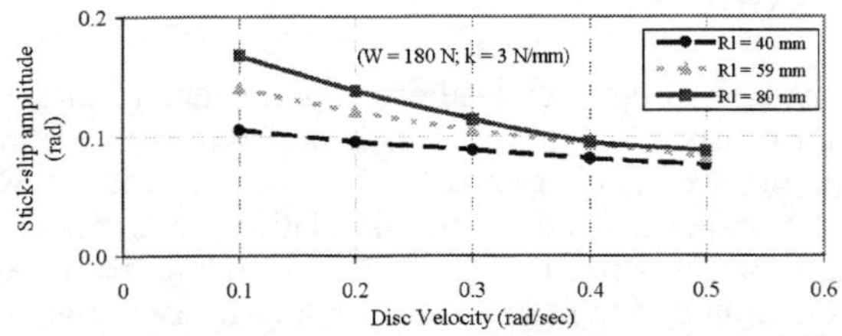

Figure 6. Simulation results for the variation of speed and loading radius

This is perhaps expected for even though the load was divided into two or four equal parts, they were positioned on the same circular track (same radial distance). The integrated frictional torque would have the same final value for the three configurations. In practice, the friction characteristic would be affected by the physical, metallurgical and surface effects that different loading configurations may incur but were not considered in the model.

One of the major concerns in studying the stick-slip phenomenon in large contact areas with multiple load points is to determine the contribution of the frictional condition of an elemental area to the overall behavior of the disc. In theory, one would expect some of these elemental areas would exhibit stick-slip mode if they were individually isolated entities. In practice, because the elements are kinematically constrained on to the same disc, the disc would behave as one component having some averaged characteristics of all the individual entities.

It has been noted earlier that the combination of contact pressure and sliding velocity at any point on the disc surface (with all other parameters being the same) could influence the probability of occurrence of stick-slip. For a given rotational speed of the disc, the velocity at any point on the disc is directly proportional to the radial distance from the center. The contact pressure decreases rapidly away from the load application point. Therefore, the frictional torque may be a function of the distance between the load point and the centre. Interestingly, they tend to have a countering effect, that is, the larger the radial distance from the centre the higher may be the localised frictional torque but also the higher is the localised sliding velocity. The former may cause an increase of the stick-slip amplitude while, at the same time, it decreases with increasing sliding velocity, Figure 6.

EXPERIMENTAL METHODOLOGY

Experimental Apparatus

Figure 7 shows a photograph of the test rig designed and built for the present study. It consists of a rotating disc (200 mm in diameter) driven by a 1/8 hp variable-speed DC motor and a slider disc of same diameter. The latter is placed on top of the driven disc and is attached to a cantilevered elastic beam, which was made from 17-4 Chromium-Nickel stainless steel (T-630) and heat-treated to H900 hardness. The discs can freely rotate about a central ball bearing. The rotational speed of the disc is controlled by a series of gears and pulleys with rubber belts, which help to isolate extraneous vibrations from the motor and gearboxes. A tachometer monitors the speed of the disc. The facing surfaces of the discs were lapped with a 1500-grit paper in order to obtain good contact between the two surfaces. Normal loads are applied by dead weights via pulleys at four load points. Four loading brackets are positioned one in each quadrant and can be moved individually along a diameter towards or away from the center thereby changing the point of the load application. An accelerometer and a LVDT are used to measure the dynamic transient motion of the upper disc. The cantilevered beam constitutes the elastic stiffness of the dynamic system and serves as a displacement and force sensor. Signal processing and data analysis are carried out in a desktop PC using LabVIEW®.

Figure 7. Laboratory experimental set-up.

Test parameters

To study the effect of normal load on the stick-slip phenomenon, the normal load using dead weights was varied from 22.5 N to 225 N in steps of 45 N while disc rotation was kept at a constant speed. The effect of load distribution on the disc was investigated using single-point, two-point and four-point loading for different load steps. Two load application radii, at 84 mm and 59 mm from the centre are chosen for the tests. In addition, five rotational speeds from 0.1 rad./sec to 0.5 rad./sec in steps of 0.1 rad/sec. and two springs with stiffness 3.0377 N/mm and 9.1864 N/mm were also tested.

Surface preparation and Test procedure

The contacting surfaces of the plates were lapped with 1500-grit paper (Grade A0814W) and cleaned with Varsol followed by isopropyl alcohol. A small amount of lubricant (Motor oil, Tellus-32) was applied to the clean surfaces to ensure that sliding would be in the boundary lubrication regime. Preliminary tests revealed that if the discs were given a run-in of 5000 revolutions a more stable interfacial condition and more consistent results could be obtained. During tests, a position sensor was placed on a selected spot of the rotating disc so that, for consistency purpose, all signals would be recorded when the disc passed the same location.

Experimental results

The results show that increasing the total load increases the stick-slip amplitude and hence the period of stick-slip, albeit not linearly. On the other hand, increasing the spring stiffness reduces the stick-slip amplitude and hence decreases the period of stick-slip. The results are shown in Figure 8.

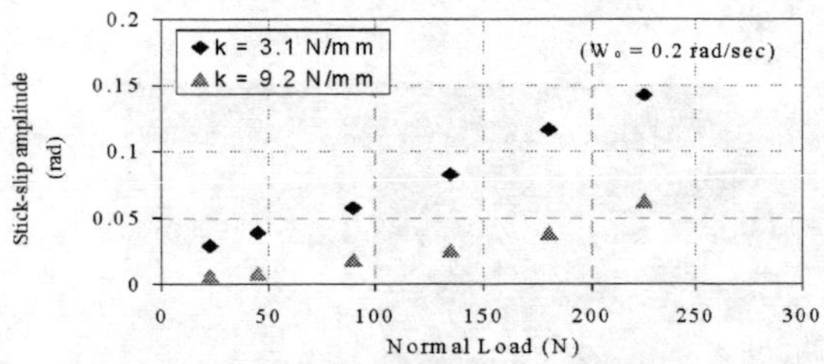

Figure 8. Experimental results for load and spring stiffness variation.

The results in Figure 9 show that stick-slip amplitude is the largest for the single-point loading followed by the two-point loading and then four-point loading of the same total load. In all three cases, the stick-slip amplitude decreases with increasing disc velocity, particularly for the case with the higher normal load. It was also observed during the tests that the critical velocity at which the stick-slip vibration disappeared was higher with the single-point loading than with the two-point and four-point loading.

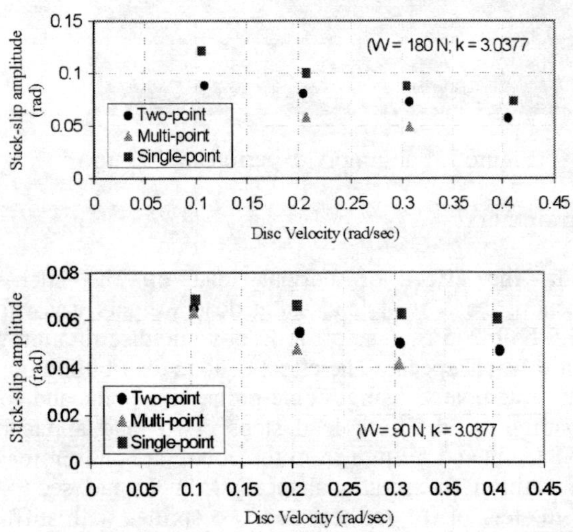

Figure 9. Experimental results for variation of speed and loading configurations.

DISCUSSION

A comparison of the simulated and experimental results, Figure 10, shows that, with the exception of loading configuration, they are in fairly good agreement. It should be noted, however, that the parameters used for the simulation were derived from an experimental point in Figure 10 by matching the simulated stick-slip amplitude to the experimentally measured amplitude for that particular set of test conditions. Nevertheless, the good agreement reached for other disc velocities and loads indicates that the model can provide good simulation of the stick-slip phenomenon.

Figure 10(a) shows the simulated and experimental results of stick-slip amplitude vs. total normal load. They show that the amplitude of stick slip increases linearly with normal load for the load range tested. The results show that the coefficient of friction is surprisingly consistent such that increasing the normal load increases the frictional torque proportionally, which in turn increases the spring beam deflection and hence the amplitude of the stick period. Although not shown in the graph, the period of the stick also increases with increasing normal load while the disc rotational speed remains the same, which would result in a decrease in the cyclic frequency of the stick-slip waveform. Reducing the spring stiffness also increases the amplitude of stick-slip and the period of stick. For the same speed and load combination, a softer spring will require a larger deflection (stick amplitude) to exert a spring force that equals or exceeds the break-away friction force.

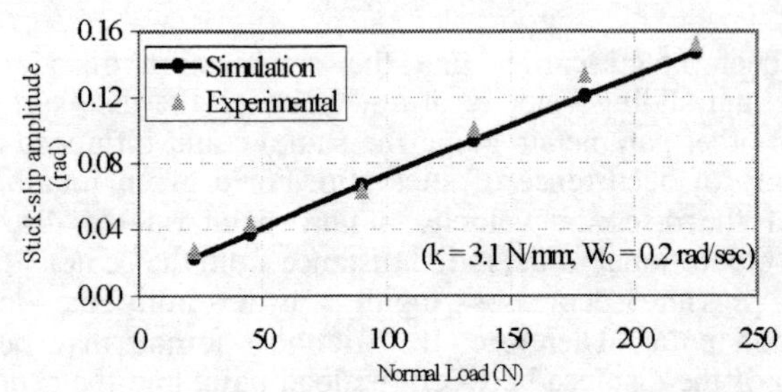

(a) Amplitude vs. normal load

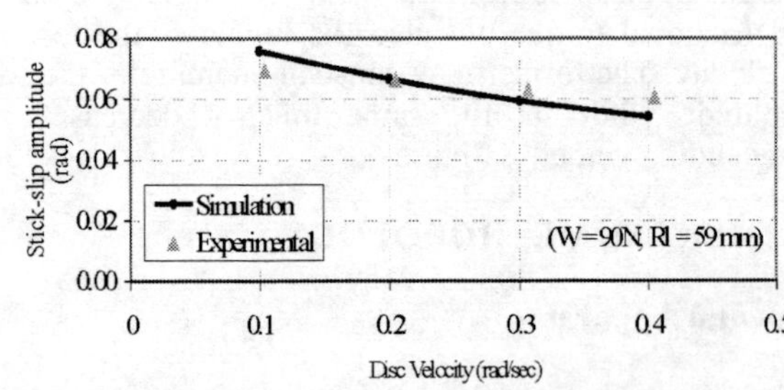

(b) Amplitude vs. rotational sliding velocity

Figure 10. Comparison of experimental and simulated results.

The simulation model is not equipped to study the effect of loading configuration. Nevertheless, the experiments reveal that loading configuration would affect the stick-slip phenomenon. If a device with large contact areas experiences stick-slip vibration, dividing and distributing the single total load to a number of load points would alleviate the stick-slip problem. The finite element model of pressure distribution has shown that even though the area of contact is large, the maximum pressure acts primary on a very small area immediately below the load point. Evidently, the actual contact area (or junction) under a large single load would be several times larger than individual contact areas under smaller multiple loads. Theoretically, the effect of the sum of these small areas (or junctions) should be the same as the single large area and hence the force needed to separate these areas would be the same. In practice, however, one or several of these smaller junctions may collapse milliseconds sooner than the others resulting in a domino effect that would reduce the period of stick and hence the amplitude of stick-slip.

A somewhat similar argument could be made about the effect of rotational speed on the stick-slip amplitude. Both simulated and experimental results, Figure 10(b), show that the stick-slip amplitude decreases with sliding speed in the low velocity range. The deflection rate of the spring beam, sometimes referred to as the rate of tangential force application [16], is related to the disc rotational velocity. As the rotational velocity is increased, the rate of tangential force application is also increased, the dynamics at the interface under the higher force rate may result in earlier gross sliding [17], hence shorter stick period and smaller amplitude of stick-slip.

The results also show smaller stick slip amplitudes when a stiffer beam was used. Again, for the same rotational disc velocity, a stiffer beam would result in a higher rate of tangential force application, that is, $k\,\omega$. Indeed, for some applications, stick-slip vibration can be prevented using a very stiff elastic system.

CONCLUSIONS

This paper presented a new formulation for modeling stick-slip friction for large contact areas as well as experimental results from a series of tests. It was observed that increasing the disc rotational speed and spring stiffness both reduced the amplitude of stick slip vibration. Increasing normal load on the other hand had the opposite effect. Specifically, with respect to large contact areas, experimental results showed that single loading is more prone to induce stick-slip vibration with larger amplitude of stick-slip vibration than the two-point or four-point loading. In general, the mathematical model provides a useful means for simulation of the stick-slip phenomenon, and is a tool that can be used in mechanical systems design analysis, and development of algorithms to control the stick-slip vibration.

REFERENCES

1. Rabinowicz, E., 1965, "Friction and wear of materials", John Wiley and Sons , NY
2. Derjaguin, B. V., Push, V. E. and Tolstoi, D. M., 1956, "A theory of stick-slip sliding of solids" , J. of Technical Physics (Moscow), 6.
3. Brockley, C. A., Cameron, R. and Potter, A. F., 1967, "Friction induced vibration", J. of Lubrication Technology, 89(2), pp. 101-108.
4. Sinclair, D. and Manville, N. J., 1955, "Frictional Vibration", Journal of Applied Mechanics, Vol. pp 207-213.
5. Gao, C., Wilsdorf, D. K. and Makel, D. D., 1993, "Fundamentals of stick-slip", Wear, Vol. 162-164, pp. 1139-1149.
6. Gao, C., Wilsdorf, D. K. and Makel, D. D., 1994, "The dynamic analysis of stick-slip motion", Wear, 173, pp. 1-12.
7. Ko, P. L., Taponat, M. C. and Rodolphe, P., 2001, "Friction-induced vibration - with and without external disturbance", Tribology International, Vol.34(1), pp.7-24.
8. Jarvis, R.B. and Mills, B., 1963, "Vibration induced by dry friction", Proceedings of the Inst. of Mech.Engs, Vol. 175(32), pp. 847-866.
9. Oden, J.T. and Martins, J.A.C., 1985, "Models and computational methods for dynamic friction phenomena", Computer methods in applied mechanics and engineering, 52, pp. 527-634.
10. Martins, J.A.C., Oden, J.T. and Simoes, F.M.F.,1990, "A study of static and kinetic friction", Int. J. Engineering Science, 28(1), pp. 29-92.
11. Armstrong-Hélouvry, B., Dupont, B. P., and Canudas, C., 1994, "A survey of Models, Analysis tools and Compensation Methods for the Control of Machines with Friction", Automatica, Vol. 30(7), pp. 1083-1138.
12. Armstrong-Hélouvry, B., 1993, "Stick-slip and control in low-speed motion", IEEE Transactions on automatic Control, Vol. 38(10), pp. 1483-1496.
13. Tseng, J. G. and Wickert, J. A., 1998, "Nonconservative Stability of a Friction Loaded Disk", Transactions of ASME., J. of Vibrations and Acoustics, Vol. 120, pp. 922-929.
14. Canudas de Wit, C., Olsson, H., Åstrom, K. J. and Lischinsky, P.,1995, "A new model for control of systems with friction". IEEE Trans. on Automatic Control, 40(3); pp. 419-425.
15. Canudas de Wit C. and Lischinsky, P., 1997, "Adaptive friction compensation with partially known dynamic friction model", Int. Journal of Adaptive Control and Signal Processing, Vol.11, pp. 65-80.
16. Johannes, V.I., Green, M.A. and Brockley, C.A., 1973, "The role of the rate of application of tangential force in determining the static friction coefficient", Wear, Vol.14, pp. 381-385.
17. Simkins, T.E., 1967, "The mutuality of static and kinetic friction", Lubrication Engineering, January, p. 26.

Proceedings of
2003 STLE/ASME Joint International Tribology Conference
Ponte Vedra Beach, Florida USA, October 26-29, 2003

2003-TRIB-0264

FRICTION AND WEAR IN THE PRESENCE OF FULLERENE CONTAINED LUBRICANTS AT BOUNDARY CONDITIONS

R. Dubrovsky
Surface Engineering Laboratory
Mechanical Engineering Department
New Jersey Institute of Technology
Newark, NJ 07102

A. Titov
Surface Engineering Laboratory
Mechanical Engineering Department
New Jersey Institute of Technology
Newark, NJ 07102

ABSTRACT

This work describes application of fullerenes as additives for liquid lubricants such as motor oils. Experiments include dynamic tracking of friction coefficient and weight measurements of samples before and after experiments. Friction coefficient measurements have been done on computer-controlled wear-friction testing machine, which has been developed at Surface Engineering Laboratory in NJIT. For the proposed series of experiments, this machine employs conforming disc-on-ring arrangement of rubbing bodies at boundary lubrication; sample couple is steel AISI 4340 and bronze SAE 40; nominal contact pressure ranges from 1.9 MPa to 3.7 MPa with sliding speed of 0.61 m/s. Tested lubricants were prepared on the basis of super duty motor oil SAE 10 by adding one weight percent of additives: pure C_{60}, mixture of C_{60} and C_{70} and graphite powder. Performance comparison of fullerene contained lubricants with pure oil and oil modified by graphite powder showed that although there is no any clear difference in friction coefficient value for tested lubricants, fullerene contained lubricants decrease wear rate of rubbing bodies.

INTRODUCTION

Additives play important role in the improvement of lubricant properties. The use of additives is inexpensive, compared to the development of synthetic lubricants, option of getting desired properties of lubricants. An additive may function in any of the following ways: protecting lubricated surfaces; improving performance; protecting the lubricant itself.

Thus, today there is a broad variety of different additives such as extreme pressure additives and rust inhibitors, viscosity index improvers and antifoaming agents, antioxidants and others. Modification of already existing and development of new additives continue constantly [1]. Now with a strong trend into micro- and nanotechnology, development of nanoscale additives attracts special attention of tribologists and lubrication engineers. Thus, fullerene C_{60} molecule has high promising perspectives for solving various tribological problems. High elasticity, strength of C_{60} crystals, weak intermolecular interactions, low surface energy, and quazispherical shape of molecules confirm this idea.

BACKGROUND

One of the possible applications of fullerenes is as additives for liquid lubricants such motor oils that give them antifrictional and wear resistant properties. There are some works [2-4], where fullerene C_{60} is used as solid film and as additive to liquid and solid lubricants. Therefore, Gruen and others [2] proposed the idea that fullerene additives in lubricants serve as precursors for developing diamond films. Ginzburg and others [3] suggested that fullerene enriched lubricants create protective films on the metal surface. Bhushan and others [4] offered opinion that fullerenes in friction process function as tiny rigid balls that are not squeezed out as the host lubricant. Today there are no consistent data of tribological behavior of fullerene-contained lubricants and detailed investigation of the effect fullerenes make is needed.

NOMENCLATURE

A_n	nominal contact area
ρ	material density
W	weight loss
J	the moment of inertia of the system
$\beta(t)$	angular velocity
f	friction coefficient
r_r	radius of the roller
β_t	angular speed of the rotor
β_o	angular speed of the roller
W_r	force of reaction acting on the beam
W_{ro}	force of reaction acting on the beam under zero applied load

d_m the distance from the rotor axis to the point of reaction

k constant of the system

P_t profile depth

R_z roughness depth

R_a mean roughness

METHODOLOGY

Wear testing methodology [5], which is developed, manufactured and implemented in Surface Engineering Laboratory at NJIT, consists of various research techniques of both quantitative and qualitative analysis of experiment output. These methods include evaluation of friction coefficient and wear rate and surface examinations before and after an experiment.

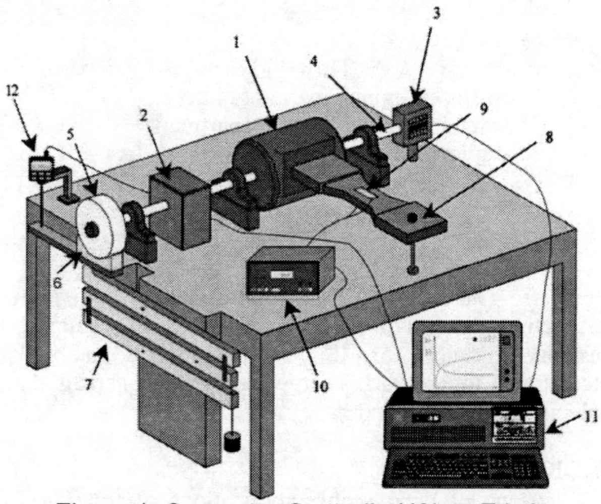

Figure 1. Computer Controlled Wear-Friction Testing Machine.

The overall design of wear-friction testing machine (WFTM) is shown in Fig. 1. It consists of electric motor (1) with suspended stator, reducer (2), cycle counter (3), driving shaft (4), rubbing couple - roller (5) and shoe (6), loading mechanism (7), strain gages beam (8), strain gages (9), strain indicator (10), computer (11) and digital linear indicator (12). For collecting and processing friction measurement data, the LabView application was developed and implemented; for wear measurement part, Visual C++ program was built and set up. The rubbing couple was selected in such a way that roller is made of AISI 4340 steel, and shoe is made of bronze SAE 40 – the couple designed in order to investigate wear and weight loss of softer material. During friction-wear test wear of soft shoe is observed and friction coefficient is recorded.

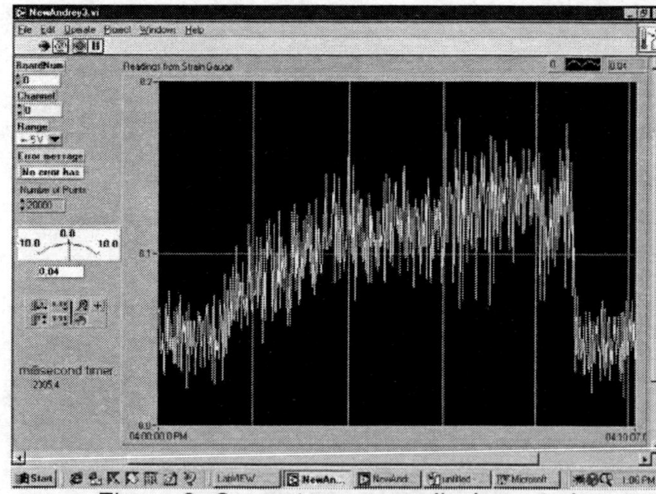

Figure 2. Computer screen displays strain gages readings.

Calculations of friction characteristics are designed on power loss principle. The stator of the suspended motor produces the power to compensate all losses in the system. Suspended electric motor placed between two bearings – this allows to attach beam by one end to the stator of electric motor and place the other end on a stopper fixed to the unmovable table. If the stator of the motor under influence of established magnetic field between rotor and stator rotates around the rotor attached to the driving shaft (4), the beam (8) will experience bending and will produce bending stresses. These stresses depend on motor's consumed power and can be measured by strain gages placed in the middle of the beam. Data from strain gages go through strain gage amplifier to the computer. Computer collects, processes and saves these data. This design makes possible to calculate motor's power changes, which are triggered by changes in friction force between shoe and roller.

Another important objective of wear-friction testing machine is that it can record online wear rate of tested material. For these purposes, wear testing machine is furnished with digital linear gage (12) (DLG), which is placed between movable shoe holder and unmovable table and can provide online recording of distance change between the center of the roller and the bed of the shoe holder. DLG has high precision of 1 micron and provide accurate readings of distance alteration recorded by the computer. This distance multiplied by the nominal contact area A_n and the material density ρ will give the weight loss W of the tested sample

$$W = l \cdot A_n \cdot \rho \quad (1)$$

For more reliable measurements of wear and weight loss, tested samples are being weighed before and after each test. Carried out experiments showed good alignment, within 5%, between online data and before-and-after test weight measurements.

For design of WFTM, an AC motor with constant speed of 1440 RPM was selected. Conversional ratio of the reducer is 6.3 and rotational roller speed of 229 RPM, which equivalent of about 0.61 m/s sliding speed. If this motor were DC motor, sliding speed could be made variable. The load mechanism consists of the lever system with transmission ratio 72 and the linear sliding-pusher, which can work in manual and automatic modes. Applied load on entrance lever from 0.5 lb to 4 lb provides 160-1280 N range on the shoe.

For acquiring data from strain gages, computer was modified by installment of Data Acquisition Card (DAS) PCI-DAS 1001. This card is multifunction analog and digital Input/Output board designed to operate in computers with PCI bus accessory slots. Because the board is fully supported by a wide variety of application software packages, LabView Virtual Instruments program was developed and installed to receive signal from Data Acquisition Card, process it and save data to a file. As shown in Fig. 2, LabView program obtains data from DAS, displays this information on the computer screen in the form of the graph and saves these data to file.

For acquiring data from digital linear gage, the gage was connected to the computer through COM1 port. Visual C++ program was developed and installed to the computer. The function of the program is to send a signal to the digital linear gage, which triggers the gage to send back its current reading, and collect the reverse signal.

Calculations of friction moment and friction coefficient are based on general form of moment balance law

$$\sum_i \vec{M}_i(t) = J \cdot \frac{d\beta(t)}{dt} \tag{2}$$

where J – the moment of inertia of the system; $\beta(t)$ – the angular velocity. Then friction coefficient f can be calculated by following expression

$$f = \frac{1}{W(1+k)} \cdot (W_r - W_{ro}) \cdot \frac{d_m \beta_t}{r_r \beta_o} \tag{3}$$

where r_r – radius of the roller, β_t and β_o – angular speed of the rotor and roller respectively, W_r – the force of reaction acting on the beam, W_{ro} – the force of reaction acting on the beam under zero applied load , d_m – the distance from the rotor axis to the point of reaction, k is a constant of the system.

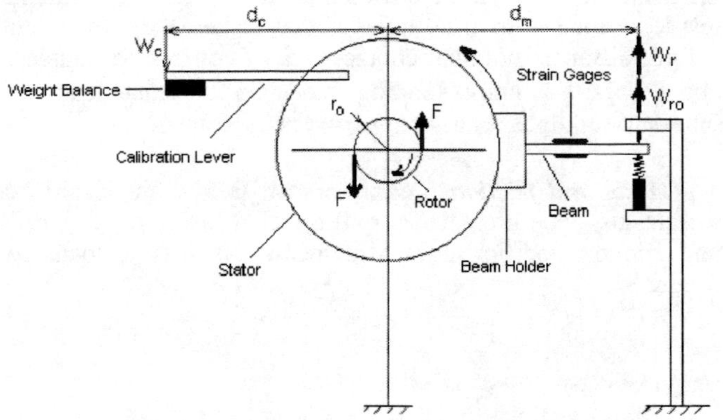

Figure 3. Scheme of the suspended motor.

After data collecting by computer, readings are processed on Microsoft Excel. Final test data include all essential characteristics of the test such as length of the test, load, normal force, strain gage readings, friction moment, friction force, power losses, friction coefficient, digital linear gage readings and weight loss. The graphs of friction coefficient and wear are superimposed on each other and dependence on time is shown. One of these graphs is shown in Fig. 4.

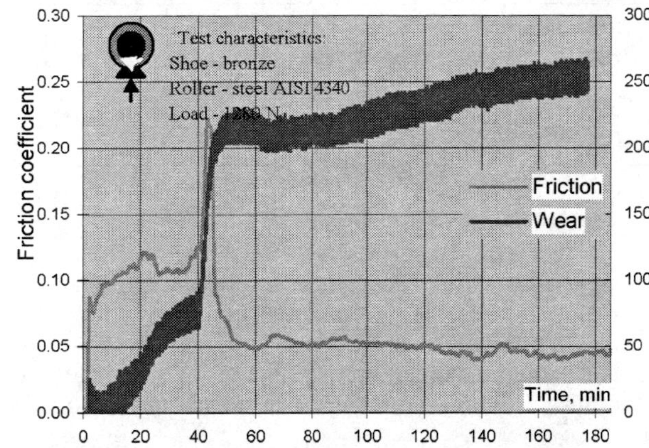

Figure 4. Dependence of friction coefficient and weight loss on time.

EXPERIMENTAL PART

Tested lubricants. Tested lubricants have been prepared on the base of heavy-duty machine oil SAE 10 by adding one weight percent of additives. This percent has been selected as the maximum solubility of fullerenes C_{60} and C_{70} in this oil. Tested lubricants are

- SAE 10 (base heavy duty motor oil)
- SAE 10 plus 1 w% of graphite powder (surface area – 80 m^2/g, bulk density – 200-230 kg/m^3)
- SAE 10 plus 1 w% of fullerene mixture C_{60} and C_{70} (3 to 1)
- SAE 10 plus 1 w% of fullerene C_{60}

The viscosity, a function of the substance's structure and its temperature, has been experimentally determined by measuring the rate of ball fall in a tube filled with a lubricant to be investigated. Test parameters are

Length of the tube, m	*2.95*10-1*
Radius of the ball, m	*5.00*10-4*
Mass of the ball, kg	*8.19*10-6*
Density, kg/m3	*890*
Ambient temperature, C	*20*

Although usually additives increase viscosity of the host lubricant, in our case, insignificant viscosity decrease is observed. Possible explanation is that after additive input and filtration by a filter paper (Whatman 1, particle retention >11 μm, medium porosity) in order to remove big insoluble particles of additives, lubricants became even more pure that the base oil. Viscosity change does not surpass an error range. Results of dynamic and kinematic viscosity measurements of the tested lubricants are shown in Table 1 (each point is based on 10 measurements).

Table 1. Dynamic and kinematic viscosity of the tested lubricants.

	Time, sec	*Velocity U, m/s*	*Dynamic viscosity, kg/m*s*	*Kinematic viscosity, m^2/s*
Oil SAE 10	3.43	0.086	0.0990	$1.11*10^{-4}$
Graphite mixture	3.35	0.088	0.0967	$1.09*10^{-4}$
C_{60}+C_{70} mixture	3.22	0.092	0.0930	$1.04*10^{-4}$
C_{60} mixture	3.24	0.091	0.0935	$1.05*10^{-4}$

Tested materials. The couple selected for this series of experiments represents common conditions of rubbing materials in mechanisms. One material, steel roller of AISI 4340, is hard and the other, bronze shoe of SAE 40, is relatively soft. In this case, only bronze shoe has experienced noticeable weight loss.

Surface analysis is applied both to roller and to shoe and assesses various parameters of examined surface texture such as profile depth P_t, roughness depth R_z, mean roughness R_a and others. For these purposes, Perthometer PGK of Mahr Federal with testing accuracy of 0.01 μm has been used. Surface texture profile and parameters for the shoe and roller are shown in Fig. 5 and 6.

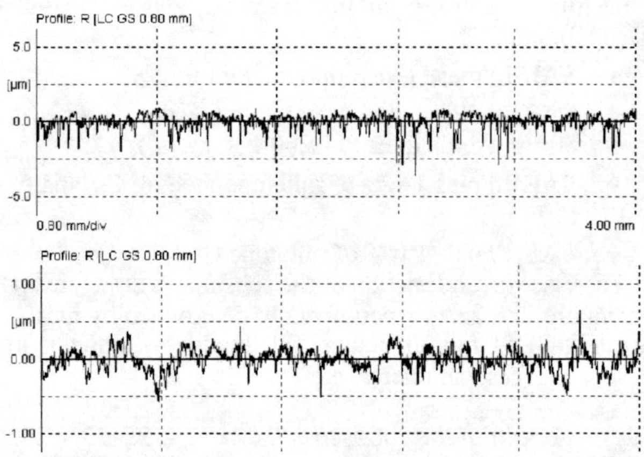

Figure 5. Surface texture profile for steel 4340 roller (a) and bronze SAE 40 shoe (b).

Parameter field for steel roller:

Ra	0.39	μm
Rz	3.54	μm
Rq	0.54	μm
Rp	0.87	μm
R3z	2.60	μm
R Sk	-1.68	
R Ku	7.42	
Rv	2.67	μm
Rv1	2.01	μm
Rv2	2.75	μm
Rv3	2.72	μm
Rv4	3.61	μm
Rv5	2.27	μm
Rv sigma	0.55	μm
Rdq	0.182	
Rk	0.98	μm
MR1	4.99	%
MR2	81.80	%
Wt	2.57	μm
Wa	0.60	μm

a)

Parameter field for bronze shoe:

Ra	0.11	μm
Rz	0.88	μm
Rq	0.14	μm
Rp	0.41	μm
R3z	0.57	μm
R Sk	-0.50	
R Ku	3.61	
Rv	0.47	μm
Rv1	0.58	μm
Rv2	0.40	μm
Rv3	0.51	μm
Rv4	0.36	μm
Rv5	0.49	μm
Rv sigma	0.08	μm
Rdq	0.035	
Rk	0.39	μm
MR1	6.76	%
MR2	87.34	%
Wt	1.26	μm
Wa	0.33	μm

b)

Figure 6. Surface texture parameters for steel 4340 roller (a) and bronze SAE 40 shoe (b). Cut-off LC - 0.25 mm, evaluating length LM – 4.00 mm.

Test conditions. Experiments have conducted at three different applied loads – 320, 480 and 640 N. For the selected nominal contact area of the shoe, nominal contact pressure ranges from 1.9 MPa to 3.8 MPa. For this series of experiments test length is set up at 60 minutes that makes 2,196 m as total sliding distance; nominal contact area of roller-shoe couple is about 1.7·10-4 mm^2; loading mode is constant load and lubricant supply is set as about 7 mg per second.

Wear and friction measurements. Below are results for weight loss measurements for all selected loads, see Figure 7, and friction coefficient measurements for 480 N load, see Figure 8.

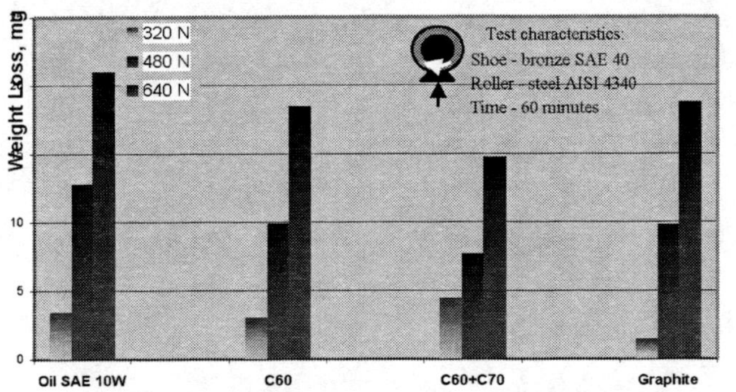

Figure 7. Weight loss of bronze shoe after 60-minute tests on different loads for different lubricants.

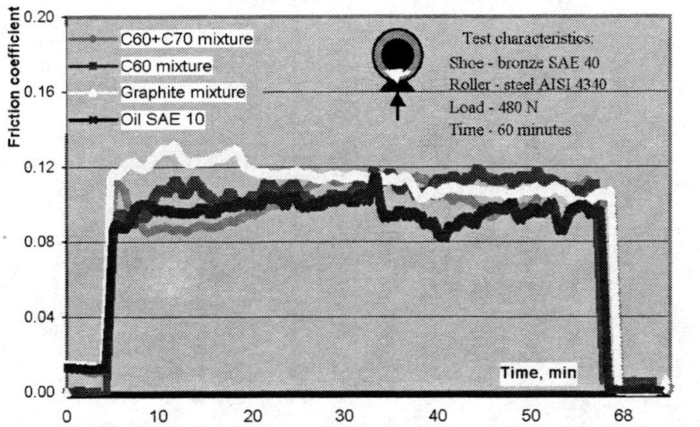

Figure 8. Friction coefficient for roller-shoe for different lubricants at 480 N load.

CONCLUSION

According to obtained experimental data fullerenes proved to affect tribological value of the host lubricant. Because of scare repeatability of friction coefficient behavior, its value clearly differs for the same conditions for the same lubricant and it is difficult to say that friction coefficient varies for the tested lubricants. Experimental data for weight loss measurements of shoe show that although data for low applied load (320 N) is inconsistent, for high loads – 480 and 640 N – there is clear dependence between weight loss of bronze shoe and the used lubricant. Weight loss for fullerene-contained lubricants is clearly lower than for pure base oil and graphite contained lubricant.

ACKNOWLEDGMENTS

Authors wish to acknowledge Foundation of New Jersey Institute of Technology for providing financial support and would like to thank all members of Surface Engineering Laboratory at NJIT and specially Dr. Valery Bezmelnitsyn for his help and inspiration.

REFERENCES

1. Sunqing, Q., Review of ultrafine particles as antiwear additives and friction modifiers in lubricating oils. Lubrication Science **1999**, 11(3), 217-226.
2. Erdemir, A., Bindal, C., Fenske, G., Zuiker, C., Krauss, A. and Gruen, D., Diamond and Related Materials **1996**, 5 (9), 923-931.
3. Ginzburg, B. M., Krasnyi V.A., Kozyrev, Yu. P., and Bulatov, V.P., Decrease of fretting wear of metals in the presence of fullerene C_{60}, Friction and Wear **1997**, 18(4), 523-526.
4. Bhushan, B., Fullerene films for solid lubrication, S T L E Tribology Transactions **1993**, 36(4), 573-580.
5. Dubrovsky, R., and Shih, I-T., Development of the wear testing methodology, Sound and Vibration Journal **1988**.
6. Kragelskii, I. V., Friction and wear: calculation methods, Pergamon Press, **1982**.
7. Dubrovsky, R., and Shih, I., Materials Research Society Symposium Proceedings, **1988**, 140, 449-454.
8. Dubrovsky R. and Titov A., Effect of Fullerene Contained Lubricants on Wear Resistance of Machine Components, Journal "Fullerenes, Nanotubes and Carbon Nanostructures", Special Issue, December **2003** (in press).

Proceedings of
2003 STLE/ASME Joint International Tribology Conference
Ponte Vedra Beach, Florida USA, October 26-29, 2003

2003-TRIB-0265

FRICTION MODELING OF A FREE-SPINNING BICYCLE WHEEL

Shuguang Huang
Mark Nagurka
Department of Mechanical and Industrial Engineering
Marquette University
Milwaukee, Wisconsin 53201-1881
Email: { huangs, nagurka }@marquette.edu

ABSTRACT

An experiment is conducted in which a free-standing bicycle wheel is given an initial angular speed and then allowed to slow down to rest. Measurements of the wheel speed, using a magnetic sensor, during the decay are compared to predictions from a model accounting for a combination of viscous and dry friction at the wheel bearing. The time history data indicate two dynamic regimes: *(i)* a higher speed phase corresponding to the first part of the motion for which a simple viscous friction model applies, and *(ii)* a slower speed phase corresponding to low speed to stop behavior for which a model involving both viscous and dry friction is proposed. A method is presented for finding the viscous and dry friction coefficients of the two phases.

1 INTRODUCTION

Anyone who has spun a bicycle wheel, with the wheel off the ground, can attest to the fact that the slowly decaying behavior has an almost mesmerizing effect. Wheel inertia keeps the wheel rotating for a long time, making the reduction in speed almost imperceptibly detectable. But, over time, the wheel speed decreases and eventually the wheel comes to rest. If the wheel is well balanced and/or if there is enough bearing friction, the wheel smoothly stops rotating. However, if the wheel is not well balanced and there is minor bearing friction, the wheel may exhibit an oscillatory tendency near the end of the motion. Prior to coming to rest, the wheel may rotate back and forth, oscillating like a pendulum, after losing most but not all of its rotational energy.

The rotational motion of a bicycle wheel about its spin axis, where the wheel has been imparted with an initial angular velocity, is an interesting and educational subject. We have introduced an experiment on bicycle wheel speed decay in course “MEEN 120: Mechanical Measurements and Instrumentation” at Marquette University. The laboratory experience requires students to study the dynamics of a free-spinning wheel using a non-contact magnetic speed sensor and a data acquisition system. The impetus for the experiment was a similar experiment introduced at the University of Washington (Seattle, WA) by Professors Joseph Garbini and William Murray, and reported in [1].

The physical system of a freely rotating bicycle wheel can be represented by a relatively simple model involving a rotational inertia and a rotational viscous damper. This linear model can be solved and predicts an exponentially decaying speed starting from the initial speed. The model can also be used to introduce the concept of time constant, which can be found experimentally. (It is the negative slope of the plotted line in a graph of the logarithm of angular speed versus time.) With the time constant, and with knowledge of, or an estimate of, the wheel rotational inertia, the value of the viscous damping coefficient can be determined.

In theory, this first-order model predicts that the wheel continues to rotate ad infinitum. A more advanced model can be posited including both viscous damping and dry friction, where the latter is introduced to ensure that the wheel stops in finite time. One challenge of this model is that there are two friction parameters that must be determined.

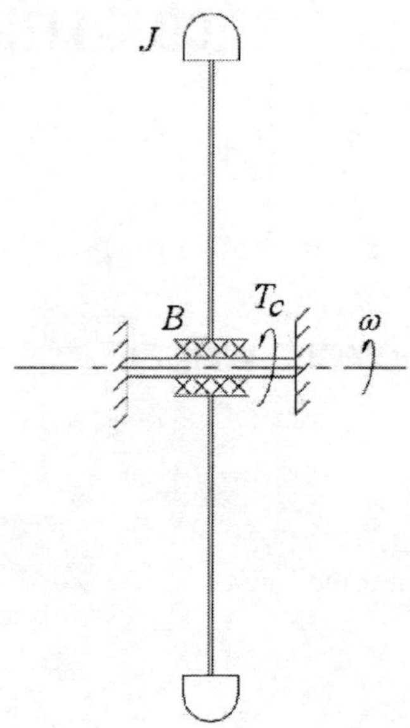

Figure 1. Lumped parameter model of bicycle wheel.

1.1 Scope

In this paper, the rotational speed decay of a free-standing bicycle wheel is investigated and an appropriate dynamic model that accounts for frictional losses is developed. The model presented includes both viscous damping and Coulombic dry friction effects. The results of simulation and experimental studies are presented.

2 DYNAMIC MODEL

Consider a bicycle wheel in a vertical plane spinning freely about its axis. It is assumed that the wheel is balanced (with its mass center at the wheel's geometric center). A dynamic model is proposed involving wheel inertia and two retarding effects of viscous and dry friction at the wheel bearing, as depicted schematically in Figure 1. The viscous friction torque is linearly related to the angular velocity, and the dry friction is constant (according to Coloumb's law).

The equation of motion is:

$$J\dot{\omega}+B\omega+T_c=0, \tag{1}$$

with initial condition

$$\omega(0)=\omega_0 \tag{2}$$

where J is the moment inertia of the wheel, B is the viscous damping coefficient and T_c is the friction torque (assumed constant). Solving gives

$$\omega(t)=\left(\omega_0+\frac{T_c}{B}\right)e^{-\frac{t}{\tau}}-\frac{T_c}{B} \tag{3}$$

where the time constant, τ, is

$$\tau=\frac{J}{B}. \tag{4}$$

The model has a physical inconsistency in that Eq. (3) is not a solution for all $t\in[0,\infty]$. The predicted terminal speed is $\omega(t\rightarrow\infty)=-T_c/B$, which is not physically consistent for a nonzero T_c.

It is possible to determine the time t^* when the wheel comes to rest. By setting $\omega(t^*)=0$ in (3),

$$\left(\omega_0+\frac{T_c}{B}\right)e^{-\frac{t^*}{\tau}}-\frac{T_c}{B}=0, \tag{5}$$

and solving yields:

$$t^*=-\tau\ln\left(\frac{T_c}{B\omega_0+T_c}\right).$$

Introducing

$$\lambda=\frac{T_c}{B}, \tag{6}$$

t^* can be written as:

$$t^*=-\tau\ln\left(\frac{\lambda}{\omega_0+\lambda}\right). \tag{7}$$

Equation (3) applies only for $t\leq t^*$. For $t\geq t^*$, $T_c=0$ and $\omega(t)\equiv 0$. Thus, the motion of the wheel can be described by:

$$\omega(t)=\begin{cases}(\omega_0+\lambda)e^{-\frac{t}{\tau}}-\lambda & \text{...if } t\leq t^* \\ 0 & \text{...if } t>t^*\end{cases} \tag{8}$$

The model is characterized by two parameters, τ and λ. τ indicates the influence of viscous friction whereas λ is the ratio of dry friction torque to viscous friction damping. From knowledge of these two parameters and a known inertia J, the viscous damping coefficient and the dry friction torque can be obtained.

For $\omega_0 \gg \lambda$, then Eq. (3) can be approximated by:

$$\omega(t) \approx \omega_0 e^{-\frac{t}{\tau}} \quad (9)$$

which is suitable for an appropriate range of $t < t^*$. However, for t less than but closer to t^*, the term λ cannot be neglected.

2.1 Two Phase Model

Based on the above, a two phase model is proposed. At higher speeds, dry friction (indicated by λ) is neglected, and the angular speed of the wheel is written in the form of Eq. (9):

$$\omega(t) = \omega_0 e^{-\frac{t}{\tau}}. \quad (10)$$

At lower speeds, dry friction must be taken into account, and the angular speed of the wheel has the form of Eq. (3):

$$\omega(t) = (\omega_0 + \lambda)e^{-\frac{t}{\tau}} - \lambda. \quad (11)$$

Let t_c is the "critical" (or "boundary") time switching between the two phases. Noting that at $t = t_c$ the angular speed must be continuous, the speed can be written:

$$\omega(t) = \begin{cases} \omega_0 e^{-\frac{t}{\tau}} & \text{...for } t \le t_c \\ \left(\omega_0 e^{-\frac{t_c}{\tau}} + \lambda\right) e^{-\frac{t-t_c}{\tau}} - \lambda & \text{...for } t_c \le t \le t^* \\ 0 & \text{...for } t > t^* \end{cases} \quad (12)$$

Another possible model accounts for the possibility that τ in Eqs. (10) and (11) may not necessarily be the same. To distinguish between the two, they are denoted by τ_1 and τ_2 corresponding to the time constant in the higher speed and lower speed phases, respectively. Thus, the angular speed can be written:

$$\omega(t) = \begin{cases} \omega_0 e^{-\frac{t}{\tau_1}} & \text{...for } t \le t_c \\ \left(\omega_0 e^{-\frac{t_c}{\tau_1}} + \lambda\right) e^{-\frac{t-t_c}{\tau_2}} - \lambda & \text{...for } t_c \le t \le t^* \\ 0 & \text{...for } t > t^* \end{cases} \quad (13)$$

3 METHOD

3.1 Optimization to Determine Parameters

In order to determine the parameters, either τ, λ and t_c for Eq. (12) or τ_1, τ_2, λ and t_c for Eq. (13), an optimization approach is suggested. The proposed objective function, F, is:

$$F = \sum_{i=1}^{n} [\omega(t_i) - \omega_i]^2 \quad (14)$$

Figure 2. Experimental setup showing bicycle wheel on front fork clamped to laboratory bench and data acquisition system.

where t_i and ω_i correspond to the time and angular speed, respectively, at the i-th data point obtained from an experiment, n is the total number of data points, and $\omega(t)$ is defined by Eq. (12) or (13).

The optimization problem is to minimize F subject to $\tau_i > 0$, $\lambda > 0$ and $0 < t_c < t^*$, where t^* is given by Eq. (7). The initial point for the optimization can be chosen arbitrarily for $\tau_{i0} > 0$, $\lambda_0 > 0$ and $t_{c0} < \tau_{i0} \ln(\frac{\lambda_0}{\omega_0 + \lambda_0})$. The first speed in the time history data is the initial speed ω_0 for both models (12) and (13). The optimization stops when a set of parameters (τ_i, λ, t_c) that minimize F is found. The optimization was performed using MATLAB and its Optimization Toolbox.

3.2 Experimental Setup

The front fork of a bicycle was clamped to a laboratory table, as shown in Figure 2. (A 27 inch wheel on a front fork assembly from a street bicycle was used in the test.) A non-contact magnetic speed sensor (CAT EYE Velo 2) was mounted to one leg of the fork. It provides a single voltage pulse as a small magnet mounted to a spoke passes, giving one pulse per revolution during the experiment. Data acquisition was accomplished using LabVIEW. The wheel was spun manually, and speed versus time data were collected.

4 RESULTS

Figure 3 compares the angular speed versus time data for an experiment with an initial angular velocity of $\omega_0 = 15.140$ rad/s to the predicted speed from the viscous model of Eq. (10). In this

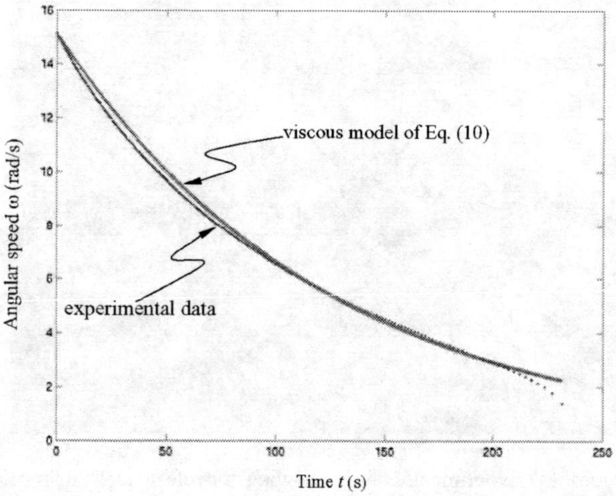

Figure 3. Angular speed ω vs. time t. Experimental data shown in blue. Viscous model of Eq. (10) shown in red.

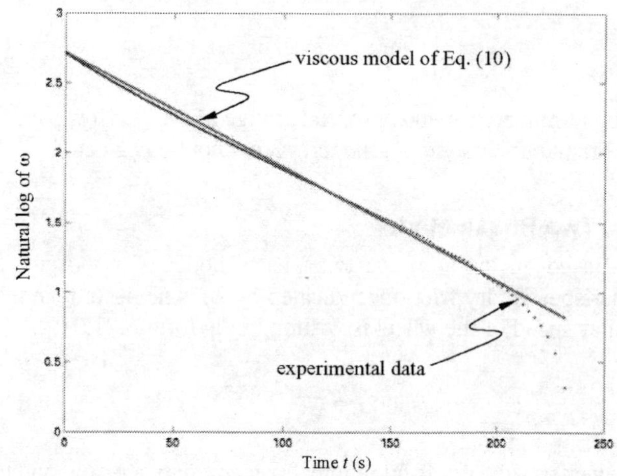

Figure 4. Natural logarithm of ω vs. time t. Experimental data shown in blue. Viscous model of Eq. (10) shown in red.

experiment, the wheel speed decayed for approximately 6 minutes. The wheel used was not perfectly balanced, and oscillated slightly prior to coming to rest. The data presented in Figure 3 is for the wheel angular speed prior to this oscillation.

Figure 4 presents the same angular speed history data in terms of the natural logarithm of the speed versus time. Also shown in the figure is a comparison to the viscous model based on the best-fit line over the time period $0 < t < 150$ s. The τ corresponding to this best-fit line was found to be 121 s, and this value was used to determine the viscous model prediction in Figure 3.

The comparison of the experimental data to the viscous model of Eq. (10) with this value of τ indicates a close match for all but the slowest speeds. The viscous model is a very reasonable approximation for the higher speeds. Figure 4 illustrates the excellent fit of the data to a straight line, with the slope of $-1/\tau$, corresponding to the log of the function $\omega(t)$ of Eq. (10). The deviation from the straight line at the slowest speeds suggests that the viscous model alone does not capture the full behavior. It is suggested below that the influence of λ cannot be neglected as the wheel comes to rest.

For the experimental data illustrated in Figures 3 and 4, the parameters resulting from the optimization were obtained as follows. For the two phase model with same viscous coefficient, i.e., the model described by Eq. (12), the parameters from the optimization are: $\tau = 117.85$ s, $\lambda = 1.515$ rad/s, and $t_c = 198.00$ s. For the two phase model with different viscous coefficients, i.e., the model described by Eq. (13), the parameters from the optimization are: $\tau_1 = 106.04$ s, $\tau_2 = 159.37$ s, $\lambda = 1.480$ rad/s, and $t_c = 42.486$ s.

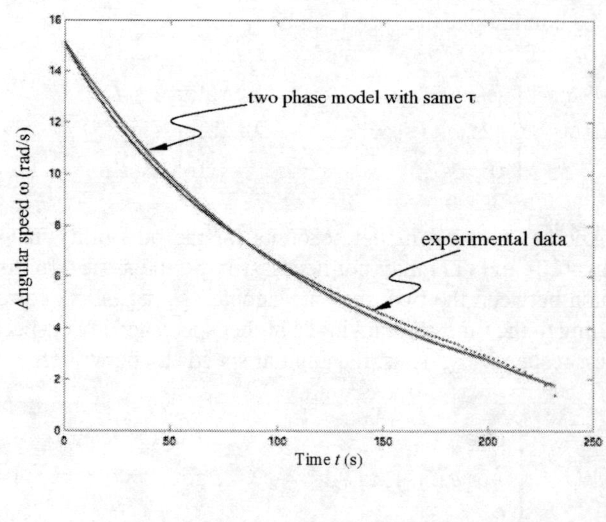

Figure 5. Angular speed time history for two phase model with same viscous coefficient. Experimental data shown in blue. Model of Eq. (12) with optimized parameters shown in red.

Figures 5 and 6 show the $\omega(t)$ curves with the above parameters for the two models, respectively, where the comparisons to the experimental data are also illustrated. The results suggest a close match in both cases, especially for the model with different viscous coefficients. It is possible to quantify how well the different models match the data by calculating the value of the objec-

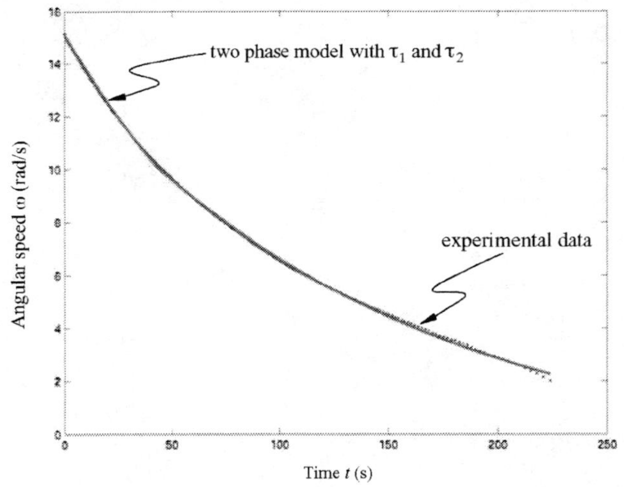

Figure 6. Angular speed time history for two phase model with different viscous coefficients. Experimental data shown in blue. Model of Eq. (13) with optimized parameters shown in red.

tive function of Eq. (14). For the one phase viscous model of Eq. (10), the value of the objective function is $F = 18.44$ (rad/s)^2. In comparison, for the two phase models the results are $F = 11.83$ (rad/s)^2 for the model of Eq. (12) and $F = 1.94$ (rad/s)^2 for the model of Eq. (13). The closest match is for the more advanced two phase model with different time constants.

5 CLOSING

This paper presents a multi-parameter model for bearing friction accounting for viscous damping and Coulomb friction that is useful in investigating the dynamics during free-spin slow down of a bicycle wheel. Measurements of the wheel speed during the decay are compared with predictions of a model accounting for a combination of viscous and dry friction at the wheel bearing. The time history data indicate two dynamic regimes, i.e., a higher speed regime corresponding to the first part of the motion for which a simple viscous friction model applies, and a slower speed regime corresponding to low speed to stop behavior for which a model involving both viscous and dry friction seems appropriate. The experiment fosters creative thinking about the modelling of a basic dynamics problem. It is well suited for undergraduate courses in "dynamics" and in "measurement and instrumentation."

ACKNOWLEDGEMENT

The authors are grateful to Mr. Norman Kirchner for setting up the bicycle wheel experiment and implementing the data acquisition system. The authors also wish to thank Messrs. Tim Dake, Casey Loughrin, and Tom Silman for their input and help with the experiment. The authors are indebted to the Kern Family Foundation for their support of the Discovery Learning Center, where the initial tests were conducted. We also gratefully acknowledge the College of Engineering Associate Dean G.E.O. Widera and the Department of Mechanical and Industrial Engineering Chair K. Kim for unwavering support of the Mechanical Measurements, Instrumentation, and Mechatronics Laboratory.

REFERENCES

[1] W.R. Murray and J.L. Garbini, "Bearing Damping: A Simple, Inexpensive Experiment in System Dynamics," Proceedings of ASME IMECE 2003, Washington, DC, November 16-21, 2003.

Proceedings of
2003 STLE/ASME Joint International Tribology Conference
Ponte Vedra Beach, Florida USA, October 26-29, 2003

2003-TRIB-0266

STICTION SUPPRESSION IN HIGH VOLUME MEMS PRODUCTS

Jack Martin
Analog Devices, Inc.
Micromachined Products Division
Cambridge, MA 02139

ABSTRACT

Analog Devices has shipped over 120 million MEMS inertial sensors (accelerometers and gyros). Most are used in automotive air bag systems so high reliability is absolutely critical. The DMD die used in the Texas Instruments' DLP image projection systems have up to 1.3 million micromirrors on each die that are designed to touch and release. The DLP product line also has a well-established record of high reliability. Clearly, both companies have solved the problem of MEMS stiction to the extent that it no longer impacts field reliability. However, stiction remains a primary cause of failure in other micromachined products that are produced on less mature processes than those employed by high volume manufacturers like ADI and TI. This paper discusses the fundamental causes of stiction and the techniques that commercial MEMS suppliers use to suppress it.

INTRODUCTION

In the early 1990's, automobile manufacturers used products like "g-switches" to detect collisions. These mechanical assemblies had a ball, a spring and an expensive electrical contact in a tube. When a collision occurred, inertial forces caused the ball to traverse the tube where it passed a detector that triggered the air bag. Unfortunately, these threshold switches could not reliably distinguish collisions from potholes so g-switches were installed at several locations in each automobile. This approach added wiring costs and reduced reliability. A better solution was required. Remember, an air bag that deploys at the wrong time ***causes*** an accident! Reliability and low cost are both paramount requirements in this application.

This dilemma was resolved when Analog Devices introduced MEMS air bag sensors. These products integrate a micromachined accelerometer and support electronics on one self-testing chip that is sealed inside a hermetic package (1-3). The resulting cost reduction allowed air bag protection to become almost universal in new vehicles. The path to high reliability was a learning experience with many surprises along the way. This paper will discuss one of the problems that had to be overcome: stiction. Since stiction is a failure mode in most classes of MEMS devices, the scope includes a discussion of the root cause as well as solution techniques that are used by MEMS suppliers.

DISCUSSION

Most MEMS products are fabricated using standard silicon IC processes. However, IC passivation processes cannot be used to seal MEMS surfaces. This is a serious concern because micromachined devices have extremely high surface-to-mass ratios – a fact that makes them very sensitive to surface effects. MEMS manufacturers had to develop designs and manufacturing processes that protect their devices from moisture and other gases that affect electrical performance, stability and reliability. They also had to solve a surface-related mechanical problem. MEMS microstructures are typically suspended on compliant springs. The elements in these devices are usually only one or two microns apart. Therefore, mechanical shock, including rough handling, can cause them to touch. This is not a problem if they separate after contact. However, if they stick, the device fails.

Why do microstructures stick when they touch? One answer is related to the clean nature of semiconductor devices. Surface atoms of clean inorganic surfaces are in a high-energy state because they cannot make normal chemical bonds to neighboring atoms in all directions. To minimize energy, they bond to whatever is available. This includes high strain bonds to nearby atoms, as well as adsorbed moisture, carbon dioxide, bacteria, dust, etc. Obviously, if a mechanical impact or an electrostatic force causes two microstructures to touch, high-energy surface atoms would attempt to bond to each other. Unless the device has adequate mechanical restoring force to pull them apart, the result will be stiction, and device failure.

Electrostatic forces also promote stiction. Avoiding electrostatic effects requires more than keeping adjacent microstructures at the same nominal potential because it is still possible for surface oxides to support charges. These charges are difficult to anticipate and measure. However, if present, they increase the likelihood of stiction.

Many techniques are used to reduce stiction. Each has advantages and challenges. Some are design-related. Others rely on materials, process or package characteristics. Suppliers must also meet another set of criteria: cost, manufacturability and reliability (4). The last item can be particularly troubling. Remember, MEMS is an enabling technology, so critical nuances of new applications may not be fully recognized until after product release.

Design

When shock or functional operation causes microstructures to touch, the support springs are expected to pull them apart. Obviously, designs with large gaps and high restoring force (relatively stiff springs) are preferred. Unfortunately, small gaps and compliant springs optimize performance characteristics (high signal-to-noise ratio, low voltage operation, low range sensor capability, etc.). Thus, trade-offs are required in order to meet both performance and reliability requirements. This invariably involves consideration of other factors such as surface modification and packaging.

In addition to the use of power and sense circuits that minimize electrostatic forces and surface charge buildup, designers also:

1. Limit contact area. For example, see reference 5.
2. Use gas-filled packages to take advantage of squeeze film damping. Note that vacuum-packaged MEMS devices are particularly susceptible to stiction because they lack gas damping to cushion microstructure motion.
3. Excite resonant behavior to enhance recovery force after contact. During the release of the TI DMD mirrors, the reset voltage is pulsed to excite a mirror resonance, and thus assist mirror release. See references 6 and 7.

Surface Modification

Many thin inorganic coatings have been proposed as stiction suppression treatments for MEMS. Examples include diamond-like carbon (DLC), inert metals (gold) and silicon nitride. Conductive films must be patterned to avoid electrical shorts – a serious challenge in 3-D microstructures. Dielectric materials must be deposited in very thin (5~50 Angstroms) layers because thicker dielectrics support static charges that cause stiction in compliant microstructures. These practical problems limit the use of inorganic surface treatments.

The primary atomic bonds in organic coatings are within the molecules. Bonds formed to the outside environment are rather weak. This leads to low energy surfaces that exhibit low stiction. However, practical deposition of organic coatings on MEMS wafers is challenging. For example, an organic thin film used in the electronics industry (parylene) is normally deposited at a thickness much greater than 1000 angstroms. The film-forming mechanism is not self-limiting so attempts to uniformly coat 3-D microstructures with 5~50 angstroms have not been successful.

Academic groups and low volume industrial applications solve wafer fab stiction by applying an organic SAM (self aligned monolayer) coating before the wafers are dried. These SAM treatments have not achieved broad commercial acceptance even though they produce low energy films that are less than 10 angstroms thick. Most are rejected for manufacturing and reliability reasons. For example, SAM materials composition is difficult to control. They also have limited storage stability. As a result, coating quality tends to be variable and particle contamination is common. Many SAM coatings also degrade in standard hermetic packaging processes or produce chlorides during the coating process. Chloride contamination is a reliability concern because even trace amounts of chloride in device metallization can cause device failure. Some devices are also susceptible to chloride ion mobility. This characteristic can lead to surface charge accumulation impacting device thresholds and functionality over time when biased.

The material in any anti-stiction process must pass through narrow passages to treat the back, top and sides of suspended microstructures with a uniform sub-50 angstrom film. If dc voltage conditions are maintained, the treatment must also act as a low defect density passivation in order to suppress corrosion and leakage. These electrical parasitics arise due to anodic oxidation of silicon if moisture is present in the package. Failure to control corrosion and leakage can cause short-term noise, long-term drift and perhaps device failure. Thus the surface treatment must be both continuous and self-limiting at a few tens of angstroms. These considerations lead to two broad conclusions:

1. Standard IC passivation processes and materials do not meet these requirements.
2. Hermetic packages with materials that do not outgas moisture over the product lifetime are essential to long-term reliability in some types of MEMS products.

Over ten years ago, Analog Devices introduced MEMS-based air bag sensors that were hermetically packaged in headers. The header package was soon replaced by lower cost cerdips and cerpacs. These ceramic packages are glass-sealed in air at about 450C. Other accelerometer suppliers followed with two chip products packaged in plastic. However, the MEMS chip inside these plastic packages is also glass-sealed at about 450C. This temperature is beyond the range of normal organic materials. Thus, there was a conflict between design/performance requirements, stiction suppression and package process temperature. Analog initially solved this problem by developing an organic anti-stiction treatment that was integrated into the package assembly process and capable of surviving the cerdip/cerpac assembly process (8, 9).

TI also adopted a package level anti-stiction process for the DMD (see reference 10 for a description of the DMD). This product has a different set of challenges. For example, the package must have an optical quality transparent cover. The solution, seam-seal, removes the high temperature constraint. This allowed the early DMD devices to be treated with a reactive fluorocarbon vapor before the package was sealed (11). Further improvements were required in order to extend the wear life and lower contamination levels so capsules are now placed inside DMD packages (12). These capsules release an organic vapor that provides stiction protection. Since the vapor persists in the package during use, it repairs wear damage. The result, when combined with spring design, use of mirror resonance and moisture control, is a high-reliability MEMS product.

The package level anti-stiction process provided satisfactory performance for Analog Devices' accelerometers. However, increasing production volumes made a wafer level anti-stiction process more attractive. Therefore, recent generations of accelerometers and gyros ("inertial sensors") have been treated in Fab before being assembled in cerdips, cerpacs, LCC and BGA packages.

Analog's wafer level anti-stiction process (13) uses standard CVD equipment and a custom-synthesized diphenyl siloxane polymer to treat 100 wafers at a time. In essence, the process infuses organic groups into and onto the surface oxide. Run-to-run thickness variation, as well as variation within a run, is only about one angstrom! The custom-synthesized siloxane was chosen because it produces antistiction performance that is both durable and well controlled. Unlike other organics, it survives 450C cerdip/cerpac furnace passes as well as the less aggressive LCC, BGA and plastic assembly processes. Its passivation qualities eliminate concerns related to bond pad corrosion. Since the coating is only about 10 angstroms thick, wire bond processes are unaffected and surface charges do not accumulate. These characteristics all contribute to the high reliability and stability that is essential to automotive safety products. Product shipments are about a million MEMS devices per week. At this scale, there was ample opportunity to identify and eliminate even low ppm anomalies before the process was released into production. Results have exceeded expectations. Anti-stiction performance and repeatability are better than the package level process, yield has increased and cost was reduced.

Package and Mount Effects

Many literature reports describe the performance of anti-stiction treatments as measured on wafers or in open packages. These measurements have limited practical value because product stiction is extraordinarily dependent on packaging. These package effects can be grouped in two categories: chemical and mechanical:

Chemical – Stiction is strongly influenced by the gas environment and by adsorbed species on the MEMS surface. This can be illustrated by using moisture as an example. Several investigators have linked stiction and wear rate to high, ***and low***, relative humidity (8, 14-19). These reports generally show that stiction/wear is significantly reduced at intermediate moisture levels (about 3000 to 10,000 ppm). This corresponds to about 15 to 40% relative humidity at room temperature (i.e., the typical lab environment). The MIL-STD-883 specification (less than 5000 ppm moisture) is an informal IC industry standard for hermetic packages. However, production products are often under 500 ppm. This difference between package humidity and room humidity is one reason why bench tests of open parts are not good predictors of stiction and wear in hermetic products.

Mechanical – In many products, full scale operation changes the microstructure element gaps by only a few percent. However, actual displacements are much larger during device testing, board level assembly and end-use operation. A MEMS product vibration model can quantify this difference. That model will consist of layers of springs, dashpots and masses. The details vary from one application to the next, but all of the models have a MEMS device mounted in the chip that is mounted in a package that is mounted on a substrate that is mounted in a module, etc. Mechanical resonances in these systems are unavoidable so unexpected shock forces that bring microstructure elements into contact in various axes are common. MEMS suppliers design their products to withstand these conditions. Unfortunately, "these conditions" is a non-specific descriptor. In essence, contact time, velocity, duration, frequency and direction are all uncontrolled variables. To deal with this situation, suppliers and large users apply math models, but also follow up with practical tests as described in reference 20.

REFERENCES

1. Core, T.A., Tsang, W.K. and Sherman, S.J., "Fabrication Technology for an Integrated Surface-Micromachined Sensor", Solid State Technology, **36**(10), 39-47 (1993).
2. Chau, K. H.-L., and Sulouff, R.E. Jr., "Technology for the High-Volume Manufacturing of Integrated Surface-Micromachined Accelerometer Products", Microelectronics J., **29**(9), 579-586, (1998).
3. Sulouff, B., "Integrated Surface Micromachined Technology", Chapter 5.2 Sensors in Automotive Technology, Vol. 4, Edited by J. Marek, Y. Suzuki, H.-P. Trah and I. Yokomori, Wiley-VCH (2002).
4. Martin, J.R., "High Volume Manufacturing and Field Stability of MEMS Products", Chapter in Nanotechnology Handbook, Edited by B. Bhushan, Springer-Verlag, Heidelberg, Germany (2004).
5. Howe, R.T., Barber, H.J. and Judy, M., "Apparatus to Minimize Stiction in Micromachined Structures", U.S. Patent 5,542,295.
6. Hornbeck, L.J. and Nelson, W.E., "Spatial Light Modulator and Method", U.S. Patent 5,096,279.
7. Douglass, M.R., "Lifetime Estimates and Unique Failure Mechanisms of the Digital Micromirror Device (DMD)", Proc. 36th Annual IEEE Int'l. Reliability Physics Symp., 9-16 (1998).
8. Martin, J.R. and Zhao, Y., "Micromachined Device Packaged to Reduce Stiction", U.S. Patent 5,694,740.
9. Martin, J.R., "Surface Characteristics of Integrated MEMS in High Volume Production", Chapter 14 in Nanotribology: Critical Assessment and Research Needs, Edited by S.M. Hsu and Z.C. Ying, Kluwer Academic Publishers (2002).
10. Mignardi, M.A., Gale, R.O., Dawson, D.J. and Smith, J.C., "The Digital Micromirror Device – A Micro-Optical Electromechanical Device for Display Applications", Chapter 4 in MEMS and MOEMS Technology and Applications, Edited by P.Rai-Choudhury, SPIE (2000).
11. Hornbeck, L.J., "Low Reset Voltage Process of DMD", U.S. Patent 5,331,454.
12. Fisher, E.C., Jascott, R. and Gale, R.O., "Method of Passivating a Micromachined Device Within a Hermetic Package", U.S. Patent 5,936,758.

13. Martin, J.R., "Process for Wafer Level Treatment to Reduce Stiction and Passivate Micromachined Surfaces and Compounds used Therefor", International Patent Application WO 01/57920 A1.

14. de Boer, M.P., Clews, P.J., Smith, B.K. and Michalske, T.A., "Adhesion of Polysilicon Microbeams in Controlled Humidity Environments", Proc. Materials Res. Symp., **518**,131-136 (1998).

15. Tanner, D.M., Walraven, J.A., Irwin, L.W., Dugger, M.T., Smith, N.F., Eaton, W.P., Miller, W.M., and Miller, S.L., "The Effect of Humidity on the Reliability of a Surface Micromachined Microengine", Proc. 37th Annual IEEE Int'l Reliability Physics Symp., 189-197 (1999).

16. Patton, S.T., Cowan, W.D. and Zabinski, J.S., "Performance and Reliability of a New MEMS Electrostatic Lateral Output Motor", Proc. 37th Annual IEEE Int'l Reliability Physics Symp., 179-188 (1999).

17. Patton, S.T., Cowan, W.D., Eapen, K.C. and Zabinski, J.S., "Effect of Surface Chemistry on the Tribological Performance of a MEMS Electrostatic Lateral Output Motor", Tribology Letters, **9**, 199-209 (2000).

18. Patton, S.T., Eapen, K.C. and Zabinski, J.S, "Effects of Adsorbed Water and Sample Aging in Air on the μN Level Adhesion Force Between Si(100) and Silicon Nitride", Tribology International, **34**(7), 481-491 (2001).

19. Patton, S.T. and Zabinski, J.S, "Failure Mechanisms of a MEMS Actuator in Very High Vacuum", Tribology International, **35**(6), 373-379 (2002).

20. Li, G.X. and Shermansky Jr., F.A., "Drop Test and Analysis of Micro-machined Structures", Sensors and Actuators, **85**, 280-286 (2000).

KEYWORDS

MEMS, Stiction, Surface treatment, Passivation, Adhesion, Package effects, High volume production.

Proceedings of
2003 STLE/ASME Joint International Tribology Conference
Ponte Vedra Beach, Florida USA, October 26-29, 2003

2003-TRIB-0267

EXTENSION OF HERTZIAN THEORY FOR BODIES WITH A SINGLE COATING

Shuangbiao Liu and Qian Wang

Department of Mechanical Engineering
Northwestern University
Evanston, IL 60208, USA
847-467-6961; liusb@northwestern.edu

ABSTRACT

The Hertzian theory is a convenient tool for analyzing counterformal bodies in mechanical contacts. However, it is limited to homogeneous materials. This paper reports the results from recent research that extends the Hertzian contact theory to layered materials. Numerical analyses are conducted to evaluate the accuracy of the formulas of the extended Hertzian theory, and the comparison with numerical solutions indicates that the formulas have sufficient accuracy.

INTRODUCTION

The classic Hertzian theory [1] reveals, in a concise form, an exact contact solution for two elastic solids in contact and has been widely applied in both academic modeling and engineering practices. From a historic point of view, the publication by Heinrich Hertz of the paper—*on the contact of elastic solids* [2] marks the beginning of the subject of contact mechanics and lays solid foundation for modern tribology.

In tribological designs, the coatings used for surface protection have a promising future. Coatings can be used for surface strengthening, structure modification, friction control, and corrosion prevention. However, the Hertzian theory generally is not applicable for analyzing the contact performance of materials with a coating layer. Numerical simulations, which require a substantial amount of code development and a certain amount of computation time, have to be used to determine contact pressure, contact area, and normal approach for coated bodies. In the numerical analysis, contact constraints have to be satisfied through an iteration scheme.

The current work extends the classic Hertzian theory to the contact of materials with one surface coating. Numerical analyses are conducted to verify the results and validate the formulas for the extended Hertzian theory.

APPROACHES

In this work, circular contact problems are investigated for materials with Young's modulus ratio of the coating to the substrate greater than one. By introducing a concept of combined Young's modulus, expressions for the maximum contact pressure, the contact radius and the normal approach (rigid body motion) in the Hertzian theory are extended to solve the contact problems for bodies covered by a single coating layer. The combined Young's modulus is expressed according to an extra part of the analytically known frequency response function [3] corresponding to the surface layer. The combined Young's modulus expression includes the Young's moduli and the Poisson's ratios of the coating and the substrate, the coating thickness and a modification parameter.

RESULTS

Substantial numerical results are obtained in order to determine the modification parameter mentioned above and validate the extended expressions for the maximum contact pressure, contact radius and approach. It is found that the absolute value of relative error between numerical results and the extended expressions is generally less than one and a half percent for a wide range of conditions, as shown in Figure 1, which presents the relative differences (in percentage) between numerical results and the solutions from extended expressions as a function of the non-dimensional coating thickness defined as the coating thickness divided by the Hertzian radius of the contact without the coating under the same contact load. The comparisons indicate that the extended expressions are accurate for engineering applications.

REFERENCES

1. Johnson, K.L., 1985, *Contact Mechanics*, Cambridge University Press.
2. Hertz, H., 1882, Über die Berührung fester elastischer Körper (on the contact of elastic solids), *J. reine und angewandte Mathematik*, **92**, 156-171.
3. Nogi, T. and Kato, T., 1997, "Influence of a Hard Surface Layer on the Limit of Elastic Contact-Part I: Analysis Using a Real Surface Model," *ASME Journal of Tribology*, **119**, pp.493 - 500.

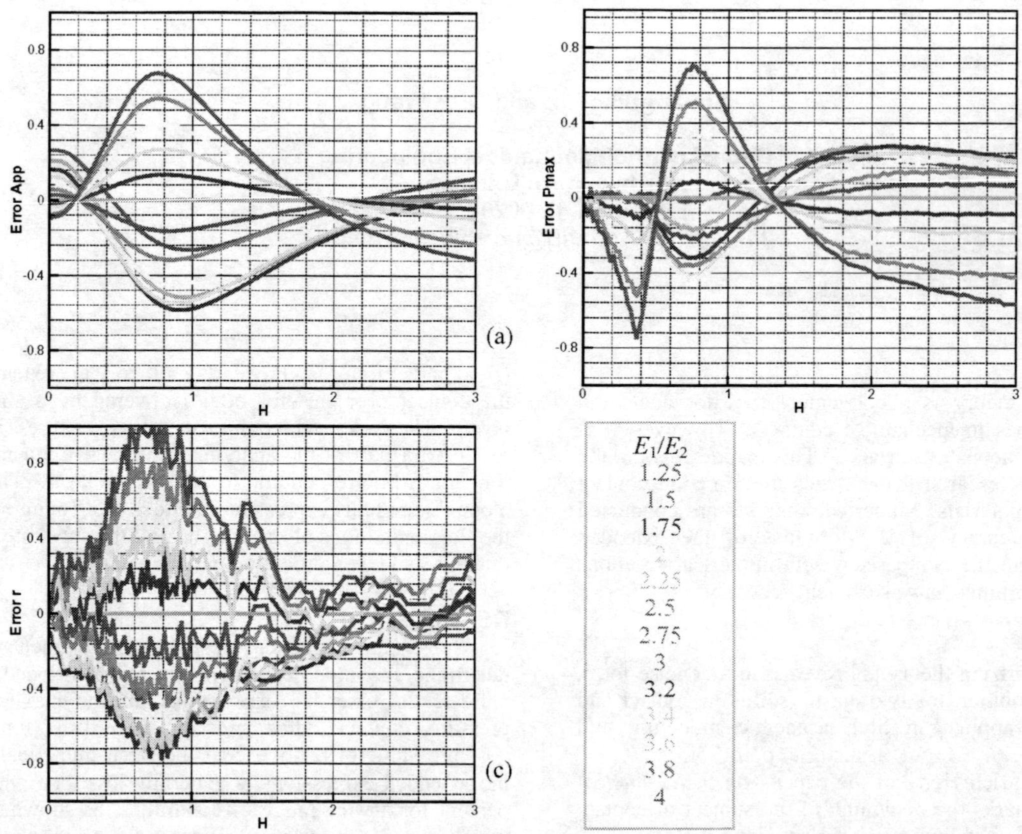

Figure 1 The relative difference (in percentage) between the numerical results and the solution from the extended expressions vs. the non-dimensional coating thickness, (a) for approach, (b) for the Hertzian pressure, and (c) for radius of contact. (E_1, E_2 - Young's moduli of the coating and the substrate, $\nu_1 = \nu_2 = 0.3$)

Proceedings of
2003 STLE/ASME Joint International Tribology Conference
Ponte Vedra Beach, Florida USA, October 26-29, 2003

2003-TRIB-0268

A FINITE ELEMENT STUDY OF ELASTO-PLASTIC HEMISPHERICAL CONTACT

Robert L. Jackson (Member, ASME)
Itzhak Green (Fellow, ASME)
George W. Woodruff School of Mechanical Engineering
Georgia Institute of Technology

ABSTRACT

This work presents a finite element study of elasto-plastic hemispherical contact. The results are normalized such that they are valid for macro contacts (e.g., rolling element bearings) and micro contacts (e.g., asperity contact). The material is modeled as elastic-perfectly plastic. The numerical results are compared to other existing models of spherical contact, including the fully plastic case (known as the Abbott and Firestone model) and the perfectly elastic case (known as the Hertz contact). At the same interference, the area of contact is shown to be larger for the elasto-plastic model than that of the elastic model. It is also shown, that at the same interference, the load carrying capacity of the elasto-plastic modeled sphere is less than that for the Hertzian solution. This work finds that the fully plastic average contact pressure, or hardness, commonly approximated to be a constant factor (about three) times the yield strength, actually varies with the deformed contact geometry, which in turn is dependant upon the material properties (e.g., yield strength). The results are fit by empirical formulations for a wide range of interferences and materials for use in other applications.

NOMENCLATURE

A = area of contact
C = critical yield stress coefficient
E = elastic modulus
H = hardness
H_G = hardness geometric limit
K = hardness factor
P = contact force
R = radius of hemispherical asperity
S_y = yield strength
a = radius of the area of contact
e_y = uniaxial yield strain, S_y/E
k = mean contact pressure factor
p_o = maximum contact pressure
z = axis of symmetry for hemisphere
ω = interference between hemisphere and surface
ν = Poisson's ratio

Subscripts

E = elastic regime
F = fit to current FEM data
c = critical value at onset of plastic deformation
o = maximum
t = transitional value from elastic to elasto-plastic behavior

Superscripts

$'$ = equivalent
$*$ = dimensionless.

INTRODUCTION

The modeling of elasto-plastic hemispheres in contact with a rigid surface is important in contact mechanics on both the macro and micro scales. This work presents a dimensionless model that is valid for both scales. In the former, e.g., rolling element bearings, load may be high and the deformations excessive. In the latter, e.g., asperity contact, a model on the micro-scale is of great interest to those investigating friction and wear. In addition, the real area of contact of such asperities will affect the heat and electrical conduction between surfaces. In either scale contact is often modeled as a hemisphere against a rigid flat.

One of the earliest models of elastic asperity contact is that of Greenwood and Williamson [1]. This model (GW) uses the solution of the contact of an elastic hemisphere and a rigid flat plane, otherwise known as the Hertzian solution, to stochastically model an entire contacting surface of asperities with a postulated Gaussian height distribution. The GW model also assumes that the asperities do not interfere with adjacent asperities. The Gaussian distribution is often approximated by an exponential distribution to allow for an analytical solution, although Green [2] has analytically solved the integrals using the complete Gaussian height distribution.

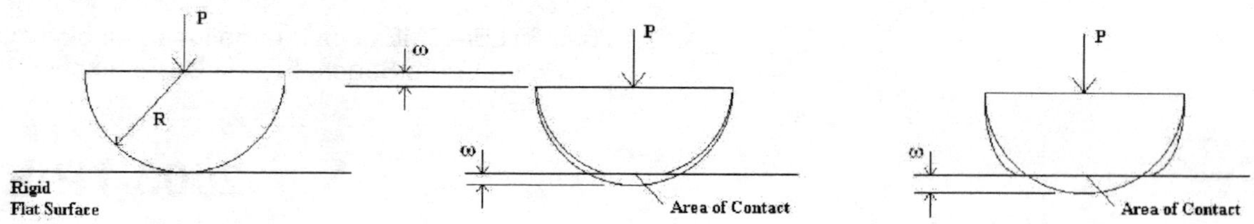

(a) (b) (c)

Figure 1: Spherical contact model before contact (a), during mostly elastic deformation (b), and during mostly plastic deformation (c).

The Hertzian solution [3] provides closed-form expressions to the deformations and stresses of two spheres in a purely elastic contact. The two spheres may have different radii and different elastic properties. However, the closed-form solutions render an equivalent case where a single elastic sphere, having an equivalent elastic modulus, E', and an equivalent radius, R, is in contact with a rigid flat (see Fig. 1, and Eqs. (1-4) that follow). The interference, ω, can be described as the distance from a point not deformed in the sphere to the rigid surface. The Hertz solution assumes that the interference is small enough such that the geometry does not change significantly. The solution also approximates the sphere surface as a parabolic curve with an equivalent radius of curvature at its tip. The resulting equations for contact radius and load from the Hertz solution are:

$$A_E = \pi R \omega \tag{1}$$

$$P_E = \frac{4}{3} E' \sqrt{R} (\omega)^{3/2} \tag{2}$$

where

$$\frac{1}{E'} = \frac{1-\nu_1^2}{E_1} + \frac{1-\nu_2^2}{E_2} \tag{3}$$

$$\frac{1}{R} = \frac{1}{R_1} + \frac{1}{R_2} \tag{4}$$

and E_1, ν_1, R_1, E_2, ν_2, R_2, are the elastic properties and radii of sphere 1 and 2, respectively.

Abbott and Firestone [4] stated that under fully plastic conditions the area of contact of an asperity pressed against a rigid flat can be approximately calculated by truncating the asperity tips as the rigid flat translates an interference, ω. For a hemisphere, this approximated fully plastic area is given by:

$$A_{AF} = 2\pi R \omega \tag{5}$$

Using Eq. (5) the contact load of the hemispherical asperity is simply the contact area multiplied by the average contact pressure, which in this case is the hardness, since the contact is assumed to be fully plastic. The approximated fully plastic contact force is thus:

$$P_{AF} = 2\pi R \omega H \tag{6}$$

From this point forward, Eqs. (5) and (6) will be referred to as the AF model.

While in the elastic regime, the stresses within the hemisphere increase with P and ω. These stresses eventually cause the material within the hemisphere to yield. The interference at the initial point of yielding is known as the critical interference, ω_c. The current work derives this critical interference analytically using the von Mises yield criterion (VM). The derivation is given in the Appendix, resulting in:

$$\omega_c = \left(\frac{\pi \cdot C \cdot S_y}{2E'} \right)^2 R \tag{7a}$$

where C is derived in the Appendix to be

$$C = 1.295 \exp(0.736\nu) \tag{7b}$$

The Poisson's ratio, ν, to be used in Eq. (7b) is that of the material which yields first. For ν=0.32, as is used in this work, Eq. (7b) results in C=1.639.

The critical load, P_c, is then calculated from the critical interference, ω_c, by substituting Eq. (7a) into Eq. (2). The resulting critical contact force at initial yielding is thus:

$$P_c = \frac{4}{3} \left(\frac{R}{E'} \right)^2 \left(\frac{C}{2} \pi \cdot S_y \right)^3 \tag{8}$$

Similarly, the critical contact area is calculated from Eq. (1) and is given by:

$$A_c = \pi^3 \left(\frac{C S_y R}{2E'} \right)^2 \tag{9}$$

These critical values predict analytically the onset of plasticity. These values are, therefore, chosen to normalize the results of all the models. The normalized parameters are:

$$\omega^* = \omega / \omega_c \tag{10a}$$

$$P^* = P / P_c \tag{10b}$$

$$A^* = A / A_c \tag{10c}$$

Normalizing the Hertzian contact area (Eq. (1)) and force (Eq. (2)), and the AF contact area (Eq. (5)) and force (Eq. (6)), by the critical values given in Eqs. (8) and (9), results in the following simplified expressions:

$$A_E^* = \omega^* \tag{11}$$

$$P_E^* = \left(\omega^* \right)^{3/2} \tag{12}$$

$$A^*_{AF} = 2\omega^* \quad (13)$$

$$P^*_{AF} = \frac{3\text{H}}{CS_y}\omega^* \quad (14)$$

Chang et al. [5] (CEB model) approximated elasto-plastic contact by modeling a plastically deformed portion of a hemisphere using volume conservation. This CEB model assumes that: (1) that the hemisphere deformation is localized to near its tip, (2) the hemisphere behaves elastically below the critical interference, ω_c, and fully plastically above that value, and (3) the volume of the plastically deformed hemisphere is conserved. Also, the critical interference used in the CEB model is given by:

$$\omega_c = \left(\frac{\pi KH}{2E'}\right)^2 R \quad (16)$$

where K is the hardness factor given by $K = 0.454 + 0.41\nu$ and the hardness is assumed $H=2.8{\cdot}S_y$. While from an engineering perspective the corresponding values given by Eq. (7) and Eq. (16) are very close, the CEB model is limited to this fixed relationship between the hardness and the yield strength. It should be noted that Eq. (7) is not limited by any such assumption. Likewise the CEB model contains a discontinuity at ω_c.

Zhao et. al. [6] devised an elasto-plastic (ZMC) model, which interpolates between the elastic and fully plastic (AF) models. The ZMC model divides the interference into three segments: (1) elastic (Hertz), (2) elasto-plastic (using a template) and (3) fully plastic (AF). A template function satisfies continuity of the function and its slope at the two transitions.

The most widely used prediction for the hardness is Tabor's [7], who calculates it to be approximately three times the yield strength for most cases. More recently, Mesarovic and Fleck [8] studied the contact of elastic-perfectly plastic spheres using the finite element analysis. They plot $P/(A{\cdot}S_y)$ as a function of a/R, i.e., the ratio of the average contact pressure to the yield strength versus the radius of the contact (or equivalently the deformation). An interesting trend then emerges. The average pressure appears to have a limiting function, now designated by H_G. The limiting function, H_G, changes from a value of approximately $3{\cdot}S_y$ (Tabor's value) to significantly lower values with an increasing ratio of a/R. Thus, the hardness is found to change in relation to the contact geometry after deformation. The current work produces precisely the same trend.

Kogut and Etsion [9] also performed a FEM analysis of the same case of an elastic-perfectly plastic sphere in contact with a rigid flat. Again in their analysis, the value of H is set to be fixed at $2.8{\cdot}S_y$. Notably, the slope of $P/(A{\cdot}S_y)$ is not zero (it still increases monotonically) at the point where full plasticity is assumed. Their work gives a very detailed analysis of the stress distribution in the contact region, and empirical expressions are provided for the contact area, the contact force, and the average contact pressure. The resulting equations have a discontinuous slope at $\omega^*=6$, and they describe the deformation only up to $\omega^*=110$, at which point full plasticity is assumed.

The current work uses the finite element method to model the case of an elastic-perfectly plastic sphere in contact with a rigid flat. The von Mises criterion defines the yielding of the material. The resulting numerical data is also fitted by continuous functions that capture deformations all the way from purely elastic to fully plastic conditions. Theses expressions, that have a relatively low statistical error, may be used in other applications whether they are on macro or micro scales. For example, a statistical model for asperity contact (such as GW [1]) can greatly benefit from such expressions.

FINITE ELEMENT MODEL

To improve upon the efficiency of computation, an axisymmetric 2-D model is used. The present study utilizes the commercial program ANSYS™, while ABAQUS™ produces the same results. Kogut and Etsion [9] also use ANSYS™. However, the mesh in the current analysis is orders of magnitude more refined as necessitated by mesh convergence.

The contact region is meshed by 100 contact elements. These are in essence very stiff springs attached between surface nodes, and activate only when penetration onset into the rigid flat is detected. This contact region varies in order to fit the expected area of contact. The contact elements thus apply forces to the nodes of the elements that are in contact.

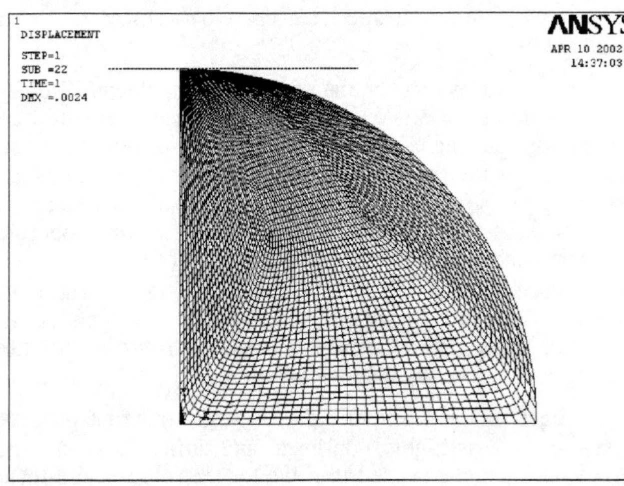

Figure 2: Finite element mesh of sphere generated by ANSYS™.

The model refines the element mesh near the region of contact to allow the hemisphere's curvature to be captured and accurately simulated during deformation. The model uses quadrilateral, four node elements to mesh the hemisphere, but the results have also been confirmed using a mesh of eight node elements. The resulting ANSYS™ mesh is presented in Fig. 2, where ABAQUS™ produces a similar mesh. The quarter-circle mesh represents the axisymmetric hemispherical body, and the straight line represents the rigid plane.

The contact force acting on the hemisphere is found from the reaction forces on the hemisphere base nodes that retain the desired interference. The radius of the contact area is determined by finding the edge of the contact, or the location of the last activated contact element.

In order to verify the validity of the model, mesh convergence must be satisfied. The mesh density was iteratively increased by a factor of two until the contact force and contact area differed by less than one percent between iterations. The resulting mesh consists of at least 11,101 elements, since the number of meshed elements will vary with the expected region of contact. The stiffness of the contact elements were also increased by an order of magnitude in successive iterations until the solution differed by no more than one percent between successive iterations.

In addition to mesh convergence, the model also compares well with the Hertz elastic solution at interferences below the critical interference. The contact force of the model differs from the Hertzian solution by no more than two percent. The contact radius differs by a maximum of 8.1%, but the average error is only 4.4%. When the contact areas are calculated from the radii, the maximum error increases to 17%. The smaller error in the contact force is attributed to overall force balance (static equilibrium) enforced by the FEM packages. However, the contact radius is obtained from a discrete mesh (which has a finite resolution). Moreover, the magnitude of the contact element stiffness also has some effect upon such radii, but not on the overall force balance. Generally, though, the differences are small enough that the FEM solution practically conforms to the Hertzian solution at interferences below critical (and even slightly above).

There are two ways to simulate the contact problem. The first applies a force to the rigid body and then computes the resulting displacement. The second applies a displacement and then computes the resulting contact force. In both methods, the displacement, stress, and strain in the elastic body can be determined, as well as the contact pressure. In this model the latter approach is used, where the base nodes of the hemisphere are displaced an interference, ω, approaching the rigid flat surface. The radial displacements of the base nodes are restricted. This method is used because the resulting solution converges more rapidly than the former.

The contact problem and the elasto-plastic material property make the analysis highly non-linear and difficult to converge. An iterative scheme is used to solve for the solution and many load steps are used to enhance solution convergence. Initially, a small interference is set of the total interference and then it is incremented after the load step converges. This continues until a converged solution is found for the desired interference.

NUMERICAL RESULTS AND DISCUSSION

The results of the described finite element hemisphere model are presented for a variety of interferences. The yield strengths cover a typical range of steel materials used in engineering [10]. While the elastic modulus and Poisson's ratio are held constant at 200 GPa and 0.32, respectively, five different material yield strengths are modeled. These are designated Mat.1 through Mat.5 corresponding to their yield strengths which are 0.210 GPa, 0.5608 GPa, 0.9115 GPa, 1.2653 GPa, and 1.619 GPa. Once the mesh is generated, computation takes from two minutes for small interferences to about one hour for large interferences on a 2.5 GHz PC.

The dimensionless contact area is normalized by the Hertz solution (Eq. (11)) and plotted as functions of ω^* in Fig. 3. The data is presented on a log scale to capture the entire range of interferences. While $\omega^* < 1.9$ the finite element model agrees well with the Hertz solution ($A^*/\omega^*=1$). This is likely because the plastic deformations are still relatively small such that the Hertz solution is not dramatically affected. As the sphere begins to plastically deform below the surface, the sphere weakens and thus does not retain its shape as well as if it were perfectly elastic throughout. Thus the area of contact eventually becomes larger in the elasto-plastic case than in the perfectly elastic case. The FEM model values for contact area continue to increase with interference even past Abbot and Firestone's fully plastic (AF) model [4] at $A^*/\omega^*=2$. Since the AF model is based on the truncation of the contacting geometries, it does not model the actual deformation of the hemisphere. It seems reasonable then that the FEM solution for contact area could continue past the AF model.

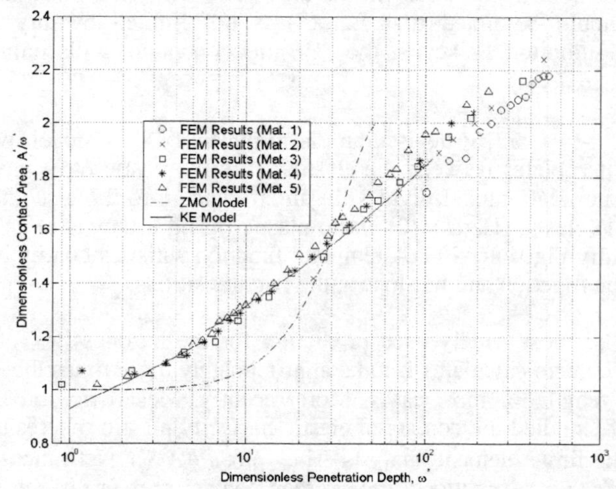

Figure 3: FEM predicted contact area.

Overall though, the FEM model generally follows the Hertz elastic solution near the critical interference and then increases past the AF model as the interference increases. Later in this work, this trend will be followed by empirical formulations fitted to the data. The FEM results also indicate a material dependence of the normalized contact area. Since the contact area is calculated by counting the number of elements in contact, and there are only a finite number of such elements, there is an inherent error in the data. The scatter in the data can be attributed mostly to this, and to the fact that the FEM is yet a discrete formulation.

For the contact area, all the models follow the same trend but differ in magnitude. The ZMC model follows the Hertz elastic solution at low and moderate interferences, but abruptly migrates to the AF model before the current model and the KE model. The KE model and the current empirical model also agree fairly well on average, except at large interferences. The KE model clearly shows a slight discontinuity at ω^*=6 and then terminates at ω^*=110. The KE model does not connect with the Hertz elastic solution at the critical interference depth. Also, the nondimensional KE model is material independent such that its contact area falls between the data of materials 1 and 5 of this work.

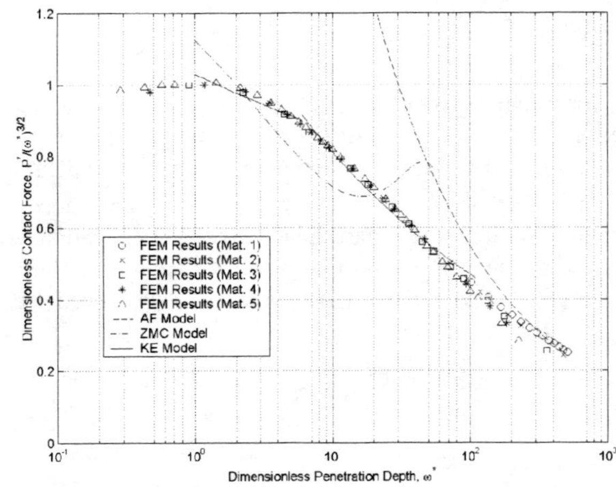

Figure 4: FEM predicted contact force.

The dimensionless contact force is normalized by the Hertz solution (Eq. (12)) and plotted as a function of ω^* in Fig. 4. This plot uses a log scale to capture the entire range of interferences. The normalized contact force $(P^*/(\omega^*)^{3/2})$ calculated from the current model follows precisely the Hertz elastic solution $(P^*/(\omega^*)^{3/2}=1)$ at small interferences. With increasing interference the current model eventually increases toward the AF model [4]. It is interesting to note that the AF model predicts higher loads at small interferences than the Hertzian solution, but eventually crosses over to become the lesser of the two. This is because the AF model assumes a constant pressure distribution, which is equal to the hardness, while the average pressure of the Hertzian solution is initially lower than the hardness. At higher interferences, the FEM data displays a material dependant behavior.

The trends of all the models are similar; however, the ZMC again crosses to the AF model prematurely. At low interferences, the KE and ZMC models predict a contact force that is greater than the elastic model. This cannot be the case, as the yield strength of the material limits the stiffness of the hemisphere. Again the KE model shows a discontinuity at ω^*=6 and then terminates at ω^*=110. Generally the KE model and the current FEM results are very similar. At about ω^*=50 the KE model crosses over the current model and continues to overestimate the contact force until ω^*=110. The KE and ZMC models also fail to capture the material dependence effects at large interferences.

The average contact pressure to yield strength ratio, $P/(A \cdot S_y)$, is calculated from the data and plotted in Fig. 5 alongside the hertz contact solution. The Firestone and Abbott [4] fully plastic (AF) model is represented by the horizontal line at $P/(A \cdot S_y)=3$. The average contact pressure should approach the hardness of the material as the contact becomes fully plastic. It is widely accepted that the hardness is approximated by $3 \cdot S_y$ [7]. It becomes evident in this plot that this is not always the case. From the data it seems that hardness is not a constant material property. The cause of this trend will be discussed later in greater detail. The work by Mesarovic and Fleck [8] also confirms this trend.

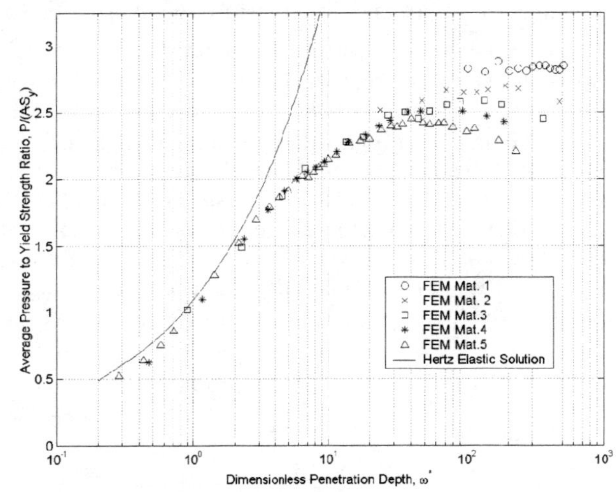

Figure 5: Average contact pressure to yield strength ratio.

EMPIRICAL FORMULATION

General empirical approximations of the FEM data are desired which can be used at any deflection and for any set of material properties. This will help designers in a variety of single contact problems, as well as it can be readily incorporated into statistical models to model rough surfaces.

As mentioned previously, the FEM solution for the area of contact continues past the AF model with increasing interference. Hence, the leading coefficient in Eq. (13) is allowed to vary when equations are fitted to the FEM data. This is reasonable, since the AF model is not an exact solution (it is based on a truncation assumption). Here a power function is used in place of this leading coefficient and fit to the numerical data. Figures 3-5 show that there are two distinct regions in the FEM data, thus a piecewise formulation is used to fit the data. At small interferences the Hertz solution is assumed and at large interferences the power function is fit to the FEM data, resulting in:

For $0 \leq \omega^* \leq \omega_t^*$

$$A_F^* = \omega^* \tag{17a}$$

For $\omega_t^* \leq \omega^*$

$$A_F^* = \omega^* \left(\frac{\omega^*}{\omega_t^*} \right)^B \tag{17b}$$

where

$$B = 0.14\exp(23 \cdot e_y) \tag{17c}$$

$$e_y = \frac{S_y}{E'} \tag{17d}$$

$$\omega_t^* = 1.9 \tag{17e}$$

The value ω_t^* represents the transition point from elastically dominant behavior to elasto-plastic behavior. The formulation follows the Hertzian solution (Eq. (17a)) for $\omega^*<1.9$. Then it transitions to the elasto-plastic case and eventually continues past the AF model for high values of ω^*. Equation (17b) is also somewhat dependant on the material properties, according to the definition in Eqs. (17c-d). Statistically, Eq. (17) differs from the FEM data for all five materials by an average of 1.3% and a maximum of 4.3%. An equation of the same form as the ZMC model fitted to the FEM data results in an average error of 43.2%. Notably, Eqs. (17a) and (17b) are continuous at ω_t^*.

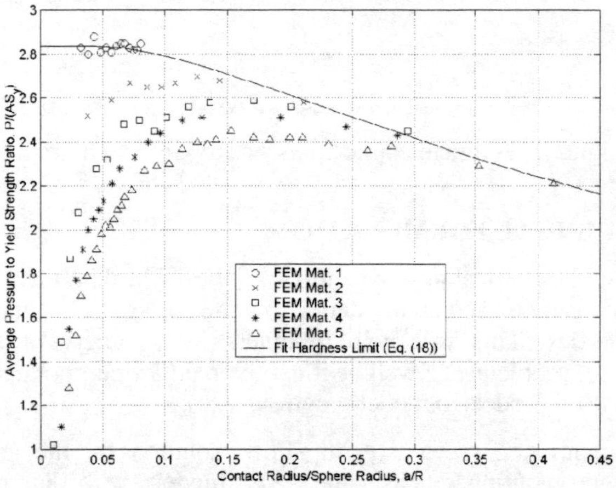

Figure 6: Average contact pressure to yield strength ratio.

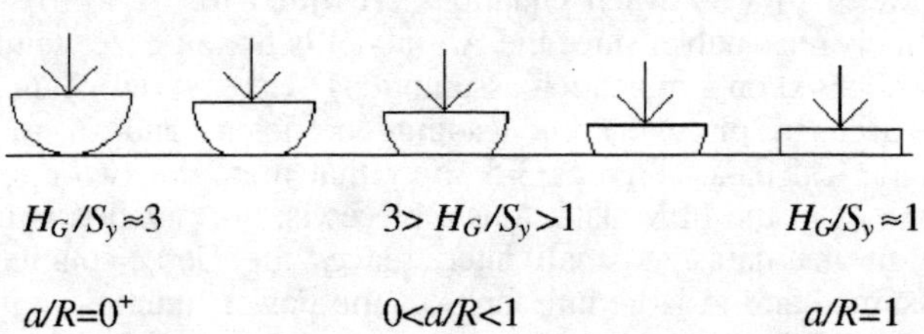

Figure 7: Diagram of progression of change in hardness with geometry.

In order to formulate a fit for the FEM contact force, the material dependant trend at high interferences shown in Fig. 4 is modeled. To assist in this model, a plot of $P/(A \cdot S_y)$ as a function of a/R in Fig. 6 reveals the cause of the material dependency. In this plot a limit appears to emerge for the fully plastic average pressure, commonly referred to as the hardness. Here the hardness appears to change as a function of a/R, or with the evolving geometry of contact. The trend may be explained by the progression schematically shown in Fig. 7. As the interference increases and the contact geometry changes, the limiting average pressure to yield strength ratio, H_G/S_y, must change from Tabor's predicted value of three to a theoretical value of one when $a=R$. The contact region when $a=R$ is essentially the case of a deformable blunt rod in contact with a rigid flat whose H_G/S_y value is theoretically one.

A Weibull function fitted to the limiting values of H_G, yields:

$$\frac{H_G}{S_y} = 2.84\left[1-\exp\left(-0.82\left(\frac{a}{R}\right)^{-0.7}\right)\right] \tag{18}$$

This formulation is plotted alongside the data in Fig. 6. Equation (18) is only valid for the range of fitted data, or $(0 < a/R \leq 0.412)$. Caution should thus be taken when using this function outside this range. This range is acceptable for many applications, particularly tribological applications where deformations above this range are either unlikely or unacceptable. From the relation $A=\pi a^2$, a is solved for and normalized by R. Then $A_c \cdot A_F^*$ is substituted for A. Equation (17b) is then substituted for A_F^*, and Eq. (9) for A_c, thus yielding:

$$\frac{a}{R} = \frac{\sqrt{A_c \cdot A_F^*}}{\sqrt{\pi} R} = \sqrt{\frac{\pi^3 (C S_y R)^2 \omega^* \left(\frac{\omega^*}{\omega_t^*}\right)^B}{4\pi (R E')^2}} = \frac{\pi C e_y}{2}\left[\omega^*\left(\frac{\omega^*}{\omega_t^*}\right)^B\right]^{\frac{1}{2}} \tag{19}$$

This substitution is valid only when $\omega^* \geq \omega_t^*$. Equation (19) can then be substituted into Eq. (18) so it may then be rewritten as a function of ω^* as follows:

$$\frac{H_G}{S_y} = 2.84\left[1-\exp\left(-0.82\left(\frac{\pi C e_y}{2}\sqrt{\omega^*}\left(\frac{\omega^*}{\omega_t^*}\right)^{\frac{B}{2}}\right)^{-0.7}\right)\right] \tag{20}$$

This results in a formulation for H_G as a function of the material properties, E, S_y, and ν (not just upon S_y as suggested by Tabor [7]). Interestingly, as a/R approaches zero, the limiting value of H_G/S_y=2.84 agrees almost precisely with the theoretical value of 2.83 (Williams [11, p.109]).

To formulate an approximation of the contact force as predicted by the FEM results, the AF model for contact force must first be corrected by way of substituting in Eq. (18) or Eq. (20) into Eq. (14), letting H_G replace H, and by allowing the AF contact area to deviate from Eq. (13) (see reasoning for Eq. (17)). This results in an equation for a corrected fully plastic model. Once again a piecewise solution is fit to the FEM data. At small interferences, the Hertz solution is assumed. The resulting piecewise equation fit to the FEM data is given as:

For $0 \leq \omega^* \leq \omega_t^*$

$$P_F^* = (\omega^*)^{3/2} \tag{21a}$$

For $\omega_t^* \le \omega^*$

$$P_F^* = \left[\exp\left(-\frac{1}{4}\left(\omega^*\right)^{\frac{5}{12}}\right)\right]\left(\omega^*\right)^{3/2} + \frac{4H_G}{CS_y}\left[1-\exp\left(-\frac{1}{25}\left(\omega^*\right)^{\frac{5}{9}}\right)\right]\omega^* \tag{21b}$$

where $\omega_t^* = 1.9$. This formulation approaches asymptotically the Hertz elastic model at small interferences, and approaches and continues past the AF model at large interferences. Statistically this formulation differs from the FEM data for all five materials by by an average error of 0.94% and a maximum of 3.5% when Eq. (20) is used for H_G.

The average pressure to yield strength ratio, $P/(A{\cdot}S_y)$, can now be modeled by combining Eq. (17) and Eq. (21). Since these equations are normalized by their critical values, the resulting formulation for the average pressure is:

$$\frac{P}{A \cdot S_y} = \frac{2}{3} C \frac{P_F^*}{A_F^*} \tag{22}$$

This ratio is shown in Fig. 8 (only the weakest and strongest materials are plotted for clarity). The largest differences between the ZMC and KE models and the current FEM model then appear. It is apparent that the KE and ZMC models do not account for material dependence in the limiting average pressure to yield strength ratio, H_G/Sy. Both ZMC and KE are monotonically increasing and truncated at some point that traditionally is considered to be the "hardness." The ZMC and KE models both estimate the average pressure in the transition from the elastic to the elasto-plastic regime fairly well. It is also apparent that these models do not intersect with the Hertzian solution at $P/(A{\cdot}S_y)=2{\cdot}C/3$. The discontinuity in the slope in the KE model at a value of 6 and in the current model at a value of 1.9 is also clearly evident (see Eqs. (17, 21)).

CONCLUSIONS

This work presents a 2D axisymmetric finite element model of an elastic-perfectly plastic hemisphere in contact with a rigid flat surface. A comparison is also made with other existing models. The material is modeled as elastic-perfectly plastic, and yielding occurs according to the von Mises criterion. A concise form is presented for the critical interference at which plastic deformation initiates within the hemisphere. It is derived from the Hertzian solution and the von Mises yield criterion. An a priori definition of the hardness is not needed.

The resulting plots indicate that the FEM results for the contact area agree closely at small interferences with the trends of the Hertzian solution. While at large interferences the FEM predicts contact areas that surpass Abbot and Firestone's fully plastic model [4] (that is based upon truncation). The ZMC model is found to differ significantly from the FEM results, where the KE model (which is also a FEM) follows more closely and has the same general trends. An empirical formulation for the contact area is also fitted to the FEM data as a function of the material properties and interference.

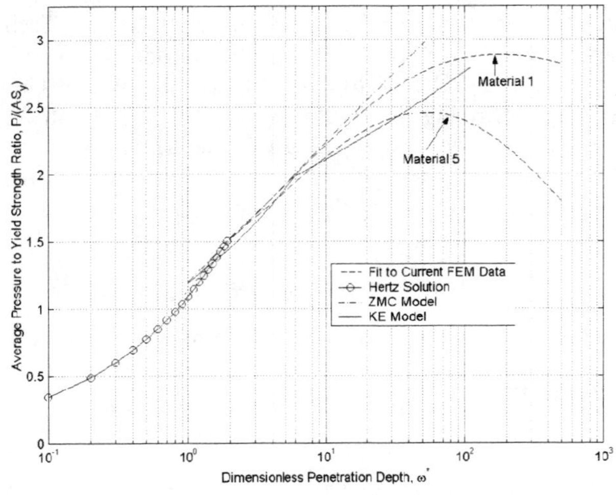

Figure 8: Predicted Average Pressure to Yield Strength Ratio for various models.

The FEM results of the contact force predict a lower load carrying capacity than the AF model for most materials and values of ω^*. This is because the AF model assumes that the average pressure distribution is simply the hardness, which is approximated by $3{\cdot}S_y$. It is found, however, that the fully plastic average contact pressure or hardness is not constant as is widely accepted. Rather, the limiting value of the fully plastic average pressure varies with the deformed contact geometry, which in turn is coupled to the material yield strength. This is accounted for in an empirical formulation for the limiting average pressure to yield strength ratio, H_G/S_y. A formulation using H_G/S_y is then fit to the FEM contact force data.

This work reveals large differences between approximate analytical models and other numerical solutions. More importantly, the contact area, force, and pressure that are found to be particularly dependent upon the deformed geometry in all regimes and effectively dependent upon the material properties (e.g., strength) in the elasto-plastic and plastic regimes. The fit-them-all equations that solely depend upon deformation, which are found in previous works, are imprecise when compared to current FEM results. For example, the average contact pressure to yield strength ratio in all previous work is shown to increase monotonically with deformation, and is assumed to terminate (or truncate) at the hardness. In this work it is shown that such a truncation is not warranted. Particularly, it is shown that the truncation model of Abbott and Firestone [4] cannot be justified. This work discovered significant geometrical and material nonlinearities, and that the hardness depends not just upon strength but also upon the modulus of elasticity, Poisson's ratio, and most importantly upon the deformation itself (i.e., hardness is not a unique or fixed material property as indicated by Tabor [7], and assumed by others after him). The results are based on the finest and adaptive mesh yet (over 11,000 elements for a single hemispherical asperity in contact with a rigid flat, and 100 contact elements) that was necessary for finite element convergence. The results were obtained by using

ANSYS™ and then independently confirmed by using ABAQUS™.

ACKNOWLEDGEMENTS

The contribution of Dawei Shen of Georgia Tech in constructing the ANSYS™ FEM model is gratefully acknowledged. The contribution of Scott Shipley of Georgia Tech in confirming the results using ABAQUS™ software is also gratefully acknowledge. Appreciation is also given to Professor Izhak Etsion of the Technion – Israel Institute of Technology for his debate and critique.

APPENDIX: CRITICAL INTERFERENCE

The Hertz solution results in the following equations for stress within the deformed sphere along the axis of revolution, z (Johnson, [12]):

$$\sigma_1 = -p_o\left(1+\left(\frac{z}{a}\right)^2\right)^{-1} \tag{A1}$$

$$\sigma_{2,3} = p_o\left\{\left[2\left(1+\left(\frac{z}{a}\right)^2\right)\right]^{-1} - (1+\nu)\left[1-\frac{z}{a}\tan^{-1}\left(\frac{a}{z}\right)\right]\right\} \tag{A2}$$

where the origin of the z-axis lies at the point of initial contact between the hemisphere and the rigid flat, and p_o is the maximum contact pressure.

The von Mises yield criterion is given as:

$$S_y = \sqrt{\frac{1}{2}\left[(\sigma_1-\sigma_2)^2+(\sigma_2-\sigma_3)^2+(\sigma_3-\sigma_1)^2\right]} \tag{A3}$$

By substituting the principal stresses given in Eq. (A1) and Eq. (A2) into Eq. (A3) and then simplifying, the following equation for the von Mises yield criterion is obtained:

$$\frac{S_y}{p_o} = \frac{3}{2}\left(1+\left(\frac{z}{a}\right)^2\right)^{-1} - (1+\nu)\left[1-\frac{z}{a}\tan^{-1}\left(\frac{a}{z}\right)\right] \tag{A4}$$

The resulting Eq. (A4), which must be positive, dictates where within the hemisphere initial yielding occurs. This is obtained by setting the derivative with respect to z to zero. Hence,

$$\frac{d}{dz}\left(\frac{S_y}{p_o}\right) = -az\left[a^2(4+\nu)+(1+\nu)z^2\right]+(1+\nu)\left(a^2+z^2\right)^2\left[\tan^{-1}\left(\frac{a}{z}\right)\right]=0 \tag{A5}$$

This equation is solved numerically for Poisson ratios between 0.01 and 0.50 to find the locations, z, at initial yielding. These locations are then substituted in Eq. (A4) to find the applied maximum contact pressure to yield strength ratio, p_{oc}/S_y. This ratio, p_{oc}/S_y, is referred to as the yield strength coefficient and designated by the symbol C. An empirical function is fitted to the final numerical data, which is given by:

$$\frac{p_{oc}}{S_y} = C = 1.295\exp(0.736\nu) \tag{A6}$$

Equation (A6) differs from the numerical solution by an average of 1.2% and by no more than 3.1%.

The interference, ω, is given as a function of p_o by the Hertz elastic solution in Johnson [12] as:

$$\omega = \left(\frac{\pi \cdot p_o}{2E'}\right)^2 R \tag{A7}$$

Thus, to find the critical interference, or the interference at the initial point of yielding, the maximum pressure when yielding first occurs, p_{oc}, is substituted into Eq. (A7) for p_o. This maximum pressure is the pressure given by the maximum contact to yield strength ratio given in Eq. (A6). The equation $p_{oc}=CS_y$ is substituted into Eq. (A7), resulting in the Eq. (7a).

A similar derivation is also given in Chang [13]. However, that derivation assumed a fixed value between strength and hardness, $S_y=0.35H$, which resulted in an equation for the hardness coefficient, $K = 0.454 + 0.41\nu$. Such an assumption is not made in this work (see discussion on H_G within).

REFERENCES

[1] Greenwood, J. A. & Williamson, J. B. P. 1966, "Contact of Nominally Flat Surfaces," *Proc. R. Soc. Lond. A,* 295, pp. 300-319.

[2] Green, I., 2002, "A Transient Dynamic Analysis of Mechanical Seals Including Asperity Contact and Face Deformation," *Trib. Trans.*, 45, 3, pp.284-293.

[3] Timoshenko, S., and Goodier, J. N., 1951, Theory of Elasticity, New York, McGraw-Hill.

[4] Abbott, E. J., and Firestone, F. A., 1933, "Specifying Surface Quality-A Method Based on Accurate Measurement and Comparison," *Mech. Engr.*, 55, pp. 569-572.

[5] Chang, W. R., Etsion, I., and Bogy, D. B., 1987, "An Elastic-Plastic Model for the Contact of Rough Surfaces," *ASME J. Tribol.*, 109, pp.257-263.

[6] Zhao, Y., Maletta, D. M., Chang, L., 2000, "An Asperity Microcontact Model Incorporating the Transition From Elastic Deformation to Fully Plastic Flow," *ASME J. Tribol.*, 122, pp.86-93.

[7] Tabor, D., 1951, The Hardness of Materials, Clarendon Press, Oxford.

[8] Mesarovic, S. Dj. and Fleck, N. A., 2000, "Frictionless Indentation of Dissimilar Elastic-plastic Spheres," *Int. J. Solids and Structures*, 37, pp.7071-7091.

[9] Kogut, L., & Etsion, I., 2002, "Elastic-Plastic Contact Analysis of a Sphere and a Rigid Flat," J. of Applied Mechanics, Trans. ASME, 69(5), pp. 657-662.

[10] Shigley, J. E., Mischke, C. R., 1989, Mechanical Engineering Design, 5th Edition, McGraw-Hill Inc., New York.

[11] Williams, J. A., 1994, Engineering Tribology, Oxford University Press, Oxford.

[12] Johnson, K. L., 1985, Contact Mechanics, Cambridge University Press, Cambridge.

[13] Chang, W. R., 1986, "Contact, Adhesion, and Static Friction of Metallic Rough Surfaces," PhD Thesis, University of California, Berkeley, pp.18-23.

Proceedings of
2003 STLE/ASME Joint International Tribology Conference
Ponte Vedra Beach, Florida USA, October 26-29, 2003

2003-TRIB-0269

A CYLINDRICAL CONTACT MODEL FOR TWO DIMENSIONAL MULTIASPERITY PROFILES

John J. Jagodnik and Sinan Müftü
Northeastern University
Department of Mechanical Engineering
Boston, MA 02115

ABSTRACT

In practice, multi-asperity contact problems are often solved as two dimensional (2D) plane problems rather than true three dimensional (3D) problems. This is accomplished by assuming that each peak on a 2D scanned profile is the pinnacle of a half sphere. Hertz contact equations are then used to solve for the radius of contact and pressure profile. In reality, the local maximum in the plane may not be the maximum in the unmeasured depth direction, creating inherent errors in the contact model. This error is shown to be significant in contact problems when estimating the area of contact and total contact force over a single asperity. The pressure corrected Sternberg-Turteltaub model is introduced, in which a cylinder is used to model a sphere in a plane. This model is shown to improve the contact area and total force estimates for a range contact parameters.

INTRODUCTION

Typically, multi-asperity contact problems are solved by assuming that each asperity is a perfect sphere. Hertz's equations for contacting spheres gives good estimates of the contact pressure profile, if the radius of the sphere is known [1]. However, the topography of the surface is often measured by using a stylus type profiler, which scans a single plane. In reality, the local maximum in the scanned plane may not be the maximum in the unmeasured depth direction. Using this radius in the spherical Hertz contact model creates inherent errors. In this paper the significance and reasoning behind these errors are introduced and corrections, including a modified form of the cylindrical contact model, are proposed to reduce the error in plane approximations.

Plane Approximations

Multi-asperity contact problems can be solved in two or three dimensions. A surface profiler can be used to determine the surface height in both the x and y directions (Fig 1). Then, the overall equilibrium of the two contacting surfaces could be found by direct modeling of each asperity as Hertz contacts [1], or by statistical techniques [2]. For these problems, each asperity is treated as a sphere and Hertz contact equations are applied. Typically, 3D solutions require significant computational time. It is also difficult to analyze complex problems involving sliding, composite layers, etc. in 3D [4].

Two dimensional (2D) "plane approximations" provide an alternative, where a surface profiler could be used to obtain the surface heights along the x direction, on a single x-z plane, as shown in Fig 1. Similar profiles can be obtained for a general surface using a random profile generator [3]. Typically each peak on the profile is treated as the pinnacle of a sphere and Hertz contact equations are applied. In this paper, we will investigate the error from ignoring the y direction effects when a 2D profile is analyzed, rather than a full 3D surface.

THEORY

Single Asperity Analysis

Here a single asperity rather than an entire surface or profile is analyzed. The asperity is treated as a perfect half sphere, whose center lies in a plane-1 as shown in Fig 2a. The sphere is subjected to interference from a flat rigid punch. It will be assumed that the surface profiler scanned the asperity along plane-2, which is parallel to plane-1, but offset by a distance y^*.

Focusing on a front view of the asperity in Fig 2b, it is

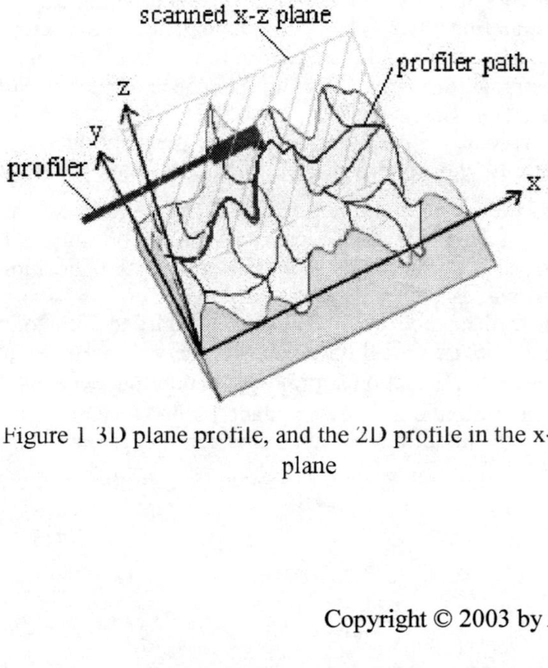

Figure 1 3D plane profile, and the 2D profile in the x-z scan plane

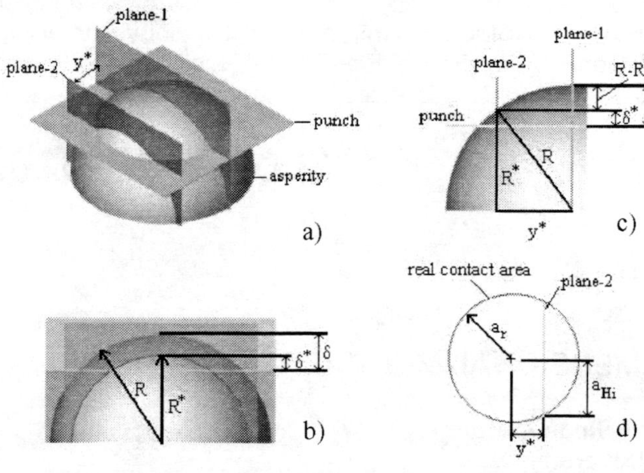

Figure 2: a) Three dimensional model of an asperity; The front b) and side c) views; and d) Details of the real contact area.

noted that if the scan plane is identical to plane-1, the true radius of the sphere (R) and the true maximum interference (δ) will be found. If y^* is not zero, a different value for the radius (R^*) and maximum interference (δ) will be observed. Clearly assuming that the observed half circle on the scanned profile is the pinnacle of a sphere will lead to some amount of error depending on the value of y^*.

Focusing on Fig 2c, the relationships between the true and observed radii and maximum interferences are found as follows,

$$\delta^* = \delta - \left(R - R^*\right) \text{ (a)}, \qquad R^* = \sqrt{R^2 - y^{*2}} \text{ (b)} \tag{1}$$

Analysis Over The Entire Asperity

In order to gain valuable information about the error associated with using the parameters obtained from a line scan, the entire asperity will be analyzed, rather than simply analyzing a single y^* value. In order to do this, n (=200) equally spaced y^* values are selected. These y^* values cover the entire range, where the punch interferes with the undeformed sphere. For each y^* value the contact problem is solved on plane-2 using three methods.

First, the full 3D model is solved using spherical Hertz contact equations with the true radius and true maximum interference from plane-1. The relevant information for plane-2 is then extracted from the 3D solution. This approach is called the Hertz (H) model.

The second method for solving the plane problem is to simply apply the spherical Hertz contact equations to the perceived radius R^* and interference δ^* value, obtained from the profile. This approach, which is based on incomplete information on surface topography is called the incomplete Hertz (Hi) model.

The third method involves using a modified form of the *cylindrical contact* model based on Sternberg and Turteltaub's (ST) work [5]. This model provides a relation between the punch penetration δ and the contact radius a, and cylinder radius R. The modified ST model is introduced later in the paper. The values of R^* and δ^* obtained from the profile are used in the ST equations. This modified cylindrical contact model is called the ST-model, here.

Equations for the Three Models

The focus of the analysis is to determine the cumulative effect of randomly chosen plane-2 (y^*) locations on the spanned contact area (A_j) and total force (F_j), using the three methods. A different subscript j used on the variables, to indicate the method. The subscript $_H$ is added for the real 3D Hertz solution. The subscripts $_{Hi}$ and $_{ST}$ are used for the incomplete Hertz model and for the cylindrical contact with Sternberg-Turtlebaum's modifications, respectively. The radius and punch penetration values in the last two models are obtained in plane-2.

Equations for the Real 3D Hertz Solution

The relation between the true contact radius a_r and the perceived contact radius a^*_{Hi} can be found from Fig. 2d as,

$$a^*_{Hi} = \sqrt{a_r^2 - y^{*2}} . \tag{2}$$

Substituting the well-known Hertz [1] relationship between contact radius and maximum interference ($a_r^2 = R\delta$), the real contact radius on any plane can be found to be,

$$a^*_{Hi} = \sqrt{R\delta - y^{*2}} . \tag{3}$$

Using the pressure distribution from Hertz contact, the equation for the pressure in the contact length (x) of the scan plane is,

$$p_H(x) = \frac{2E\delta}{\pi\left(1-\nu^2\right)a_r}\sqrt{1-\frac{x^2+y^{*2}}{a_r^2}} . \tag{4}$$

The force per unit width on the scan plane $\left(F'_H\right)$ is found as,

$$F'_{Hr} = \frac{E\delta}{1-\nu^2}\left(1-\frac{y^{*2}}{a_r^2}\right) , \tag{5}$$

by integrating p_H over the real contact length for the scan plane-2.

The spanned contact area (A_H) is found by numerically integrating the contact length $2a^*_r$ through the y direction with,

$$A_H = 2\Delta y^* \sum_{i=1}^{n} \sqrt{R\delta - y_i^{*2}} . \tag{6}$$

where n is the number of line scan locations.

Similarly, the total force applied to the asperity is found by numerically integrating F'_H through the y^* direction with,

$$F_r = \frac{E\delta\Delta y^*}{\left(1-\nu^2\right)} \sum_{i=1}^{n}\left(1-\frac{y_i^{*2}}{a_r^2}\right) . \tag{7}$$

In these summations, Δy^* is the spacing between scan planes in the y direction. It should also be noted that the Hertz equations can be applied directly to find the true contact area and force supported by the asperity without numerical integration. These equations can be used to verify that a small enough Δy^* value has been selected.

Equations for the Incomplete Hertz Model

For the Hertz sphere solution, the radius of contact on any scan plane is given as,

$$a^*_{Hi} = \sqrt{R^*\delta^*}. \tag{8}$$

where R^* and δ^* are given in Eqns (1a) and (1b).

The contact area is found by integrating in the y^* direction in the same manner that was followed for the 3D solution, giving,

$$A_{Hi} = 2\Delta y^* \sum_{i=1}^{n} \left(R^*\delta^*\right)^{1/2}. \tag{9}$$

The pressure at a given x value in the contact length on a scan plane is,

$$p_{Hi}(x) = \frac{2E\delta^*}{\pi(1-\nu^2)a^*_{Hi}}\sqrt{1-\left(\frac{x}{a^*_{Hi}}\right)^2}. \tag{10}$$

The force per unit width on the scan plane can be found by integrating p_{Hi} over the contact length for the scan plane. Performing this integration leads the force per unit width, given as,

$$F'_{Hi} = \frac{E\delta^*}{1-\nu^2}. \tag{11}$$

Numerical integration through the entire asperity in the y direction leads to the total force on the asperity, given as,

$$F_{Hi} = \frac{E\Delta y^*}{1-\nu^2}\sum_{i=1}^{n}\delta^*_i. \tag{12}$$

Equations for the Cylindrical Contact with Sternberg-Turteltaub Modifications

Sternberg and Turteltaub analyzed an infinitely long cylinder with a finite radius R compressed a distance 2δ between two flat rigid punches [5]. From this analysis the following equations were obtained under the assumption that radius of the cylinder R is much greater than the half-contact length a,

$$\frac{\delta}{R} = \frac{\gamma^2}{2}\left(\ln\left[\frac{4}{\gamma}\right]-\frac{1}{2}\right)+\frac{9\gamma^4}{64}$$
$$\frac{a}{R} = \gamma - \frac{\varphi\gamma^2}{4} \tag{13a,b,c}$$
$$\upsilon p^s_o = \gamma + \frac{3\gamma^3}{16}$$

where p^s_o is the maximum contact pressure, $\gamma = \left(2\upsilon P/\pi R\right)^{1/2}$, $\upsilon = 2\left(1-\nu^2\right)/E$, and $\varphi = (1-2\nu)/2(1-\nu)$.

Using symmetry about the mid-plane between the two punches allows the problem to be treated as a half cylinder. In this work Eqns (13a) and (13b) are solved numerically by using $\nu = 0.3$, and a curve fit relationship between a, R, and δ is obtained,

$$a = C_s R^{1-m}\delta^m \tag{14}$$

with $C_s = 0.962$ and m = 0.557. Note that this equation is valid in the range $1.15\times10^{-6} \le \delta/R \le 2.25\times10^{-3}$. Substituting eqns (1a) and (1b) yield the half-contact length,

$$a^*_{ST} = 0.962R^{*0.443}\,\delta^{*0.557}. \tag{15}$$

The contact area is obtained, by numerically integrating through the y direction,

$$A_{ST} = 1.924\Delta y^* \sum_{i=1}^{n} R^{*0.413}\delta^{*0.557}. \tag{16}$$

The pressure at any point in the contact length on a scan plane is represented by,

$$p_{ST}(x) = \frac{E}{1-\nu^2}\left(\eta+\frac{3}{4}\eta^3\right)\left[1-\left(\frac{x}{a^*_{ST}}\right)^2\right]^{1/2} \tag{17}$$

where,

$$\eta = \frac{1}{\varphi}\left[1-\left(1-\varphi\frac{a^*_{ST}}{R^*}\right)^{1/2}\right]. \tag{18}$$

The force per unit width on the scan plane for the ST cylinder is found in the same manner as the Hertz analysis,

$$F'_{ST} = 0.481\pi\frac{E}{1-\nu^2}R^{*0.443}\delta^{*0.557}\left(\eta+\frac{3}{4}\eta^3\right). \tag{19}$$

The total force on the asperity is found as,

$$F_{ST} = 0.481\pi\frac{E\Delta y^*}{1-\nu^2}\sum_{i=1}^{n}R^{*0.443}\delta^{*0.557}\left(\eta_i+\frac{3}{4}\eta_i^3\right). \tag{20}$$

Non-dimensional Equations

All of the variables involving a length are scaled with a_r,

$$\bar{a}^*_j = \frac{a^*_j}{a_r},\ \bar{R}^* = \frac{R^*}{a_r},\ \bar{R} = \frac{R}{a_r},\ \bar{\delta}^* = \frac{\delta^*}{a_r},\ \bar{\delta} = \frac{\delta}{a_r},$$
$$\bar{y}^* = \frac{y^*}{a_r},\ \bar{F}'_j = \frac{F'_j}{p_o a_r},\ \bar{F}_j = \frac{F_j}{p_o a_r^2},\ \bar{p}_j = \frac{p_j}{p_o} \tag{21}$$

where the subscript j takes the values H, Hi and ST as appropriate, and the maximum pressure is defined as,

$$p_o = \frac{2E\delta}{\pi a_r\left(1-\nu^2\right)}. \tag{22}$$

This non-dimensionalization results in only three independent parameters that control the problem, ν, $\bar{R}$ and $\bar{y}^*$.

RESULTS

Contact Pressure

The contact pressure distribution for the three models are plotted for three different $\bar{y}^*$ values and $\bar{R}$ =110 in Fig 3. This figure shows that the Hi- and the ST-models approach the Hertz pressure distribution as $\bar{y}^* \to 0$ and $\bar{y}^* \to 1$, respectively. This indicates that, as the line-scan moves further away from the pinnacle of the asperity, the match between the ST and Hertz models gradually improves.

Contact Radius

Contact radius $\overline{a}$ is plotted for different $\overline{y}^*$ locations and $\overline{R}$ values in Figure 4a. The contact radius $\overline{a}_H$ for the Hertz solution falls on a unit circle as expected. However, the contact radius of the Hi-model ($\overline{a}_{Hi}$) overestimates and the contact radius of the ST model ($\overline{a}_{ST}$) underestimates the Hertz solution. It is interesting to note that the effect of asperity radius $\overline{R}$ on the Hi-model is so small that it can not be detected in Fig. 4a. On the other hand the ST-model further underestimates the contact radius with increasing $\overline{R}$.

Contact Area Span

The probability of scanning a given asperity at a $\overline{y}^*$ location is the same for all $\overline{y}^*$ values. Therefore, if we integrate the area under the $\overline{a}$ curve, given in Fig 4a, we get an estimate of the contact area span A_j that the three models are capable of estimating.

A_j is obtained numerically from eqns (6), (9) and (16). Figure 4b displays the A_j values for all three methods. Note that the spanned area for the Hertz model is the constant π, and is equal to the real contact area of the Hertz contact model with the non-dimensional contact radius of 1. The spanned area for the Hi-model is nearly constant with 4.53. This value is 44% greater than π. Finally, the spanned area of the ST-model is variable as shown in Fig 4b. The error between spanned area estimates of the Hertz model (π) and the ST-model varies between 3% and 30%, where the error becomes smaller at smaller values of $\overline{R}$.

When using plane approximations, the Hi- and ST-models wrongly assume the sphere or the cylinder is centered on the scan plane and thus the first point to touch the flat punch is the peak as seen in the scan plane. This reasoning explains the overshoot area for the Hi- and ST-models. The remaining error in the ST-model is simply attributed to modeling a portion of a sphere with a cylinder.

While it may not be apparent as to why the ST cylindrical model is being considered for modeling spheres in a plane, it is interesting to note that the this model provides a better approximation to the true contact area in the range of asperity radii ($15 \leq \overline{R} \leq 250$) considered in Fig. 4b. This is a result of the underestimation of the contact radius near the center of the sphere, which to some extent corrects the overestimation in the overshoot area. This correction is lacking in the Hi-model where the contact radius is overestimated at all points except the center where the solution is precise.

Force per Width

The force per width $\overline{F}_j'$ expressions, for the three models, are given in equations (5), (11) and (19). These expressions are plotted in Fig 5a as a function of scan location $\overline{y}^*$, for three different values of $\overline{R}$. This figure shows that the force per width is overestimated by the Hi-model as compared to the Hertz model; and, it is underestimated by the ST-model for a large section of the $\overline{y}^*$ range. The Hi-model is insensitive to changes in $\overline{R}$, whereas the estimate of the ST-model becomes worse as $\overline{R}$ increases.

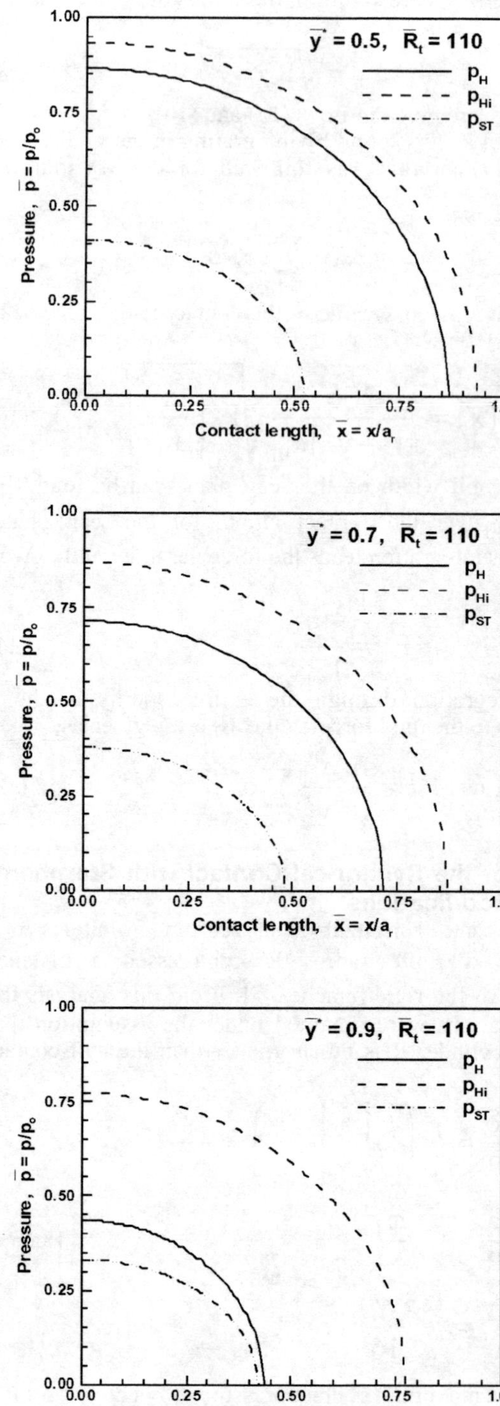

Figure 3 Contact pressure distribution for $\overline{y}^*$ =0.5, 0.7, 0.9 and $\overline{R}$ =110.

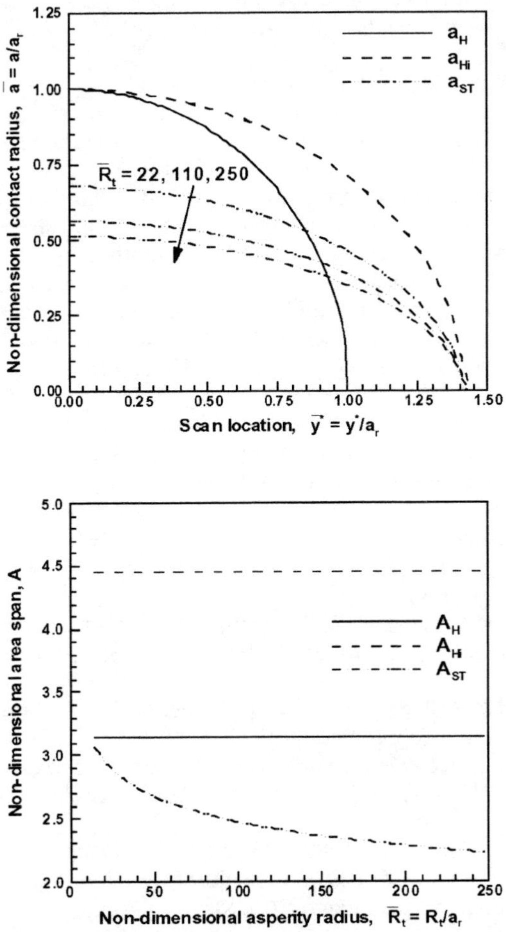

Figure 4 a) Non-dimensional contact radius for different $\overline{R}$ values. b) Contact area span variation with $\overline{R}$ for the three models.

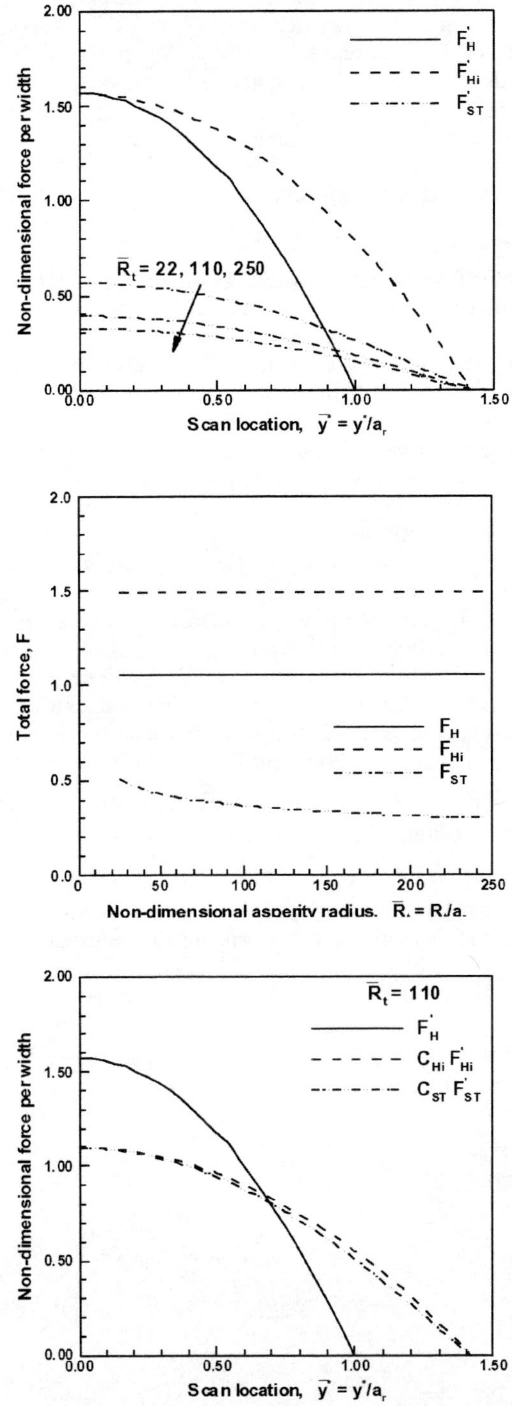

Figure 5 a) Dimensionless force per width as a function of scan location y^*. b) Total spanned asperity force as a function of asperity radius. c) Estimates of the Hi and ST models with correction factors C_{Hi} and C_{ST}.

Total Force

As was the case for the spanned contact area, we can view the effects of asperity radius $\overline{R}$ on the total spanned force $\overline{F}_j$. Eqns (6), (13) and (20), which give the $\overline{F}_j$ expressions for the three models, are plotted in Fig 5b. This figure shows that $\overline{F}_H$ = 1.05 for the Hertz model, and that it is independent of $\overline{R}$. The total force for the Hi-model is nearly constant with the value $\overline{F}_{Hi}$ = 1.49. However, the total force for the ST-model varies with $\overline{R}$ as shown in Fig 5b. The error between the ST-model and the Hertz solution varies between 52 and 71%.

The error between the total force estimates can be reduced by introducing correction factors for the pressure estimates given by eqns (10) and (17) for the Hi- and ST-models, respectively. Fig 5b suggests that a constant correction factor of $C_{Hi} = 0.7$ should be used for the Hi-model. On the other hand the correction factor C_{ST} for the ST-model depends on the value of $\overline{R}$. For example, we found that for $\overline{R}$ =110 the correction factor should be $C_{ST} = 2.8$. The value of $\overline{R}$ =110 represents, for example, R = 121 μm and δ = 10 nm, which are parameters typical of magnetic tape applications [4].

Fig 5c shows the effect of using the correction factors indicated above on the estimates of force per width of the asperity. Both the Hi- and ST-models follow very similar trends, and the areas under the two modified curves nearly equals that under the H-model.

It is also important to note that the pressure correction factors do not effect the contact area analysis stated above. Therefore, the errors in contact area estimation remain as they are given in Fig 4b. As the area estimate errors are smaller for the ST-model, as compared to the Hi-model use of the ST-model may be an attractive choice.

DISCUSSION

Qualitative Analysis of Multi-asperity Contact

In the previous sections contact over a single asperity was analyzed. While this analysis provides insight into the reasoning behind the use of the pressure corrected ST model, the real purpose for using either model is to analyze a single plane from a multi-asperity surface.

In the previous section a single asperity was broken into 200 thin strips. Each strip was analyzed and then combined to find the resulting area of contact and total force approximations. The behavior of a plane from a multi-asperity contact problem closely parallels the single asperity analysis.

Let us consider a scan plane, which intersects n contacting asperities, and assume the y coordinates for the centers of each asperity are random and uncorrelated. If this is the case there is an equal probability for any one of the 200 thin strips of a given asperity falls within the scan plane. This means that if n is large, we expect to see n/200 strip 1s, n/200 strip 2s, etc. This is close to analyzing n/200 single asperities, with the only difference being the variation of $\overline{\delta}/\overline{R}$ from asperity to asperity due to the random nature of the surface. Neglecting the effects of the variation in $\overline{\delta}/\overline{R}$, the conclusions reached for a single asperity can be directly applied to the multi-asperity contact case. While this qualitative analysis appears reasonable, a more detailed quantitative analysis of multi-asperity plane approximations will be the subject of a future paper.

SUMMARY AND CONCLUSION

While the accuracy of the Hertz contact equations are uncontested in a 3D analysis, often the equations are used with information based on 2D plane scans. In these analyses, the local peaks are assumed to be the pinnacles of a sphere, when in fact they may not be the pinnacle had the depth direction been considered. The errors for this assumption have been analyzed for a single asperity and shown to be substantial when estimating the contact area and total force over the asperity. Correction factors for the contact pressure are discussed. In particular, a modified (ST) cylindrical contact model is introduced. This model is shown to be better in predicting contact area. A more robust model for determining the pressure correction factor and a more vigorous multi-asperity contact analysis will be the subject of future papers.

REFERENCES

1. Johnson, K.L. (1985) *Contact Mechanics*, Cambridge University Press.
2. Greenwood, J.A. and Williamson, J.B.P. (1966) "Contact of Nominally Flat Surfaces," *Proceedings, Royal Society*, **A295**, 300.
3. Patir, N. (1978) "Numerical Procedure for Random Generation of Rough Surfaces," *Wear*, v 47, n 2, 263.
4. Bhushan, B. (1996) *Tribology and Mechanics of Magnetic Storage Devices*, Springer- Verlag New York Inc., New York (Ch. 2 and 3).
5. Gladwell, G.M.L. (1980) *Contact Problems in the Classical Theory of Elasticity*, Sijthoff & Noordhoff Int. Publishers, The Netherlands (Ch. 8).

Proceedings of
2003 STLE/ASME Joint International Tribology Conference
Ponte Vedra Beach, Florida USA, October 26-29, 2003

2003-TRIB-0270

CONTACT RESISTANCE AND ADHESION IN A MEMS MICROSWITCH

Sumit Majumder
Radant MEMS, Stow, MA 01775, USA
Email: smajumder@radanttechnologies.com

Nicol McGruer
Department of Electrical and Computer Engineering
Northeastern University, Boston, MA 02115, USA
Email: mcgruer@ece.neu.edu

George G. Adams
Department of Mechanical, Industrial and Manufacturing Engineering
Northeastern University, Boston, MA 02115, USA
Email: adams@neu.edu

ABSTRACT

A multi-asperity model of the contact resistance in a MEMS microswitch has been developed. This model includes the effects of elastic and plastic deformation, adhesion, and scale-dependent constrictive resistance. The number of asperities in contact is small enough that a discrete, rather than a continuous, distribution of asperity heights is used. Due to the combined effects of plasticity and adhesion, the surface profile changes with repeated load and unload cycles. Furthermore, adhesion produces significant hysteresis in the contact resistance vs. contact force characteristics. Measurements of the contact resistance as a function of actuation voltage show good qualitative agreement with the model.

INTRODUCTION

MEMS switches and relays have a number of properties that make them promising candidates to replace conventional relays or solid-state switches in a number of low-power applications. MEMS switches are smaller and can switch faster than the smallest existing mechanical relays (reed relays). They also have lower on-resistance and lower parasitic capacitances than transistor switches. Unlike mechanical relays and PIN diode switches, electrostatically actuated MEMS switches consume practically zero steady-state power. Potential uses of MEMS switches include RF applications such as in cell phones, phase shifters and smart antennas (Rebeiz and Muldavin, [1]), as well as relay-replacements in Automatic Test Equipment (ATE) and industrial and medical instrumentation.

Most MEMS switches reported in the literature use one of two actuation mechanisms – electrostatic or electromagnetic. Electrostatic switches generally consume less power than electromagnetic switches; electromagnetic switches usually draw steady-state power in the on-state. Also, electrostatic switches are generally smaller and faster, and easier to build. On the other hand, electromagnetic switches have higher contact forces, and therefore have lower and more reliable contact resistance, particularly when operated in air.

A fundamental constraint in electrostatic microswitches is the low available actuation force which is typically at most a few mN, and more often several hundred μN. Part of the available actuation force deflects the beam to close the switch, whereas the rest of the actuation force provides the contact force. Therefore, both the contact force and the spring force are limited to at most several hundred μN. These forces are generally 2-3 orders of magnitude higher in reed relays, and 1-2 orders of magnitude higher in MEMS electromagnetic relays. There are two important implications of this low contact force. First, it is difficult to obtain a low and stable contact resistance – at the least, an inert, hermetic environment is required. Secondly, because of the necessarily low spring force, adhesion between the contact surfaces becomes an important consideration. In the extreme case, the adhesive force between the contacts exceeds the restoring spring force, holding the beam in its "on" position even when the drive voltage is removed, and causing a "stuck-closed" failure.

These two problems (low contact force and low restoring force) are inter-related. Within the constraint of a certain available actuation force, the designer may try to obtain a higher contact force - thereby reducing the restoring force, and also creating a more intimate contact, which in turn tends to make the switch more susceptible to the adhesion force. Also, a low available contact force tends to encourage the choice of gold as a contact material because it is soft (resulting in a larger contact area and more intimate contact for a given contact force), and inert (resulting in a clean, truly metallic contact surface). However, a large contact area and a clean-metal contact increase the adhesion force, thereby increasing the possibility of contact stiction.

There is relatively little literature dealing specifically with microswitch/microrelay contacts. Investigations of microswitch contacts with and without adhesion have been reported by Zavracky et al. [2] and Majumder et al. [3-5]. Other authors have used simpler contact models, generally from the viewpoint of device design, e.g. Kruglick and Pister [6] and Lafontan et al. [7]. In those cases, the reported contact resistance was high (1 Ω or greater), and a clean-metal model may not have been applicable. There has also been some work in looking at contact resistance in the force range applicable to microswitches, using an experimental set-up instead of real devices, i.e. Beale and Pease [8], and Bromley and Nelson [9]. In these papers, the effect of contact cycling on resistance and adherence was not reported.

The objective of this paper is to develop a basic understanding of adhesion in microswitch contacts, and how it affects microswitch performance. More details of this investigation can be found in the dissertation of Majumder [10]. To support this work, we have used measurements of a microswitch with gold-on-gold contacts (Figure 1), originally developed at Northeastern University (Zavracky et al, [2]).

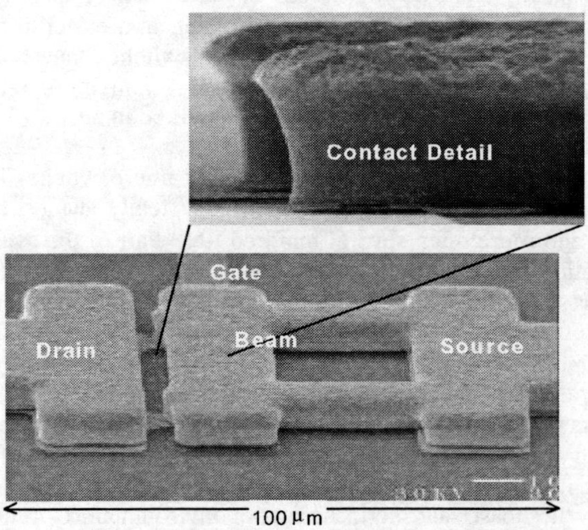

Figure 1. SEM micrographs of the Northeastern University microswitch.

ADHESION MODELING

Two of the most widely used models for adhesion in elastic contacts of elastic spheres are those of Johnson, Kendall and Roberts [11] know as the JKR model, and that of Derjaguin, Muller and Toporov [12] known as the DMT model. It was shown by Tabor [13] that these theories are each applicable for two different limits of a dimensionless parameter related to the ratio of the elastic deformation to the range of surface forces. We seek to develop a multi-asperity model (Majumder, [10]) in which a typical asperity radius is estimated to be R = 100 nm, the work of adhesion for a clean gold-on-gold contact is w = 2.5 J/m^2, and Young's elastic modulus is E = 91 GPa. These parameters imply that the contacts are well within the JKR regime. During the unloading phase of the JKR model, adhesion causes a finite adherence force ($F_{adh} = 1.5\pi wR$) to be necessary to separate the bodies. Separation occurs at a finite contact radius, referred to as the minimum stable JKR contact radius (a_{min}). For our case a_{min} = 11.3 nm.

In order to include the effect of plasticity, we use the approach of Maugis and Pollock [14], which is based on a deformation model given by Roy Chowdhury and Pollock [15]. In the elastic regime, the deformation is assumed to be governed by the JKR model. The transition from elastic to elasto-plastic behavior is assumed to occur at a contact radius (a_e) at which the Hertz load becomes equal to the load given by the Studman [16] elasto-plastic model for the onset of plastic deformation. Using the value for hardness H = 2.2 GPa, we have a_e = 4.5 nm. Note that this contact radius is smaller than the minimum stable contact radius for the elastic JKR model, which implies that for our parameters, contact is never completely elastic.

In both the elasto-plastic and plastic regimes, the stress distribution and contact geometry are assumed to be those for a purely plastic contact between a sphere and flat with no adhesion. The mean contact pressure p_m is assumed to vary with contact radius according to a modified version of the Studman model (following the approach of Maugis and Pollock [14]), i.e.

$$p_m = Y(1.1 + 0.69 \ln \frac{Ea}{3.9YR}), \tag{1}$$

with the yield stress $Y = H/3$. Since the contact geometry is assumed to be that for a purely plastic contact without adhesion, the contact force is related to the contact area by

$$F = \pi a^2 p_m - 2\pi Rw . \tag{2}$$

Eventually, the contact pressure becomes equal to the hardness H; this occurs at a critical contact radius $a_p \approx 60YR/E$ (with our parameters a_p = 48 nm). Thereafter, the deformation is purely plastic, and the contact pressure remains constant, with

$$F = \pi a^2 H - 2\pi Rw . \tag{3}$$

Due to plastic deformation, the effective radius of curvature after deformation (R_{eff}) is different than its value (R) before. Since the effect of the adhesive forces on the stress distribution decreases progressively with further loading into the plastic regime, R_{eff} is assumed to be the same as it would be in the absence of adhesion (Johnson [17]). Therefore

$$R_{eff} = \frac{2a_f E}{3\pi p_m (1-\nu^2)} \tag{4}$$

in the elasto-plastic regime and

$$R_{eff} = \frac{2a_f E}{3\pi H (1-\nu^2)} \tag{5}$$

in the plastic regime, where a_f is the maximum value of the contact radius.

The asperity penetration δ is assumed to be related to the contact radius as

$$\delta = \frac{a^2}{R} - \sqrt{\frac{8\pi a w}{3K}} \tag{6}$$

for elastic contact and

$$\delta = \frac{a^2}{2R} + \frac{\delta_e}{2} - \frac{1}{2}\sqrt{\frac{8\pi a_e w}{3K}} \tag{7}$$

for the elasto-plastic and plastic regimes, where δ_e is the penetration at the onset of plastic deformation.

If the loading is elastic, the unloading will also be elastic [11]. However if the loading is elasto-plastic or plastic, three different unloading modes are possible (Maugis and Pollock, [14]). When the contact is gradually unloaded starting from the maximum contact force F_f, the initial unloading is equivalent to the elastic unloading of a flat punch. This initial regime persists until one of the following three conditions is met: (a) the average tensile axial stress becomes equal to the hardness – the contacts separate and the separation is called *ductile*; (b) the contact force reaches the JKR equilibrium value F_m, for which the contact radius $a_f < a_{min}$ – and the contacts separate in what is referred to as a *brittle* mode; (c) the contact force reaches F_m, and $a_f > a_{min}$ – subsequent unloading of the contact follows the JKR model until the JKR equilibrium becomes unstable, and the contacts separate in a brittle manner. Note that it is possible for $a_f < a_{min}$ because a_{min} increases as a result of plastic deformation i.e. $R_{eff} > R$.

During subsequent load-unload cycles *without adhesion*, plastic deformation only occurs on the first load cycle. In the elasto-plastic/plastic deformation models *with adhesion*, there need not be purely elastic deformation on subsequent load cycles. There does not appear to be published work on analytical modeling of deformation under repeated loading-unloading of asperities in the presence of adhesion. There are serious difficulties in predicting the deformed asperity radius after a load-unload cycle. Clearly, if the separation of the contacts is ductile, and therefore does not necessarily occur at the contact interface, the asperity shape defined by the maximum load may not be preserved after the contacts separate.

MULTI-ASPERITY CONTACT RESISTANCE MODEL

The single asperity adhesive contact model discussed in the previous section will now be applied to the multiple-asperity system consisting of the contact tip (Figure 2) and the drain.

Figure 2. SEM micrograph showing a contact bump.

Consider N asperities on the contact tip, each with an end radius of curvature R, and heights $z_1 > z_2 > \ldots > z_N$, and let the roughness of the drain be neglected. The separation between the reference planes (the mean of asperity heights for the tip, and the flat surface for the drain) be d for a given contact force F, such that $z_n > d > z_{n+1}$. Asperities $1,2,\ldots,n$ come into contact, and the penetration of asperity i is given by

$$\delta_i = z_i - d \tag{8}$$

This procedure is similar to that of Greenwood and Williamson [18] with the exception that we use a finite number of asperities and include the effects of elastic and plastic deformation as well as adhesion. For a given δ_i, the contact radius a_i is given by the appropriate model: elastic, elasto-plastic, or plastic. Recall, however, that any finite contact with a 100 nm radius asperity results in elasto-plastic or plastic deformation. The force required to produce this deformation, F_i, is given by Eqns. (2) and (3) in the elasto-plastic and plastic regimes respectively. The total contact force is the sum of the loads on the individual asperities

$$F = \sum_{i=1}^{n} F_i \, . \tag{9}$$

For a given separation between the reference planes, the force on each asperity, and the radius of each of the corresponding contact spots can be obtained using Eqns. (2)-(7). Model calculations were performed for a surface covered with 100 asperities with a Gaussian distribution and a standard deviation of asperity heights of $\sigma = 10$ nm. For a contact force between 10-300 μN, most asperities are in elasto-plastic contact.

After the contact force has reached the desired maximum, the unloading and separation of the asperities are modeled, by increasing the separation between the reference planes. If an asperity has only undergone only elastic deformation (not applicable for the case studied here), its contact radius and load are governed by the JKR model during unloading. If an asperity has been plastically deformed, it has a new effective radius of curvature given by Eqns. (4),(5). Unloading proceeds according to the three different unloading modes [14] discussed in the previous section.

The remaining step is to determine the contact resistance, given a certain number of contact spots of known radii. First consider the constrictive resistance of a single spot of radius a, separating two semi-infinite bodies of resistivity ρ. If the radius a is small compared to the electron mean free path length l_e of the material, the resistance of the contact spot is dominated by the Sharvin mechanism (Jansen, [19]), in which electrons are projected ballistically through the contact spot without being scattered. In this case

$$R_{con} = \frac{4\rho l_e}{3\pi a^2} \quad (10)$$

On the other hand, if the contact radius is much larger than the mean free path length, the resistance is dominated a diffuse scattering mechanism, as given by the Maxwell spreading resistance formula (e.g. Holm, [20])

$$R_{con} = \frac{\rho}{2a} \quad (11)$$

Wexler [21] has given a simple interpolation formula which accounts for the transition between the Maxwell and Sharvin regimes:

$$R_{con} = \frac{4\rho l_e}{3\pi a^2} + \lambda \frac{\rho}{2a} \quad (12)$$

where λ is a slowly varying function of the ratio l_e/a, with $\lambda(0)=1$, and $\lambda(8)=0.694$.

In general, when multiple asperities come into contact, the result is multiple contact spots of varying sizes. The effective contact resistance arising from the contact spots depends on the radii of the spots (given by the contact area model discussed previously) and the distribution of the spots on the contact surface. A lower bound on the contact resistance can be obtained by assuming that contact spots are independent and conduct in parallel. Denoting the constrictive resistance of spot i as $R_{con,i}$,

$$1/R_{con,lb} = \sum_i 1/R_{con,i} \quad (13)$$

An upper bound on the contact resistance can be obtained by computing the resistance of a circular spot of radius a_{eff}, and of area equal to the total area of all the individual contact spots combined, i.e.

$$a_{eff} = \sqrt{\sum_i a_i} \quad (14)$$

This upper bound is equivalent to the exact solution when all the conducting spots become large enough to merge into a single conducting spot.

The model can now be used to calculate the contact resistance characteristic of a contact interface with multiple asperities, during loading and unloading. The model calculations are shown Figure 3 during loading. Comparison with the model without adhesion shows, as expected, that adhesion results in a significantly lower contact resistance at small contact loads, say less than 10 μN.

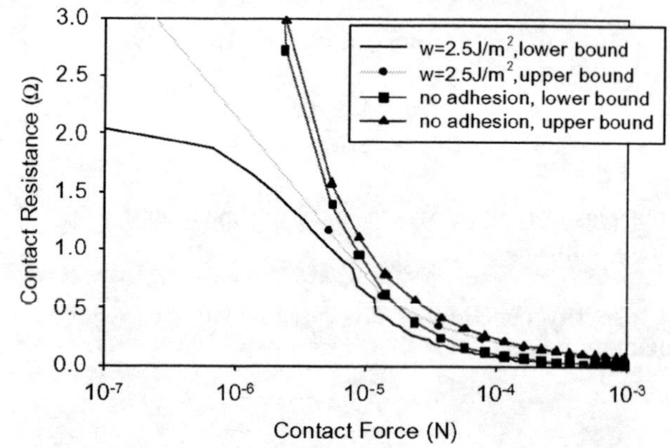

Figure 3. Modeled contact resistance as a function of load, with adhesion (w = 2.5 J/m^2), and without adhesion.

Figure 4 shows the contact resistance during a load-unload cycle, with a maximum load of 100 μN. The contact resistance remains nearly constant (within a few mΩ) while unloading, because the radii of the contact spots remain nearly constant (i.e. mode (b) unloading) reaching a negative value (i.e. the adherence force) before breaking free.

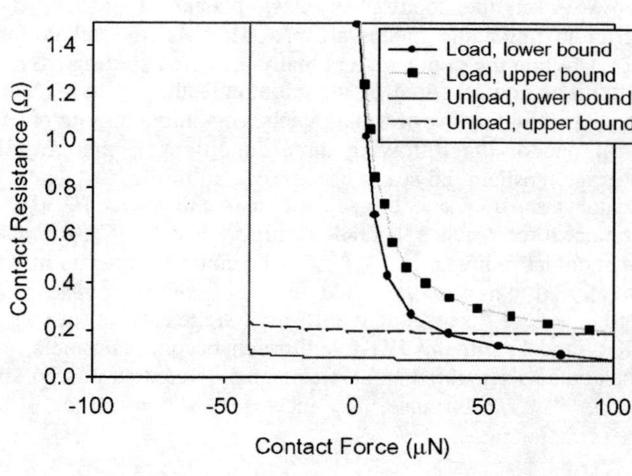

Figure 4. Modeled contact resistance as a function of load, while loading and unloading, with a maximum load of 100 μN.

The effects of σ and R on the adherence force are shown in Figure 5, once again for a maximum contact force F_f= 100 μN. The roughness σ is seen to have a strong effect. There are two reasons for the higher adherence of smoother surfaces. For a single asperity, the dependence of the adherence force on the ultimate contact load is weak. Therefore, a larger number of contacting asperities at a given maximum load results in a larger adherence force, in spite of a smaller maximum load on individual asperities. Secondly, during unloading of contacts with rougher surfaces, when the shortest asperities are separating, the tallest asperities (which suffer the most deformation) are still under compressive load, and "help" the short asperities to separate by pushing the contact surfaces apart. This effect becomes more pronounced with higher values of maximum load. If the contact surfaces are smooth, nearly all the asperities are under tensile load during separation. The effect of R on the adherence force is weak.

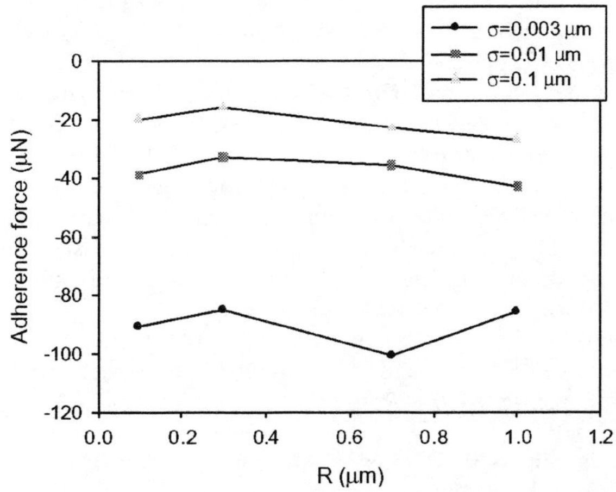

Figure 5. Predicted adherence force at a 100 μN

The dependence of the contact resistance lower and upper bounds on R and σ is shown in Fig. 6. As surface roughness decreases, the total contact area at a given load increases, as discussed earlier. Therefore, the upper and lower bounds on the contact resistance decreases as surface roughness decrease.

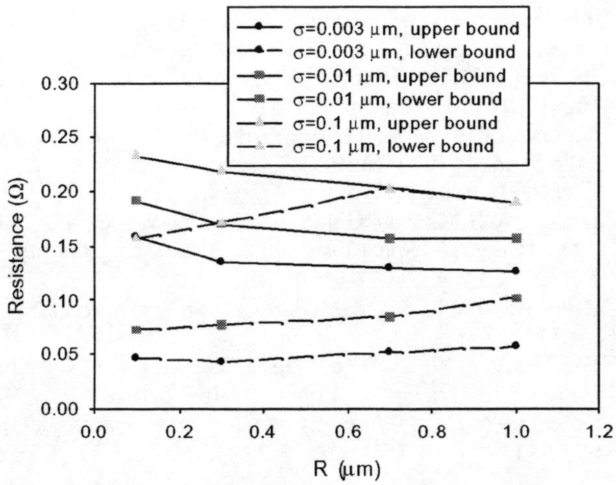

Figure 6. Predicted contact resistance bounds at a load of 100 μN.

MEASUREMENTS

Hysteresis in the measured load-unload contact resistance of a microswitch was used to motivate this study of adhesion effects in electrical contacts. Contact resistance during load and unload of a new switch has been measured as a function of actuation voltage and is shown in Fig. 7. A model of contact force vs. actuation voltage [2] has then been used in order to determine contact force vs. actuation voltage which, when combined with Fig. 7 produces a plot of contact resistance vs. contact force (Fig. 8). The effect of hysteresis is clearly evident in Figs. 7-8 in which the separation of the contact during unloading takes place at a lower voltage than needed to initiate contact during loading. The hysteresis due to electrostatic snap-through accounts for about 1 V of hysteresis. Figure 8 also shows that a tensile (pull-off) force is needed to break the contact. In [10] Majumder further shows that the pull-off force increases and the contact resistance decreases as a result of cycling the switch up to 10^4 times. Commonly, switches with a mechanical restoring force of 100 μN stick permanently closed when cycled 10^4–10^7 cycles with 50 μN or greater contact force. Comparison of the models with measured contact resistance is complicated by large variations in the above aspects of the measured characteristics.

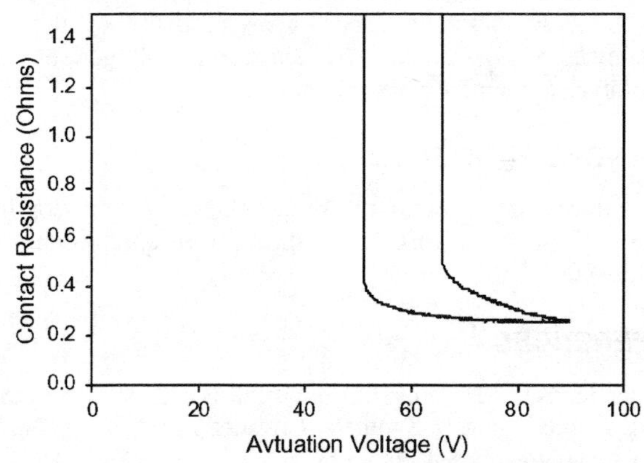

Figure 7. Contact resistance of a microswitch, measured as a function of gate voltage while loading and unloading.

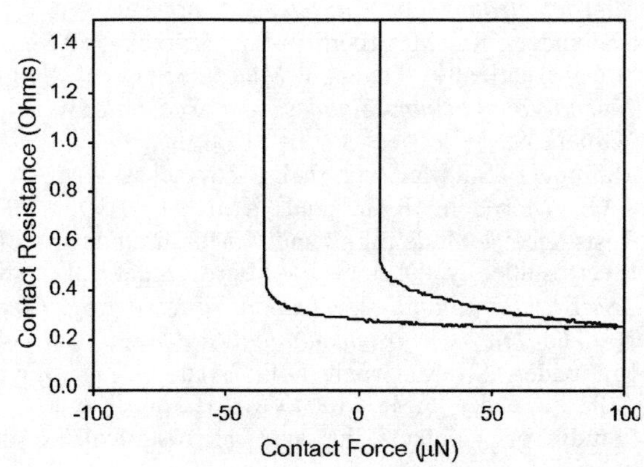

Figure 8. Contact resistance of a microswitch, measured as a function of contact force while loading and unloading.

The evolution of the measured contact resistance characteristics was studied quantitatively [10], specifically to determine the effects of the contact force and the number of cycles on the evolution of the contact resistance and the adherence force. The cause of the evolution in contact characteristics may be due to changes in the surface topology with cycling, i.e. the roughness of the surface and the radii of the asperities. Additionally, the surface properties (specifically, the surface energy) may change with cycling. Progressive material transfer between the contact bodies may occur which would tend to remove contaminant films and cause clean metallic surface to be exposed.

CONCLUSIONS

The contact resistance characteristics of a MEMS microswitch with gold-on-gold contacts have been studied analytically and experimentally. The multi-asperity analytical model includes the effects of elastic and plastic deformation, adhesion, and scale-dependent constrictive resistance. The number of asperities in contact is small enough that a discrete distribution of asperity heights is used. Due to the combined effects of plasticity and adhesion, the surface profile evolves with repeated load and unload cycles. Also the effect of adhesion is to produce significant hysteresis in the contact resistance vs. contact force characteristics. Measurements of the contact resistance as a function of actuation voltage show good qualitative agreement with the model

ACKNOWLEDGMENTS

The authors are grateful to Analog Devices, Inc. for having supported most of this work through research contracts to Northeastern University.

REFERENCES

1. Rebeiz, G.M., and Muldavin, J.B., 2001, "RF MEMS Switches and Switch Circuits", *IEEE Microwave Magazine*, pp. 59-71.
2. Zavracky, P.M., Majumder, S., and McGruer, N.E., 1997, "Micromechanical Switches Fabricated Using Nickel Surface Micromachining," *Journal of Microelectromechanical Systems*, **6**, pp. 3-9.
3. Majumder, S., McGruer, N.E., Zavracky, P.M., 1997, "Electrostatically Actuated Micromechanical Switches," *Journal of Vacuum Science and Technology, A15*, pp. 1246-1249.
4. Majumder, S., McGruer, N.E., Zavracky, P.M., Adams, G.G., Morrison, R.H., and Krim, J., 1998, "Contact Resistance Modeling and Measurements of an Electrostatically Actuated Micromechanical Switch," *ASME International Mechanical Engineering Congress*, **DSC-66**, *Microelectromechanical Systems*, pp 341-346.
5. Majumder, S., McGruer, N.E., Adams, G.G., Zavracky, A.P., Zavracky, P.M., Morrison R.H., and Krim, J., 2001, "Study of Contacts in an Electrostatically Actuated Microswitch," *Sensors and Actuators A*, **93**, pp. 19-26.
6. Kruglick, E.J.J., and Pister, K.S.J., 1999, "Lateral MEMS Microcontact Considerations," *Journal of Microelectromechanical Systems*, **8**, pp. 264-271.
7. Lafontan, X., Dufaza, C., Robert, M., Pressecq, F., and Perez, G., 2001, "Concepts, Characterization and Modeling of MEMS Micro-Switches with Gold Contacts in MUMPS", *Proceedings of SPIE – The International Society for Optical Engineering*, **4408**, pp. 381-390.
8. Beale, J., and Pease, R.F., 1994, "Limits of High-Density, Low-Force Pressure Contacts," *IEEE Transactions on Components, Packaging, and Manufacturing Technology Part A*, **17**, pp. 257-262.
9. Bromley, S.C., and Nelson, B.J., 2001, "Performance of Microcontacts Tested with a Novel MEMS Device," *Electrical Contacts, Proceedings of the Annual Holm Conference on Electrical Contacts*, pp. 122-127.
10. Majumder, S., 2003, *Contact Properties of a MEM Switch with Gold-on-Gold Contacts*, Doctoral Dissertation, Department of Electrical and Computer Engineering, Northeastern University, Boston, MA.
11. Johnson, K.L., Kendall, K., and Roberts, A.D., 1971, "Surface Energy and the Contact of Elastic Solids," *Proceedings of the Royal Society of London*, **A324**, pp. 301-313.
12. Derjaguin, B.V., Muller, V.M., and Toporov, Y.P., 1975, "Effect of Contact Deformation on the Adhesion of Particles", *Journal of Colloidal and Interface Science*, **53**, pp. 314-326.
13. Tabor, D., 1976, "Surface Forces and Surface Interactions", *Journal of Colloidal and Interface Science*, **58**, pp. 1-13.
14. Maugis, D., and Pollock, H.M., 1984, "Surface Forces, Deformation and Adherence at Metal Microcontacts", *Acta Metallurgica*, **32**, pp. 1323-1334.
15. Roy Chowdury, S.K., and Pollock, H.M., 1981, "Adhesion Between Metal Surfaces: The Effect of Surface Roughness," *Wear*, **66**, pp. 307-321.
16. Studman, C.J., Moore, M.A. and Jones, S.E., 1976, "On the Correlation of Indentation Experiments", *Journal of Physics D, Applied Physics*, **10**, pp. 949-956.
17. Johnson, K. L., 1985, *Contact Mechanics*, Cambridge University Press, Cambridge, MA.
18. Greenwood, J. A., and Williamson, J.B.P., 1966, "Contact of Nominally Flat Surfaces", *Proceedings of the Royal Society (London)*, **A295**, pp. 300-319.
19. Jansen, A.G.M., van Gelder, A.P., and Wyder, P., 1980, "Point-Contact Spectroscopy in Metals", *Journal of Physics*, **C13**, pp. 6073-6118.
20. Holm, R., 1967, *Electric Contacts*, Springer-Verlag, New York.
21. Wexler, G., 1980, "The Size Effect and the Non-Local Boltzmann Transport Equation in Orifice and Disk Geometry", *Proceedings of the Physical Society*, **89**, pp. 927-941.

Proceedings of
2003 STLE/ASME Joint International Tribology Conference
Ponte Vedra Beach, Florida USA, October 26-29, 2003

2003-TRIB-0271

ADHESION ANALYSIS FOR MEMS BASED ON ELECTRICAL CONTACT RESISTANCE MEASUREMENTS*

L. Kogut, Post-Doctoral Fellow, Mem. ASME **K. Komvopoulos, Professor, Fellow ASME**

Department of Mechanical Engineering **University of California** **Berkeley, CA 94720**

*This article is based on a full-length paper accepted for publication in Journal of Applied Physics

Because adhesion forces are especially important at the submicron scale,[1] they play a dominant role in several fields of nanotechnology, such as biology, atomic force microscope (AFM) imaging, magnetic disk drives, and microelectromechanical systems (MEMS). The profound importance of adhesion forces in MEMS has been the principal theme of several studies.[2] A common approach[3] for measuring the surface energy is based on balancing the elastic energy stored in microcantilever beams partially adhered to substrates with the work of adhesion, assumed equal to the surface energy multiplied by the apparent area of the attached beam length. However, because the apparent contact area is significantly larger than the real contact area and the elastic energy stored in the deformed asperity microcontacts is neglected, this traditional method may greatly underestimate the interfacial adhesion energy. Consequently, the objective of this study was to develop a method for determining indirectly adhesion forces and adhesion energies form relatively simple *in situ* electrical contact resistance (ECR) measurements. The method presented herein is based on a theoretical treatment of the ECR encountered during contact of isotropic, conductive, rough surfaces,[4] using multi-scale fractal description of the equivalent surface topography,[5] constitutive contact relations for elastic-perfectly plastic asperity microcontacts,[6] and size-dependent constriction resistance of microcontacts.[7] Results are presented for the adhesion force and adhesion energy in terms of ECR for different surface topographies.

Figure 1 shows a schematic with the various forces arising at contact interfaces. The external force F_e is produced by electrostatic actuation. The total adhesion (attractive) force F_a may be due to various adhesion mechanisms, e.g., capillary, electrostatic, and van der Waals attractive forces and hydrogen bonding.[2] A restoring force F_r may be encountered in the case of microstructures suspended over substrates by flexure beams. This force can be obtained by multiplying the stiffness of the flexure beams by the vertical displacement of the microstructure from its free-standing position to contact. The contact load P is the sum of the contact (repulsive) loads P_i due to deformation of the asperity microcontacts. The contribution of the gravitational force is neglected as several orders of magnitude less than the external force. Force equilibrium of the system shown in Fig. 1 yields

$$F_a = P + F_r - F_e . \tag{1}$$

Since F_r and F_e can be easily obtained, the adhesion force F_a can be determined indirectly through measurement of the contact load P and use of Eq. (1).

The contact load can be estimated from relatively simple ECR measurements.[4] Since the contact resistance R is a function of contact load, surface topography, and mechanical and electrical properties of the contacting surfaces,[4] it provides a convenient means of determining the contact load when the geometrical, mechanical, and electrical properties of the interacting surfaces are known. Specifically, it is necessary to determine *a priori* the apparent contact area A_a, fractal roughness G and fractal dimension D (for fractal description of the surface topography), hardness of the softer surface H (assumed equal to KY, where Y is the yield strength of the softer material and K is typically equal to 2.8[8]), reduced elastic modulus E, defined as $E = \left[(1-\nu_1^2)/E_1 + (1-\nu_2^2)/E_2\right]^{-1}$, where ν_1, ν_2, and E_1, E_2 are the Poisson's ratios and elastic moduli of the two surfaces, respectively, specific resistivities of the contacting surfaces ρ_1 and ρ_2, respectively, and the average electron mean free path λ, i.e., $\lambda = (\lambda_1 + \lambda_2)/2$. For convenience, the following dimensionless parameters are defined $G^* = G/A_a^{1/2}$, $\lambda^* = \lambda / A_a^{1/2}$, $F_a^* = F_a/(A_a E)$, $P^* = P/(A_a E)$, $F_r^* = F_r/(A_a E)$, $F_e^* = F_e/(A_a E)$,

$R^* = RA_a^{1/2}/(\rho_1+\rho_2)$, and $A^* = A/A_a$, where the real contact area A is the sum of the microcontact areas a_i (Fig. 1).

The adhesion energy is defined as $\Delta\gamma = \gamma_1+\gamma_2-\gamma_{12}$, where γ_1 and γ_2 are the surface energies of the contacting surfaces and γ_{12} is the surface energy of the interface. The adhesion energy can be determined from an energy-based approach. In the absence of adhesion forces, the contact load is equal to $F_e^*-F_r^*$ (Eq. (1)). The actual contact load P^* that includes the adhesion force can be determined from ECR measurements, as discussed above. The dimensionless surface interference distance $\delta^* = \delta/A_a^{1/2}$ versus contact load is obtained from a numerical procedure.[6] The mechanical work W_m due to the adhesion force is the portion of the area below the δ^*-P^* curve between $F_e^*-F_r^*$ and P^* given by

$$W_m^* = \frac{W_m}{E(A_a)^{3/2}} = \int_{F_e^*-F_r^*}^{P^*} \delta^*(P^*)dP^*. \tag{2}$$

This mechanical work is attributed to the work of adhesion

$$W_a = A\,\Delta\gamma, \tag{3}$$

where the real contact area is determined from ECR measurements.[4] Since for lightly loaded contacts the ECR is dominated by the Sharvin mechanism,[9] the majority of the microcontact areas are smaller than the electron mean free path λ.[4] For the Sharvin case, the inverse dependence of ECR on microcontact area yields a simple relation between dimensionless real contact area and dimensionless contact resistance[4]

$$A^* = \frac{1}{2}\left(\frac{\lambda^*}{R^*}\right). \tag{4}$$

According to Eq. (4), when the Sharvin solution is applicable, the real contact area can be directly obtained from relatively simple ECR measurements that are independent of topography (fractal) parameters, mechanical properties, contact load, and deformation mode of asperity microcontacts. By substituting Eq. (4) in Eq. (3), the dimensionless work of adhesion is obtained as

$$W_a^* = \frac{W_a}{E(A_a)^{3/2}} = \frac{1}{2}\left(\frac{\lambda^*}{R^*}\right)\Delta\gamma^*, \tag{5}$$

where $\Delta\gamma^*$ is the dimensionless adhesion energy, defined as $\Delta\gamma^* = \Delta\gamma/[E(A_a)^{1/2}]$. Equating the mechanical work (Eq. (2)) with the work of adhesion (Eq. (5)) yields

$$\Delta\gamma^* = 2\left(\frac{R^*}{\lambda^*}\right)\int_{F_e^*-F_r^*}^{P^*} \delta^*(P^*)dP^*. \tag{6}$$

In this energetic approach, variations in the elastic strain energy stored in the suspension beams and electrostatic energy stored in the actuation electrodes of a MEMS device are negligible due to the small interference distances occurring during surface contact, typically a few nanometers.

In the general case that accounts also for elastic-plastic deformation, the constitutive laws for the contact area and contact load of a single microcontact are complex,[6] and it is impossible to obtain analytical solutions for the adhesion force. Therefore, a numerical scheme[4] was used to obtain solutions for R^* versus P^* for lightly loaded contacts. These numerical results can be used to determine P^* in terms of the measured R^* for given surface topography parameters D and G^* and material properties E/Y and λ^*. Then, the magnitude of the adhesion force can be obtained from Eq. (1) in terms of F_e and F_r. To study the surface topography effect on the adhesion force-ECR relation, three-dimensional fractal surfaces were generated[4] with D = 2.3 and various values of G^* ($10^{-10} \le G^* \le 10^{-7}$). Contact simulations were performed for conductive surfaces with E/Y = 106 and λ^* = 0.03. In the following, the results are limited to contacts possessing $A^* < 10^{-2}$, i.e., small average size of microcontacts relative to their average spacing in order for interaction effects to be negligible as secondary. For the sake of numerical solutions, it is assumed that $F_e^*-F_r^* = 10^{-5}$; thus, the minimum contact load (in the absence of adhesion) is $P^* = F_e^*-F_r^* = 10^{-5}$ (Eq. (1)). Since only adhesive contacts are considered in the present study, results are presented for $P^* > 10^{-5}$.

Figure 2 shows the adhesion force versus electrical contact resistance for different values of the fractal roughness parameter. The upper and lower bounds of the adhesion force data shown in Fig. 2 (as well as in following figure) correspond to the minimum and maximum ECR obtained for $A^* = 10^{-2}$ and $P^* > 10^{-5}$, respectively. For fixed fractal parameters, the adhesion force decreases with increasing ECR. This is because higher ECR values result from lower contact loads,[4] which, for fixed applied force $F_e^*-F_r^*$, yield lower adhesion forces (Eq. (1)). For a given ECR, the adhesion force increases with the fractal roughness parameter. Because higher G^* values are associated with rougher surfaces, higher contact loads must be applied in order to maintain a constant ECR value,[4] thereby resulting in higher adhesion forces.

To analyze the effect of the surface topography on the dependence of the adhesion energy on ECR, simulations were performed for conductive surfaces with the same parameters used to obtain the adhesion force results (Fig. 2). By curve fitting the numerical solutions of the surface interference distance versus contact load, the δ^*-P^* response is expressed as

$$\delta^* = a(P^*)^b, \tag{7}$$

where a and b are constants depending on the mechanical properties and surface topography (fractal) parameters. Substituting Eq. (7) into Eq. (6) and performing the integration yields

$$\Delta\gamma^* = 2\frac{a}{b+1}\left(\frac{R^*}{\lambda^*}\right)\left[(P^*)^{b+1} - (F_e^*-F_r^*)^{b+1}\right], \tag{8}$$

where P^* can be determined from ECR measurements as discussed earlier.

Figure 3 shows the adhesion energy versus electrical contact resistance for various values of the fractal roughness parameter. For fixed fractal parameters, the adhesion energy decreases with increasing ECR. As mentioned earlier, high ECR values correspond to low adhesion forces and, therefore, less mechanical work (Eq. (2)). Consequently, the balance between mechanical and adhesion work yields low adhesion energy. For a given ECR, the adhesion energy increases with the fractal roughness parameter. Higher G^* values are associated with rougher surfaces, characterized by higher adhesion forces for the same ECR and, in accordance, higher adhesion energies. Determination of the adhesion energy of contacting bodies with known surface energies γ_1 and γ_2 provides a means of estimating the surface energy of the interface γ_{12}. Moreover, the same approach can be used to obtain the surface energy of two identical conductive surfaces ($\gamma_1 = \gamma_2$ and $\gamma_{12} = 0$).

ACKNOWLEDGMENTS

This research was supported by the National Institute of Standards and Technology under Grant No. 60NANB1D0078. The financial support provided to the first author (L.K.) by a Fulbright Fellowship and a New-England (Technion) award is also gratefully acknowledged.

REFERENCES

[1] K. Kendall, Science **263**, 1720 (1994).
[2] K. Komvopoulos, J. Ashes. Sci. Technol. **17**, 477 (2003).
[3] C. H. Mastrangelo and C. H. Hsu, J. Microelectromech. Syst. **2**, 33 (1993); **2**, 44 (1993)
[4] L. Kogut and K. Komvopoulos, J. Appl. Phys. **94**, (2003, in print).
[5] W. Yan and K. Komvopoulos, J. Appl. Phys. **84**, 3617 (1998).
[6] K. Komvopoulos and N. Ye, J. Tribol. **123**, 632 (2001).
[7] A. Mikrajuddin, F. G. Shi, H. K. Kim, and K. Okuyama, Mater. Sci. Semiconductor Processing **2**, 321 (1999).
[8] D. Tabor, *The Hardness of Metals* (Clarendon Press, Oxford, UK, 1951).
[9] Y. V. Sharvin, Sov. Phys. JETP **21**, 655 (1965).

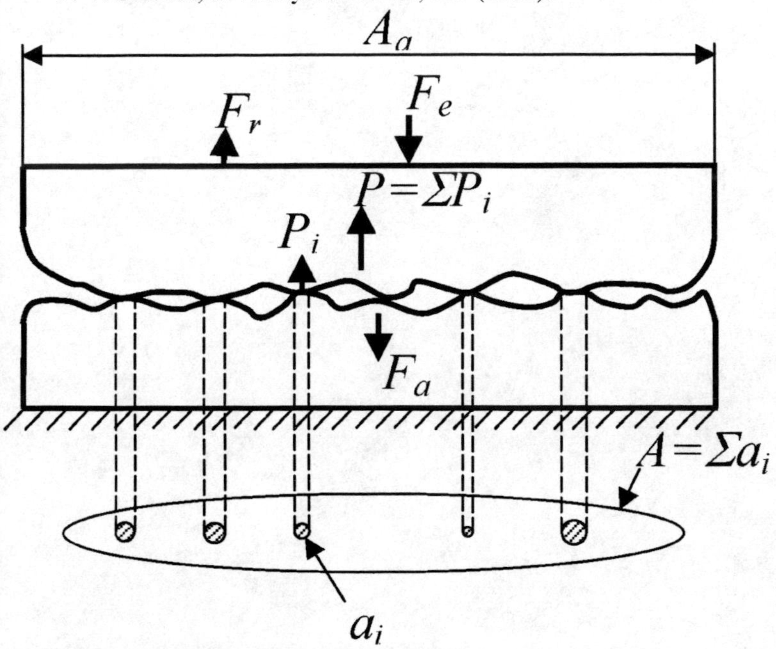

FIG. 1. Schematic of contact interface in MEMS devices and pertinent nomenclature.

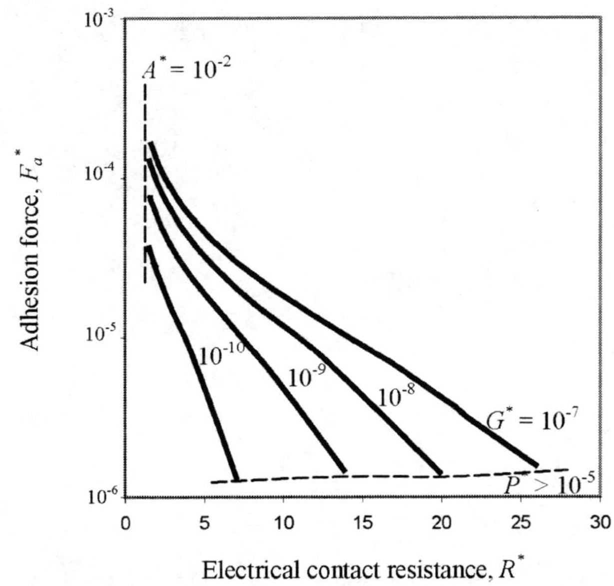

FIG. 2. Dimensionless adhesion force F_a^* versus dimensionless electrical contact resistance R^* for contacting rough surfaces with $F_e^* - F_r^* = 10^{-5}$, E/Y = 106, λ^* = 0.03, D = 2.3, and various values of G^*.

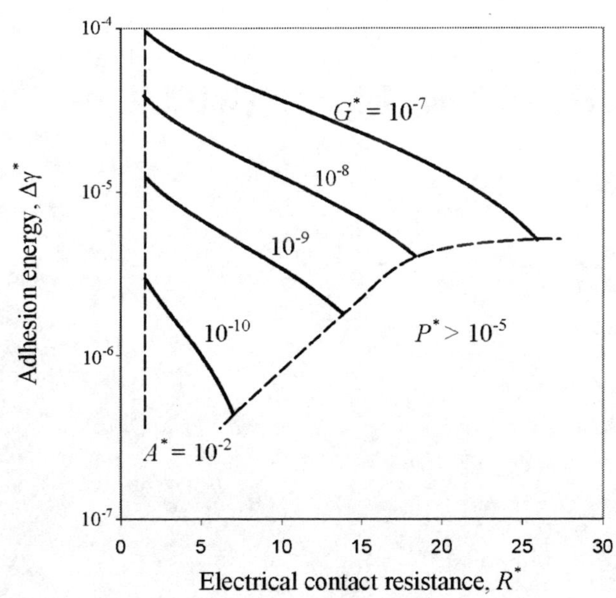

FIG. 3. Dimensionless adhesion energy $\Delta\gamma^*$ versus dimensionless electrical contact resistance R^* for contacting rough surfaces with $F_e^* - F_r^* = 10^{-5}$, E/Y = 106, λ^* = 0.03, D = 2.3, and various values of G^*.

Proceedings of
2003 STLE/ASME Joint International Tribology Conference
Ponte Vedra Beach, Florida USA, October 26-29, 2003

2003-TRIB-0272

DETERMINATION OF REAL MATERIAL PROPERTIES FROM NANOINDENTATION EXPERIMENTS*

L. Kogut, Post-Doctoral Fellow, Mem. ASME **K. Komvopoulos, Professor, Fellow ASME**

Department of Mechanical Engineering **University of California** **Berkeley, CA 94720**

*This article is based on a full-length paper submitted for publication in ASME Journal of Applied Mechanics

In recent years, due to extensive development of depth-sensing indentation techniques, nanoindentation has been used to evaluate the mechanical properties of surface layers and thin films of different materials. However, current nanoindentation procedures are based on simplified assumptions about the material behavior during unloading and empirical relations of the contact area (e.g., Oliver and Pharr, 1992) with little input from analytical and numerical solutions. Therefore, it is unclear what properties can be measured using instrumented nanoindentation techniques and what is the validity of the present procedures for measuring reduced elastic modulus and material hardness. Thus, the main objective of the present study was to analyze the validity of the current approaches for determining material properties and to propose an alternative approach for measuring the reduced elastic modulus, yield strength, and material hardness.

Figure 1 shows a rigid sphere of radius R indenting an elastic-perfectly plastic half-space. The indentation depth, δ, and contact radius, r, correspond to an indentation load, P. The most frequently measured mechanical properties using load and depth sensing nanoindentation techniques are the hardness and reduced elastic modulus. In a commonly used method (Oliver and Pharr, 1992), indentation load versus depth data are obtained for one complete load/unload cycle of arbitrary maximum load, P_{max}. The material hardness is then determined as

$$H = P_{max}/a, \tag{1}$$

where a is the contact area corresponding to P_{max}. Since a cannot be measured during loading, methods for estimating the contact area from the indenter shape function have been proposed (e.g., Oliver and Pharr, 1992). In the case of a relatively small indentation depth, a spherical indenter can be approximated as a paraboloid of revolution and, according to the elastic analysis of Sneddon (1965) and Oliver and Pharr (1992), the contact depth, h_c (Fig. 1), is given by

$$h_c = \delta - 0.75\frac{P_{max}}{S}, \tag{2}$$

where S is the experimentally measured stiffness at the inception of unloading. From geometric considerations, the contact area, a_s (hereafter referred to as the shape function of the indenter), is

$$a_s = \pi h_c\left(2R - h_c\right). \tag{3}$$

The analysis of the unloading data is based on the assumption that unloading is fully elastic. For this case, Hertz theory (Johnson, 1985) can be used to determine the reduced elastic modulus

$$E = \frac{\sqrt{\pi}}{2}\frac{S}{\sqrt{a}}, \tag{4}$$

where a in Eq. (3) is the contact area corresponding to the arbitrary maximum indentation load, P_{max}, and E is the reduced elastic modulus, given by $E = \left[\left(1-\nu_1^2\right)/E_1 + \left(1-\nu_2^2\right)/E_2\right]^{-1}$, in which E_1, E_2 and ν_1, ν_2 are the elastic moduli and Poisson's ratios of the two materials, respectively. In the case of a rigid indenter, $E_2 \to \infty$.

A finite element model was built and verified in order to analyze the validity of the current approaches for determining material properties and to propose an alternative approach for measuring the reduced elastic modulus, yield strength, and material hardness. In order to generalize the numerical solutions and eliminate the dependency of the results on input parameters, the global contact parameters and material properties are presented in dimensionless form, i.e., p_m/Y, δ / r', and E/Y, where Y is the material yield strength and r' is the radius of the truncated contact area given by

$$r' = \sqrt{\delta(2R - \delta)}. \tag{5}$$

The dependence of the mean contact pressure, p_m/Y, on indentation depth, δ / r', is shown in Fig. 2 for different material properties in the range of $11 \le E/Y \le 450$. Figure 2 shows that yielding commences when $p_m/Y = 1.07$, i.e., $\delta / r' = 1.78/(E/Y)$, as predicted by the Hertz theory. This is the lower bound of the elastic-plastic deformation regime. The upper bound of the elastic-plastic deformation regime is defined by the inception of fully plastic deformation, determined by the indentation depth at which the mean contact pressure reaches its maximum value for the first time, i.e., the material hardness. Therefore, applying an arbitrary indentation load in order to measure the hardness could result in measuring an arbitrary contact pressure that is smaller than the actual material hardness. In order to measure the actual material hardness, it is required to apply a specific indentation load that corresponds to the boundary between elastic-plastic and fully plastic deformation regimes. By curve fitting the numerical results of the maximum contact pressure, the hardness can be expressed in terms of the reduced elastic modulus and yield strength as,

$$\frac{H}{Y} = 0.201 \ln\left(\frac{E}{Y}\right) + 1.685 , \tag{6}$$

while the corresponding indentation depth at the inception of fully plastic deformation is given by

$$\frac{\delta}{r'} = \frac{1}{1 + 0.037(E/Y)} . \tag{7}$$

For the case of a spherical rigid indenter and a rigid-perfectly plastic half-space, the hardness is given by the similarity solution and is equal to $3Y$, i.e., independent of the elastic material properties. According to Eq. (6), this condition is reached when $E/Y=700$ and for materials possessing $E/Y>700$ the relation $H/Y=3$ should be used. However, the hardness of materials with $E/Y < 270$ is much less than $3Y$. The deviation from the similarity solution increases as E/Y decreases, demonstrating a significant effect of elastic deformation on the indentation behavior of these materials.

The present finite element analysis is also used to examine the conditions under which the assumptions used in traditional nanoindentation approaches for measuring the reduced elastic modulus and hardness are valid. The evaluation comprises indentation loading up to the inception of fully plastic deformation, where the material hardness should be measured, followed by full unloading. As discussed earlier, a major assumption in nanoindentation hardness and reduced elastic modulus measurement is that the contact area determined from the indenter shape function, a_s (Eqs. (2) and (3)). Equation (2) is based on the assumption of purely elastic deformation and small surface indentation (linear elasticity). However, the indentation depth and plastic region at the inception of fully plastic deformation may be large, depending on the magnitude of E/Y. Simulation results show that the error between a_s and a is less than 5.7% when $E/Y < 50$ and, therefore, the approximation of the spherical indenter by a paraboloid of revolution is acceptable, even for a large indentation depth distance. However, for $E/Y > 50$, the accuracy of a_s decreases, probably due to extension of the plastic region beyond the contact edge to the free surface of the half-space. It should be noted that a_s is used for estimating the contact area both for hardness (Eq. (1)) and reduced elastic (Eq. (4)) determinations.

The validity of the procedure for determining the reduced elastic modulus (Eq. (4)) is also examined, using the contact area at the inception of unloading, a, and the contact area obtained from the shape function of the spherical indenter, a_s (Eq. (3)), all corresponding to the inception of fully plastic deformation. Equation (4) assumes purely elastic unloading and, consequently, uses the contact area predicted by the Hertz solution (through h_c, Eq. (2)). Nevertheless, the contact area at the inception of unloading is different from that predicted by the Hertz theory, particularly for materials with high E/Y values that are dominated by plastic deformation. When using the contact area, a, the accuracy of Eq. (4) increases with decreasing E/Y, i.e., material response is dominated by the elastic response. Significantly lower error in determining the reduced elastic modulus is produced by using the shape function, a_s, in Eq. (4). The reduced elastic modulus can be predicted from Eqs. (4) and (5) with accuracy better than 6.3% when the unloading is initiated from a maximum load in the elastic or elastic-plastic deformation regimes. However, unloading from indentation depth distances well within the fully plastic deformation regime increases the error for all materials in the range of $11 \le E/Y \le 450$ due to extensive plastic deformation.

As discussed earlier, applying an arbitrary maximum indentation load for measuring the material hardness may yield a mean contact pressure less than the actual hardness. The dimensionless indentation depth corresponding to the real material hardness is given by Eq. (7) and the corresponding indentation load can be calculated from the following equation derived by curve fitting the simulation results

$$\frac{P}{a'Y} = \frac{0.839 + \ln\left[\left(\frac{E}{Y}\right)^{0.656}\left(\frac{\delta}{r'}\right)^{0.651}\right]}{2.193 - \ln\left[\left(\frac{E}{Y}\right)^{0.394}\left(\frac{\delta}{r'}\right)^{0.419}\right]} , \tag{8}$$

where a' is the truncated contact area. However, in order to use Eqs. (7) and (8), it is necessary to measure first the reduced elastic modulus and yield strength. As explained earlier, Eq. (4) yields fairly accurate estimates of the reduced elastic modulus over a wide range of E/Y. However, it is preferred to calculate the reduced elastic modulus from the elastic indentation response using the Hertz solution for a single load path,

$$E = \frac{3}{4}\frac{P}{R^{1/2}\delta^{3/2}} . \tag{9}$$

Equation (9) may also be used to find the indenter radius if the material elastic properties are known *a priori*. Then, the hardness may be determined from an iterative procedure involving an initial assumption for the value of Y, calculation of the dimensionless indentation depth, δ / r', at the inception of the fully plastic deformation regime (Eq. (7)), and estimation of the corresponding dimensionless indentation load, $P / a'Y$ (Eq. (8)). Next, the new value of the indentation depth distance is measured experimentally based on the estimated indentation load, and the new value of Y is calculated from Eq. (7). This procedure can be repeated until a convergence to a specific tolerance is reached. Finally, the material hardness can be determined by substituting the obtained values of the reduced elastic modulus and yield strength in Eq. (6).

ACKNOWLEDGMENTS

This research was supported by the Computer Mechanics Laboratory at the University of California at Berkeley. The financial support provided to the first author (L.K.) by a Fulbright Fellowship and a New-England (Technion) award is also gratefully acknowledged.

REFERENCES

Johnson, K. L., 1985, *Contact Mechanics*, Cambridge University Press, Cambridge, UK, pp. 90-95 and 170-184.

Oliver, W. C., and Pharr, G. M., 1992, "An Improved Technique for Determining Hardness and Elastic Modulus Using Load and Displacement Sensing Indentation Experiments," Journal of Materials Research, 7, pp. 1564-1583.

Sneddon, I. N., 1965, "The relation Between Load and Penetration in the Axisymmetric Boussinesq Problem for a Punch of Arbitrary Profile," International Journal of Engineering Science, **3**, pp. 47-57.

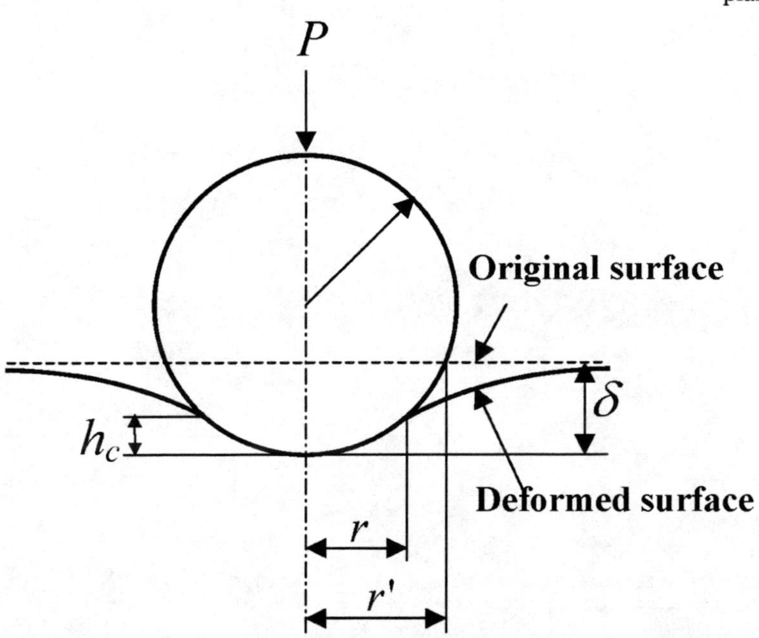

Fig. 1. Schematic of spherical nanoindentation and pertinent nomenclature.

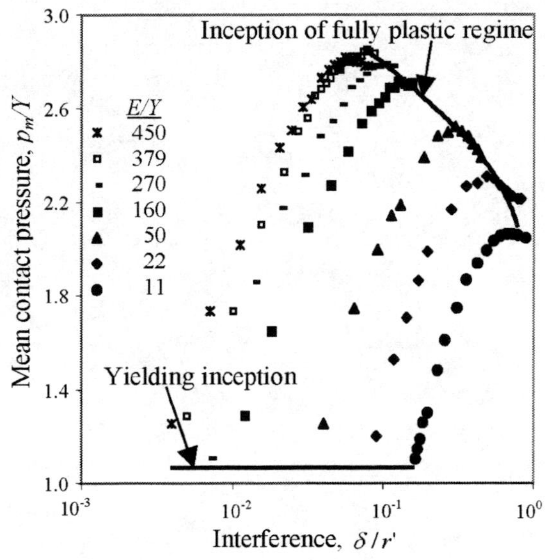

Fig. 2. Dimensionless mean contact pressure, p_m/Y, versus dimensionless interference, δ/r', in the elastic-plastic and fully plastic deformation regimes for different material properties, E/Y.

Proceedings of
2003 STLE/ASME Joint International Tribology Conference
Ponte Vedra Beach, Florida USA, October 26-29, 2003

2003-TRIB-0273

A TRANSIENT CONTACT STRESS ANALYSIS FOR MIXED-EHL OF ROUGH SURFACES, STRESS FILED ANIMATION, AND A VIRTUAL TEXTURING TECHNOLOGY

D. Zhu[1], Q. Wang[2], S. Liu[2] and A. Martini[2]

1 Eaton Corporation
Innovation Center
Southfield, MI 48037

2 Northwestern University
Mechanical Engineering
Evanston, IL60208

ABSTRACT

Understanding subsurface stresses that occur during the mixed-EHL (elastohydrodynamic lubrication) contact of rough surfaces is fundamental to mechanical design and failure analysis. This is driven by the fact that most of the critical tribological components of heavy-duty machinery are subjected to asperity contact. Moreover, the trend towards increasingly compact design further necessitates incorporating surface roughness. The deviation of the subsurface stress field for rough surfaces from that of smooth surfaces can be used to identify the effect of roughness on stress distribution and variation. This study presents a transient contact stress analysis for rough surfaces in mixed-EHL problems. The solutions are illustrated using animation of the results, and the method of model-based simulation is summarized for a surface virtualization technology.

KEYWORDS

EHL, Stress, Model-Based Simulation

INTRODUCTION

Analysis of mixed-EHL (elastohydrodynamic lubrication) contact is more complicated than that of dry contact (1-3) due to the presence of both fluid-surface and surface-surface interactions (4-6). Furthermore, mixed-EHL for rough surfaces is transient, even under steady state operating conditions, because of asperity interactions that occur when the two surfaces are in relative motion. The transient asperity contact that occurs during mixed-EHL creates transient pressure which, in turn, causes the subsurface stress to vary. It is important to address these complexities because most of the critical tribological components of heavy-duty machinery are subjected to asperity contact. This issue is accentuated by the trend towards increasingly compact design. Surface design should consider subsurface stress and ensure that the components satisfy performance and life requirements.

This paper presents a transient contact stress analysis for contacting surfaces in mixed-EHL. The results of this analysis are animated and then summarized as a comparison of critical geometric and mechanical quantities, such as surface roughness, film thickness, pressure, temperature, friction, and stress. The resulting trends can then be applied to surface design, which is directly linked to surface mechanics, lubrication, and performance evaluation. The mixed-EHL and contact stress visualization methods presented in this paper lay the groundwork for a model-based simulation approach to surface design and virtual surface texturing technology. This approach may find wide application in future design of mechanical components operating under severe conditions.

MODEL-BASED SIMULATIONS

In this study, the mixed-EHL problems are modeled using the Zhu-Hu unified micro mixed lubrication formulation (5-6), and the subsurface elastic stresses are analyzed with the Liu-Wang discrete convolution and fast Fourier transform (DC-FFT) approach (3). These models are used to analyze several cases of two contacting surfaces in mixed-EHL: Smooth surfaces with the presence of a single asperity, surfaces with sinusoidal asperities, and several different machined surfaces. The models are summarized in the following sections.

Mixed lubrication

Zhu and Hu's micro-EHL model [5-6] for the mixed lubrication of rough surfaces is applied and briefly mentioned here for clarity. This model considers that the lubricant film is continuous in non-contacting regions and that asperity contact occurs when the local film is sufficiently thin. It computes the film thickness from the deformed average gap and treat flows and asperity contact in one universal model. In the regions of hydrodynamic lubrication, the pressure is governed by the Reynolds equation expressed as follows:

$$\frac{\partial}{\partial x}\left(\frac{\rho}{12\eta^*}h^3\frac{\partial p}{\partial x}\right)+\frac{\partial}{\partial y}\left(\frac{\rho}{12\eta^*}h^3\frac{\partial p}{\partial y}\right)=U\frac{\partial(\rho h)}{\partial x}+\frac{\partial(\rho h)}{\partial t} \tag{1}$$

Here, p is the pressure, h the film thickness, ρ density of the lubricant, and η the viscosity. The x-coordinate has been chosen to coincide with the motion direction. The instantaneous lubricant film thickness, h, or the gap between two rough surfaces, is calculated by the geometric equation given below:

$$h=h_0(t)+B_x x^2+B_y y^2+V(x,y,t)+\delta_1(x,y,t)+\delta_2(x,y,t) \tag{2}$$

where B_x and B_y are constants related to the original geometry of contacting bodies, δ_1 and δ_2 denote the roughness amplitude of surface 1 and 2, respectively. V is the surface deformation, including the elastic deformation caused by pressure and frictional shear, u^P and u^s.

$$V(x_1,x_2)=u^p(x_1,x_2)+u^s(x_1,x_2) \tag{3}$$

An effective viscosity η^* has been introduced in Equation (5) to describe the non-Newtonian lubricant properties, which may be a function of the shear stress, τ_l, temperature, T, and pressure, p (τ_0 is a reference shear stress):

$$\frac{1}{\eta^*}=\frac{1}{\eta}\frac{\tau_0}{\tau_1}\sinh\left(\frac{\tau_1}{\tau_0}\right) \tag{4}$$

$$\text{and } \eta=\eta_0\exp(\alpha p) \tag{5}$$

The variation of the lubricant density is a function of pressure [7]:

$$\rho=\rho_0\left(1+\frac{0.6\times10^{-9}p}{1+1.7\times10^{-9}p}\right) \tag{6}$$

The unified lubrication-contact approach in mixed lubrication developed by Zhu and Hu [5-6] uses the same equation systems consistently in both hydrodynamic and contact regions. When $h\approx 0$, the left-hand side of Equation (1), which represents the lubricant flow due to the hydrodynamic pressure gradient, vanishes because of the third power of the film thickness. As a result, the Reynolds Equation is reduced to the following form:

$$U\frac{\partial h}{\partial x}+\frac{\partial h}{\partial t}=0 \quad at \quad h\le\varepsilon \tag{7}$$

Mises stress

Liu and Wang's DC-FFT approach (3) by means of influence coefficients is used for the stress analysis. The response to the loading of a Dirac delta function is a Green's function, g. Influence coefficients, D, are determined by integrating the product of a Green's function and a shape function over each discrete element, and are usually evaluated on each grid point. The stress field can be expressed in the following equation,

$$\sigma_{qr}^{ijl}\equiv\sigma_{qr}(x_i,y_j,z_l)=\sum_\xi\sum_\eta[p_{\xi\eta}D_{N\,qr}^{i-\xi,j-\eta,l}+s_{\xi\eta}D_{S\,qr}^{i-\xi,j-\eta,l}]$$

$$q,r=x,y,\text{or } z \tag{8}$$

where the influence coefficients are (the distribution of the shape function over an element is assumed to be unity, a rectangular impulse)

$$D_{N\,qr}^{i-\xi,j-\eta,l}=\int_{x_i-x_\xi-(\Delta_1/2)}^{x_i-x_\xi+(\Delta_1/2)}\int_{y_j-x_\eta-(\Delta_2/2)}^{y_j-x_\eta+(\Delta_2/2)}g_{N\,qr}(x,y,z_l)dxdy \tag{9}$$

By defining a new function, $T_{N\,qr}(x,y,z)\equiv 2\pi\int\int g_{N\,qr}(x,y,z)dxdy$ Eq. (9) can be rewritten as

$$D_{N\,qr}^{i-\xi,j-\eta,l}=[T_{N\,qr}(x_+,y_+,z_l)+T_{N\,qr}(x_-,y_-,z_l)-T_{N\,qr}(x_-,y_+,z_l)-T_{N\,qr}(x_+,y_-,z_l)]/2\pi \tag{10}$$

where

$x_+=x_i-x_\xi+\frac{\Delta_1}{2},\ x_-=x_i-x_\xi-\frac{\Delta_1}{2},$

$y_+=y_j-y_\eta+\frac{\Delta_2}{2},\ y_-=y_j-y_\eta-\frac{\Delta_2}{2}$

The influence coefficients obviously depend on the z coordinate, the intervals, (Δ_1,Δ_2), and the relative coordinates in the x and y directions. For an uncoated elastic halfspace, the Green's functions for the elastic field due to pressure and shear tractions are well known as the Boussinesq and Cerruti solutions [8], respectively. Based on the formulae derived by Ahmadi [9], Kalker [10], and Hills et al. [11], the influence coefficients for stresses due to the pressure and the shear traction may be expressed as,

$$T_{Nxx}(x,y,z)=-2\nu\tan^{-1}\frac{xy}{Rz}+2(1-2\nu)\tan^{-1}\frac{x}{R+y+z}-\frac{xz}{R(R+y)}$$

$$T_{Nyy}(x,y,z)=T_{Nxx}(y,x,z)$$

$$T_{Nxz}(x,y,z)=-z^2/[R(R+y)]$$

$$T_{Nzz}(x,y,z)=-\tan^{-1}\frac{xy}{Rz}+\frac{xz}{R(R+y)}+\frac{yz}{R(R+x)}$$

$$T_{Nxy}(x,y,z)=(2\nu-1)\log(R+z)-\frac{z}{R}$$

$$T_{Nyz}(x,y,z)=-z^2/[R(R+x)] \tag{11}$$

$$T_{Sxx}(x,y,z)=z(1-2\nu)\left[\frac{y}{R(R+z)}+\frac{z}{R(R+y)}\right]+2\ln(R+y)-\frac{2\nu x^2}{R(R+y)}$$

$$T_{Syy}(x,y,z)=2\nu\ln(R+y)-z(1-2\nu)\frac{y}{R(R+z)}-\frac{2\nu y}{R}$$

$$T_{Szz}(x,y,z)=-z^2/[R(R+y)]$$

$$T_{Sxy}(x,y,z)=\ln(x+R)-z(1-2\nu)\frac{x}{R(R+z)}-\frac{2\nu x}{R}$$

$$T_{Sxz}(x,y,z)=-\frac{xz}{R(R+y)}-\tan^{-1}\frac{xy}{Rz}$$

$$T_{Szy}(x,y,z)=-z/R \tag{12}$$

The stresses, σ_{qr}, are numerically calculated through the DC-FFT algorithm [3,12] by using the influence coefficients in terms of Eqs. (11-12), and the FFT brings no additional error. The von Mises stress is defined by

$$\sqrt{J_2} = \sqrt{[(\sigma_1 - \sigma_2)^2 + (\sigma_2 - \sigma_3)^2 + (\sigma_3 - \sigma_1)^2]/6} \quad (13)$$

Mises stresses are calculated with respect to the steps of each mixed-lubrication problem modeled, and animations for each case are created for the entire process of the stress variation as asperities pass through the contact region. Such animations help visualize the characteristics of the subsurface stress field for each surface texture design. They also reveal the position and the migration of critical spots and the level of stress severity, as well as their variation in the process of the mixed-lubricated contact.

SIMULATION RESULTS

The model-based simulations are used to analyze mixed-lubrication problems of surfaces with a single asperity, sinusoidal asperities, and machined roughness. Operating conditions are consistent for all cases. Both surfaces are steel and the material properties are E' = 219.8 GPa, η_o = 0.01119 Pa·s, and α = 14.94 GPa^{-1}. The geometric parameters are R_x = R_y = 19.005 mm. The applied load is fixed at w = 1600 N. Such conditions yield a Hertzian contact radius of a = 0.5953 mm and a maximum Hertzian pressure of P_h = 2.155 GPa. The rolling speed and the slide-to-roll ratio are U = 7.5 m/s and S = 0.25, respectively.

Single asperity

A single, semi-spherical asperity is created on the lower surface of an otherwise smooth EHL contact. Figure 1 presents snapshots of the pressure and film thickness distributions and the solutions for pressure, film thickness, and stress along the centerline (the x-axis). Animations of these results clearly show the variations of pressure, film thickness, and stress as the asperity moves through the contact region. The film thickness variation reflects the shape of the asperity. The pressure distribution shows a pressure peak corresponding to the film thickness minimum as the asperity is engaged in the contact and mixed lubrication process at the inlet. This pressure peak gets smaller relative to its value at the inlet as the asperity moves towards the outlet of the contact. However, the pressure wave caused by the pressure peak is stronger as the asperity gets closer to the outlet. The stress contour shows that the Mises stress is also larger at the inlet corresponding to the pressure variation.

Sinusoidal asperities

Two surfaces with identical sinusoidal roughness are engaged in an EHL contact. Figure 2 presents snapshots of the pressure and film thickness distributions and the solutions for pressure, film thickness, and stress along the centerline. Animation of these transient results shows the variations of pressure, film thickness, and stress as the asperities move through the contact region. The occurrence of peak-peak and peak-valley interactions is clearly visible in the film thickness distributions. The resultant pressure distribution shows the same pattern. As expected, the Mises stress experiences a corresponding stressed-released cycle as the phase difference of the asperities of the two surfaces varies.

Machined surfaces

Eight types of machined/polished surfaces are analyzed in the present work in order to study the effect of surface finish/machining processes on the contact and lubrication characteristics. In each case, both surfaces are of the same type, but not identical because the two simulation surfaces were taken from measurements at different locations on a real surface. Results are included in this paper for the ground surface case as an example. Figure 3 presents the film thickness and pressure distributions, the solutions for pressure, film thickness, and stress along the centerline, and an image of a sample ground surface. The film thickness and pressure distributions both show a strong preference along the direction of the roughness orientation. The asperity-induced stress varies with pressure. The animation can be used to follow highly stressed regions (indicated by a red color) and relate it to the surface irregularities. The results for all of the machined surface cases, as well as for the smooth-surface reference case, are summarized in Figure 4. The cases are compared in terms of composite surface roughness, film thickness/roughness ratio (λ), maximum pressure, maximum stress, contact load ratio, coefficient of friction, and maximum flash temperature.

CONCLUSIONS

This paper presents a model-based simulation approach that enables direct observation of the effects of asperity interaction in mixed-EHL contact. The results show that different surface roughness/finishing processes have a relatively limited influence on the average film thickness, but greatly affect the λ ratio, the pressure peak height, the subsurface stress, the load shearing, the friction, and the flash temperature. By analyzing these trends, the simulation can be used as a powerful tool for design optimization and virtual texturing.

ACKNOWLEDGMENTS

The authors would like to express their sincere gratitude for the support of the US National Science Foundation IGERT Program and Office of Naval Research, Eaton Corporation, and the Co-Center for Surface Engineering and Tribology at Georgia Institute of Technology and Northwestern University.

REFERENCES

1. Liu, S.B., and Wang, Q., 2001, "A Three-Dimensional Thermomechanical Model of Contact Between Non-Conforming Rough Surfaces," ASME Journal of Tribology, 123, pp. 17 - 26.
2. Ju, Y., and Farris, T.N., 1996, "Spectral Analysis of two-Dimensional Contact problems," ASME Journal of Tribology, 118, pp. 320 - 328.
3. Liu, S., and Wang, Q., 2002, "Studying Contact Stress Fields Caused by Surface Tractions with a Discrete Convolution and Fast Fourier Transform Algorithm," ASME Journal of Tribology, Vol. 124, pp. 36-45
4. Lubrecht, A. A., ten Napel, W. E., and Bosma, R., 1988, "The Influence of Longitudinal and Transverse Roughness on the Elastohydrodynamic Lubrication of Circular

Contacts," ASME Journal of Tribology, Vol. 110, pp. 421-426.

5. Zhu, D., and Hu, Y.Z., "A Computer Program Package for the Prediction of EHL and Mixed Lubrication Characteristics, Friction, Subsurface Stresses and Flash Temperatures Based on Measured 3-D Surface Roughness," Tribology Transactions, 44, pp.383-390 (2001).
6. Zhu, D., and Hu, Y.Z., "Effects of Rough Surface Topography and Orientation on the Characteristics of EHD and Mixed Lubrication in Both Circular and Elliptical Contacts," Tribology Transactions, 44, pp.391-398 (2001).
7. Dowson, D., and Higginson, G. R., 1966, "Elastohydrodynamic Lubrication," Pergamon Press.
8. Johnson, K.L., 1985, Contact Mechanics, Cambridge University Press.
9. Ahmadi, N., 1982, "Non-Hertzian Normal and Tangential Loading of Elastic Bodies in Contact," Ph.D. dissertation, Northwestern University, Evanston, IL.
10. Kalker, J.J., 1986, "Numerical Calculation of the Elastic Field in a Half-space," Communications in Applied Numerical Methods, 2, pp.401 - 410.
11. Hills, D.A., Nowell, D., Sackfield, A., 1993, Mechanics of Elastic Contacts, Butterworth Heinemann Ltd., Oxford.

Liu, S., Wang, Q., and, Liu, G., 2000, "A Versatile Method of Discrete Convolution and FFT (DC-FFT) for Contact Analyses," Wear, Vol. 243 pp. 101-111.

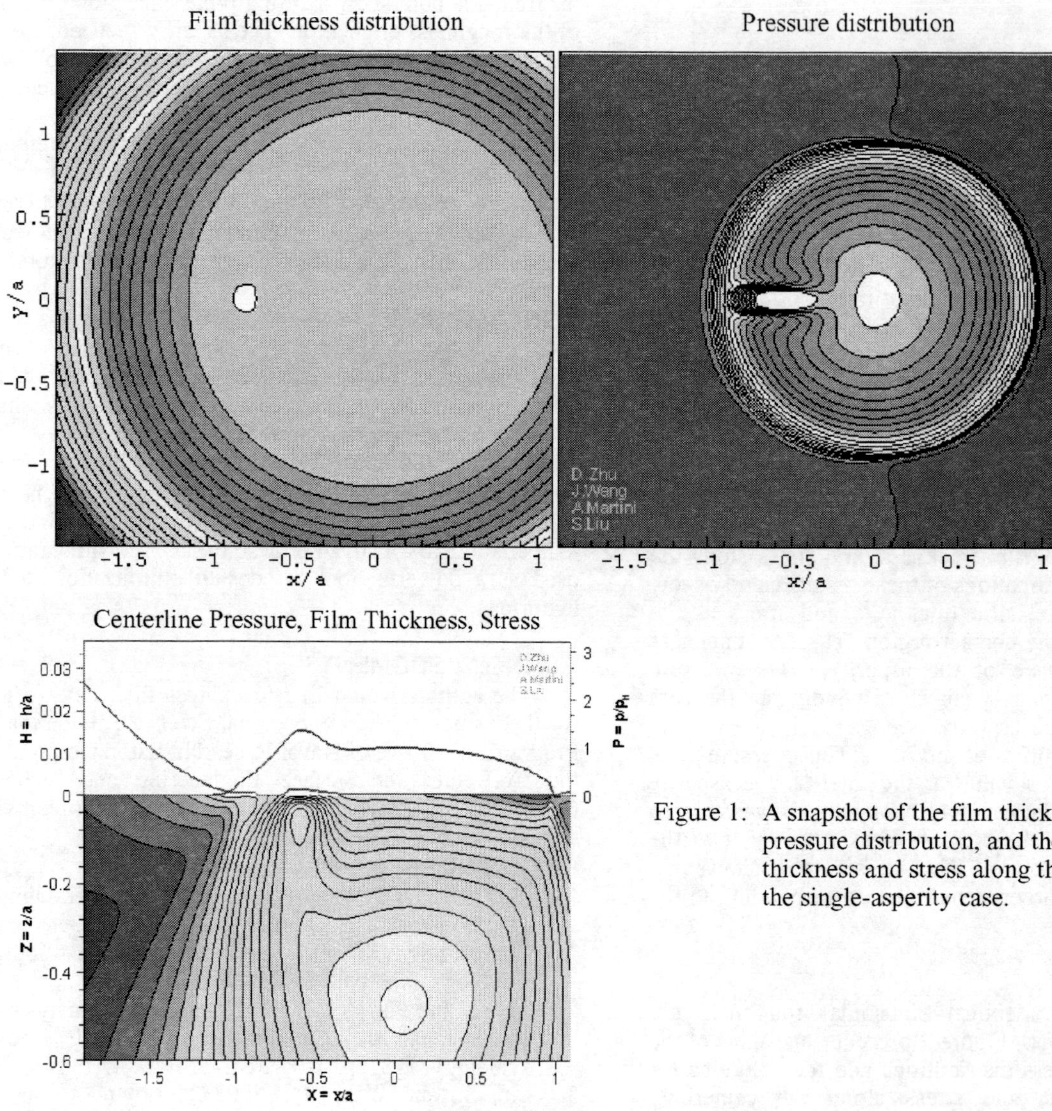

Figure 1: A snapshot of the film thickness distribution, pressure distribution, and the pressure, film thickness and stress along the centerline for the single-asperity case.

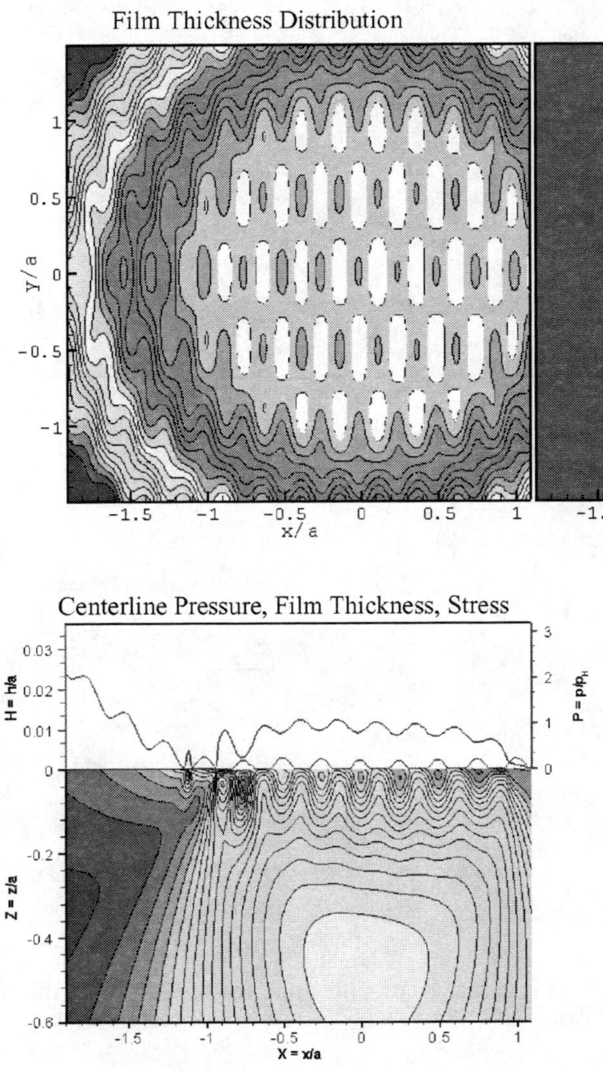

Figure 2: A snapshot of the film thickness distribution, pressure distribution, and the pressure, film thickness and stress along the centerline for the sinusoidal-asperity case.

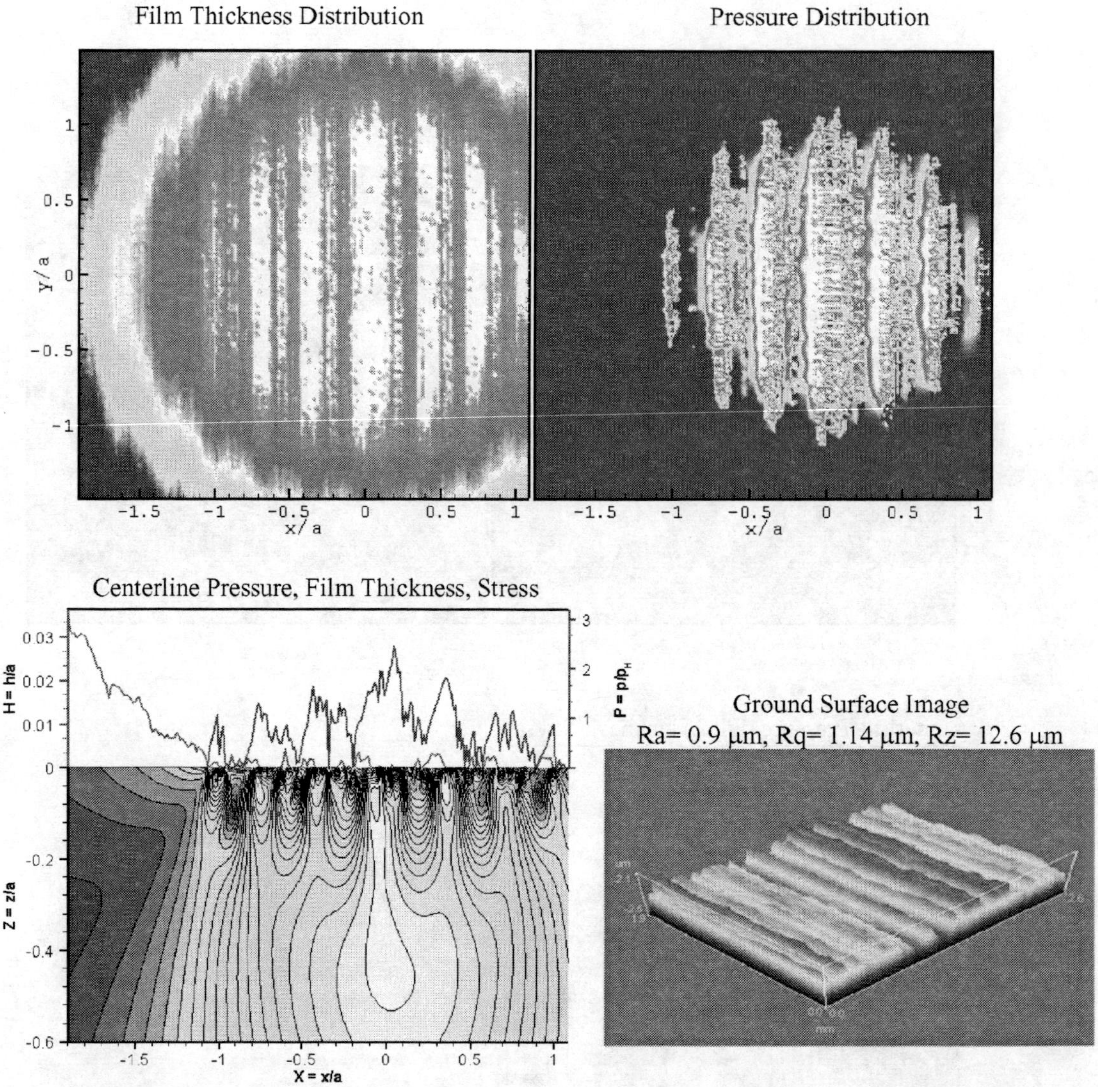

Figure 3: A snapshot of the film thickness distribution, pressure distribution, and the pressure, film thickness and stress along the centerline for the ground case. Also a sample image taken of a ground surface.

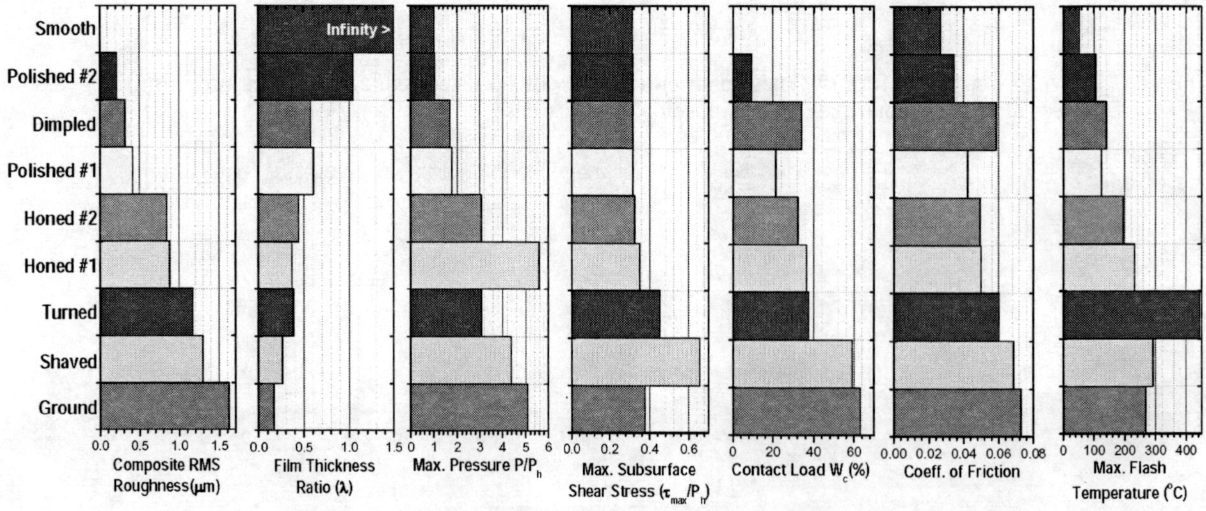

Figure 4: Summary of the results of eight different rough surfaces as compared to those for two contacting

Proceedings of
2003 STLE/ASME Joint International Tribology Conference
Ponte Vedra Beach, Florida USA, October 26-29, 2003

2003-TRIB-0274

PRESSURE DISTRIBUTION IN A HERTZ CIRCULAR CONTACT AFFECTED BY A CENTRAL, CIRCULAR DENT

Emanuel N. Diaconescu/University of Suceava, ROMANIA

Delia F. Cerlinca/University of Suceava, ROMANIA

ABSTRACT

Rolling contact fatigue is strongly influenced either by subsurface imperfections or by surface defects. The latter tend to become the major factor influencing contact fatigue. The first step to assess the effect of these defects upon contact fatigue life is to find the pressure distribution alterations they generate.

This paper investigates the pressure riser effect around a circular dent placed centrally in a Hertz circular contact. The solutions of Erdogan are used as a start point. The superposition principle for the half-space is applied to these solutions to find the pressure distribution. The involved integrals are evaluated numerically. The dent affects the Hertz pressure only in its immediate neighborhood. It produces significant pressure risers near the outer edge of the dent and a central circular region free of pressure inside. In comparison to the case of smooth surfaces, the outer radius of the contact extends by a very small amount dependent on dent geometry. Calculations evidence the influence of dent geometry upon pressure riser effect.

Keywords: circular elastic contact, surface defects, circular dent, pressure distribution.

INTRODUCTION

It is well known that surface defects, such as furrows, scratches, dents or bumps significantly reduce rolling contact fatigue life by surface nucleated failures. This is due to the fact that when entering the contact area, these defects alter the Hertz contact pressure and generate considerable pressure risers. In its turn, this pressure generates into the contacting bodies a modified internal stress field, which is characterized by local stress risers in the neighborhood of surface distress. Finally, the correlation between the variable stress state and the fatigue life allows to assess the life reducing effect of the considered surface defect.

The literature offers a limited amount of information upon this subject. The most known is the effect of furrows, treated by Chiu and Liu [1], Lubrecht, Dwyer-Joyce and Ioannides [2], Gao, Dwyer-Joyce and Beynon [3], Diaconescu and Cerlinca [4], of bumps and surface roughness as shown by Lubrecht and Ioannides [5]. Gao, Dwyer-Joyce and Beynon [3] investigated experimentally the effect of surface dents for a rather untypical dent geometry having conical or various pyramidal shapes.

This paper aims to deal theoretically with more realistic dents, which have a circular cross-section and a parabolic profile in the neighborhood of the initial contact. Such dents can occur by an impact between a ball and a flat surface. The solution is found by applying the superposition principle for the half-space to the solution of Erdogan for a circular ring of surface tractions. The integral equation obtained this way is then digitized and transformed into a system of linear algebraic equations. A program was developed under the Mathcad to solve the problem and to find the radial distribution of contact pressure. The effect of two geometrical dent parameters, namely the curvature radius of dent shoulder and the diametric extension of the dent upon the pressure risers is assessed numerically and plotted graphically.

NOMENCLATURE

- G: shear modulus;
- Q: normal load;
- R: radius of equivalent punch;
- c: radius of dent shoulder;
- q: line load applied on a circle in the plane xy of a half-space;
- w: normal displacement in the surface points.

METHOD OF SOLUTION

A circular Hertz contact can be modeled by a rigid sphere of radius R pressed against an equivalent elastic half-space made of a material that incorporates the elastic parameters of the materials of both contacting bodies. The circular dent can be considered placed either on this rigid sphere, either on the half-space. It is more convenient to consider that the defect belongs to the spherical surface because its geometry does not change by loading. The simplest situation is considered when the dent is placed centrally on the contact area. In this case, the profile of the contact is shown in Fig. 1.

It is assumed that the dent contacts initially the elastic half-space at a radius c measured from contact center and its curvature radius in the contacting points is r_0. Under an axial load, Q, the half-space deforms elastically and an annular contact area occurs. An inner circle of radius r_i and an outer circle of radius r_e bound this area.

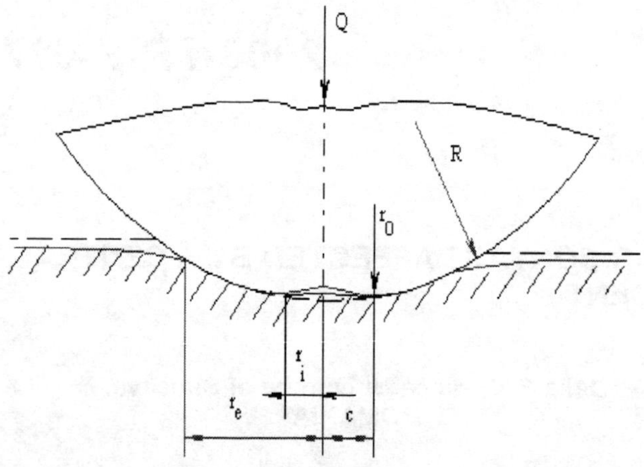

Fig. 1. Model of circular contact with a central dent

By assuming that the profile of the rigid dent is parabolic in the neighborhood of the initial circular contact, its radial profile is described by following equation:

$$z(r) = \begin{cases} \dfrac{r^2}{2R} + \dfrac{(c-r)^2}{2r_0} & \text{if } r < c; \\ \dfrac{r^2}{2R} & \text{if } |r| \geq c. \end{cases} \quad (1)$$

The problem is axi-symmetric and this feature simplifies the procedure of solution derivation. This symmetry suggests that it seems convenient to use as a start point the solutions supplied in 1965 by Erdogan for a half-space subjected to a ring of concentrated load on its bounding plane. These solutions can be found in the book of Hills, Nowell and Sackfield [6]. For the simpler case when the traction reduces to its normal component, the basic problem is illustrated in Fig. 2.

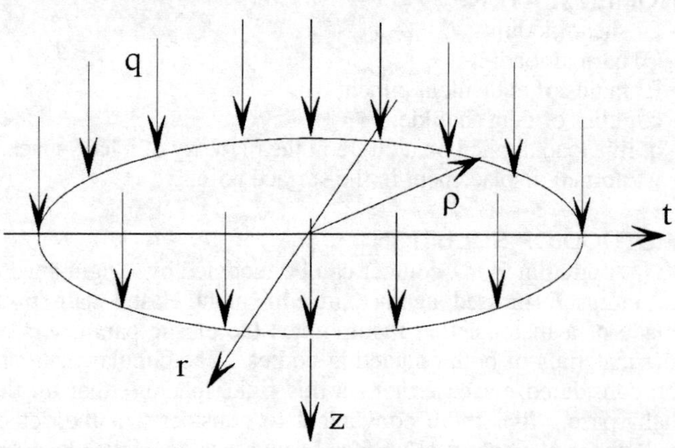

Fig. 2. The half-space loaded by a ring of concentrated normal tractions

In this case the surface displacements along z axis, in its turn chosen as the inner normal of the half-space, are expressed by following equation:

$$w(r) = \begin{cases} \dfrac{2}{\pi G}(1-\nu)\, q\, K\left(\dfrac{r}{\rho}\right) & \text{if } r < \rho; \\ \dfrac{2}{\pi G}(1-\nu)\, q \dfrac{\rho}{r} K\left(\dfrac{\rho}{r}\right) & \text{if } r > \rho. \end{cases} \quad (2)$$

where r is the position radius of the point in which the displacement w is calculated, ρ is the radius of the traction ring and $K(x)$ is the complete elliptic integral of the first kind. As known, this integral has the form:

$$K(x) = \int_0^{\frac{\pi}{2}} \frac{d\theta}{\sqrt{1 - x^2 \sin^2\theta}} \quad (3)$$

To pass from this solution to the discussed problem, the principle of effect superposition for the half-space must be applied to the solution (2). According to this principle, if the displacement $w' = w'(r)$ generated at radius r by a ring of normal traction q applied at a radius ρ is expressed as $w' = q\,w'(r,\rho)$, then the displacement $w = w(r)$ generated by an axi-symmetrical pressure distribution $p = p(r)$, applied over a circular annulus bounded by an inner circle of radius r_i and an outer one of radius r_e can be expressed as follows:

$$w(r) = \int_{r_i}^{r_e} p(\rho)\, w'(r-\rho,\rho)\, d\rho\,. \quad (4)$$

If Eq. (4) is applied to the solution (2), the following relations yield for the displacement $w(r)$:

$$w(r) = \begin{cases} \dfrac{4\eta}{\pi} \displaystyle\int_{r_i}^{r_e} K\left(\dfrac{r}{\rho}\right) p(\rho)\, d\rho & \text{if } r < r_i; \\ \dfrac{4\eta}{\pi} \displaystyle\int_{r_i}^{r_e} \dfrac{\rho}{r} K\left(\dfrac{\rho}{r}\right) p(\rho)\, d\rho & \text{if } r > r_e; \\ \dfrac{4\eta}{\pi} \left[\displaystyle\int_{r_i}^{r} \dfrac{\rho}{r} K\left(\dfrac{\rho}{r}\right) p(\rho)\, d\rho + \int_{r}^{r_e} K\left(\dfrac{r}{\rho}\right) p(\rho)\, d\rho \right] & \text{if } r \in [r_i, r_e], \end{cases} \quad (5)$$

where $\eta = (1-\nu_1^2)/E_1 + (1-\nu_2^2)/E_2$ is the elastic constant of involved materials, defined in terms of their Poisson ratios and Young moduli.

On the other hand, the integral condition of deformation over the contact area can be written under the form:

$$w(r) = \delta - z(r) \quad \text{for } r \in [r_i, r_e], \quad (6)$$

where δ is the penetration of the half-space and $z(r)$ is the initial ordinate of the ball with respect to the initial half-space. This equation is used further down to find the unknown pressure distribution.

Because the elliptic integral is, in its turn, defined by an integral, there is no chance to find an analytical solution for the problem. Therefore, the application of a numerical method is required. To this end, the contact must be digitized. The contact area is divided in $N+1$ concentric annuli of a width, depending on the radial position. These are numbered increasingly towards the outer radius. Let w_{ij} be the normal displacement generated by the pressure applied on element j in the points of central circle of the element i. The widths of these two annuli are, respectively, $2\delta_i$ and $2\delta_j$. Over each annulus, the pressure is assumed to be constant, namely p_i and p_j. Consequently, by having in mind the Eqs. 5 and the displacement w_{ij} is given by following equation:

$$w_{ij} = \begin{cases} \frac{4\eta}{\pi} p_j \int_{r_j-\delta_j}^{r_j+\delta_j} K\left(\frac{r_i}{x}\right) dx & \text{if } i<j; \\ \frac{4\eta}{\pi} p_j \int_{r_j-\delta_j}^{r_j+\delta_j} \frac{x}{r_i} K\left(\frac{x}{r_i}\right) dx & \text{if } i>j; \\ \frac{4\eta}{\pi} p_j \left[\int_{r_j-\delta_j}^{r_i} \frac{x}{r_i} K\left(\frac{x}{r_i}\right) dx + \int_{r_i}^{r_j+\delta_j} K\left(\frac{r_i}{x}\right) dx \right] & \text{if } i=j, \end{cases} \quad (7)$$

where r_i and r_j are the radii of central circles of the annuli i and j, respectively. The total displacement at radius r_i, w_i is the sum of all displacements $w_{i,j}$ when j takes values between 0 and N:

$$w_i = \sum_j w_{i,j}. \quad (8)$$

Expressions 6 and 8 yield the following linear algebraic equation with pressures p_j as unknowns:

$$\sum_j c_{ij}\, p_j = \frac{\pi}{\eta}(\delta - z_i) \quad (9)$$

where the coefficients c_{ij}, called influence coefficients, are given by following integrals:

$$\int_0^{\frac{\pi}{2}} \left(\sqrt{(r_j+\delta_j)^2 - r_i^2 \cdot \sin^2\theta} - \sqrt{(r_j-\delta_j)^2 - r_i^2 \cdot \sin^2\theta} \right) d\theta \qquad \text{if } j>i;$$

$$\int_0^{\frac{\pi}{2}} \frac{1}{\sin^2\theta} \left[\sqrt{r_i^2 - (r_j-\delta_j)^2 \sin^2\theta} - \sqrt{r_i^2 - (r_j+\delta_j)^2 \sin^2\theta} \right] d\theta \qquad \text{if } j<i;$$

$$\int_0^{\frac{\pi}{2}} \left[\sqrt{(r_i+\delta_i)^2 - r_i^2 \sin^2\theta} - r_i\cos\theta + \frac{1}{\sin^2\theta}\sqrt{r_i^2 - (r_i-\delta_i)^2 \sin^2\theta} - r_i \frac{\cos\theta}{\sin^2\theta} \right] d\theta \qquad \text{if } j=i. \quad (10)$$

If Eq. 9 is written for every value of i, an algebraic system of $N+1$ equations is obtained. This has $N+2$ unknowns, namely $N+1$ values of the pressures p_j and the penetration δ. A supplementary equation is needed. This is the condition of force balance equation on z direction, which can be written as:

$$4\pi \sum_j r_j\, p_j\, \delta_j = Q. \quad (11)$$

In order to have a more detailed insight in the region of the dent, the half-widths of the elementary annuli are chosen in a geometrical series, decreasing towards the inner radius of the contact. To solve the system when the radial extension of the contact is not known initially, one has to use for a start guessed values for the radii r_i and r_e. Then, a computer program adjusts these values until the calculated ones coincide with those assumed. To this end, the soft is developed under a mathematics program, like the Mathcad.

NUMERICAL RESULTS

A typical numerical result for the diametric profile of pressure distribution for $\overline{c} = 0.3$ is shown by continuous line in Fig. 3a in comparison with the Hertz profile in the absence of the defect, traced by dotted line. A 3-D plot of the same pressure distribution is illustrated in Fig. 3b. A very large dent of a radius of .3 of Hertz radius was chosen for illustration, in order to better evidence the particular aspects of pressure distribution.

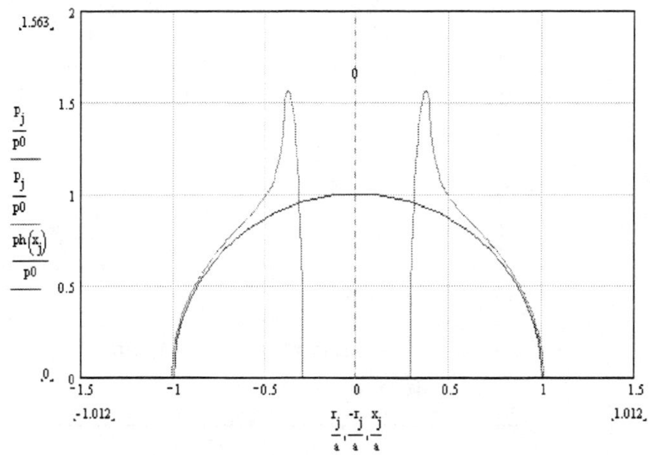

a: diametric profile

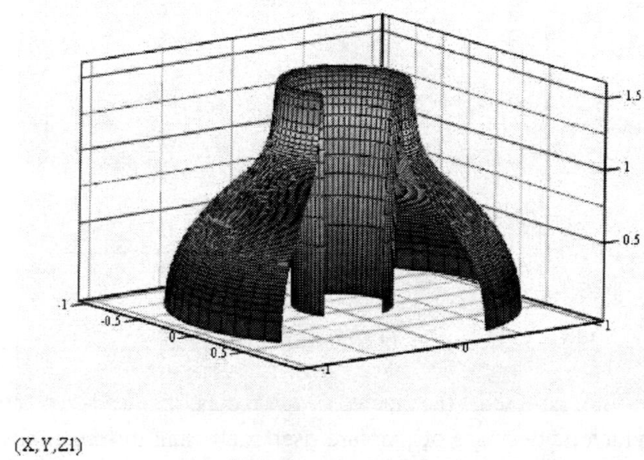

b: 3-D representation

Fig. 3. Pressure distribution in the presence of a circular dent

It is obvious that a small area free of pressure occurs around the centre of the contact. On a radial direction the pressure increases abruptly from the inner circle of contact area, reaches a maximum value in the region of dent shoulder and then tends asymptotically towards the hertz profile. Eventually, if the dent is large, the outside radius of contact area exceeds by a very small amount the value of Hertz radius.

A pressure riser coefficient due to the presence of the dent is defined as the ratio between the calculated maximum pressure and the corresponding maximum Hertz pressure in the case there is no dent:

$$k_p = \frac{p_{max}}{p_0}. \quad (12)$$

The numerical results obtained for various parameters of the dent geometry allow to assess the effect of dimensionless curvature radius $\bar{r}_0 = r_0 / a$ of the dent and of its initial dimensionless contact diameter, $\bar{c} = c / a$, upon the pressure riser effect, *a* being the calculated Hertz radius of the contact. These results are synthesized in Figs. 4 and 5.

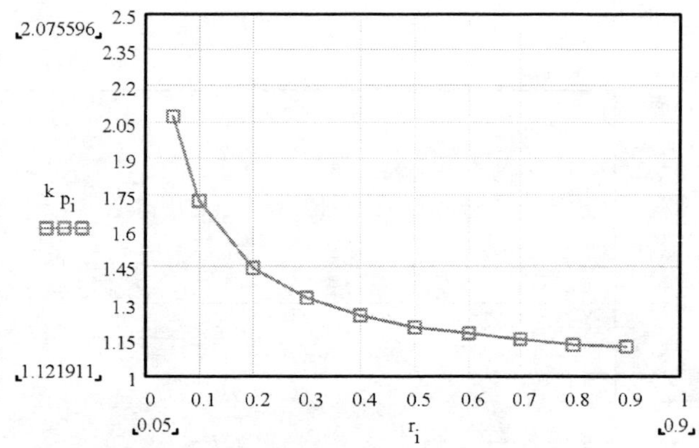

Fig. 4. Effect of radius r_0 upon pressure riser

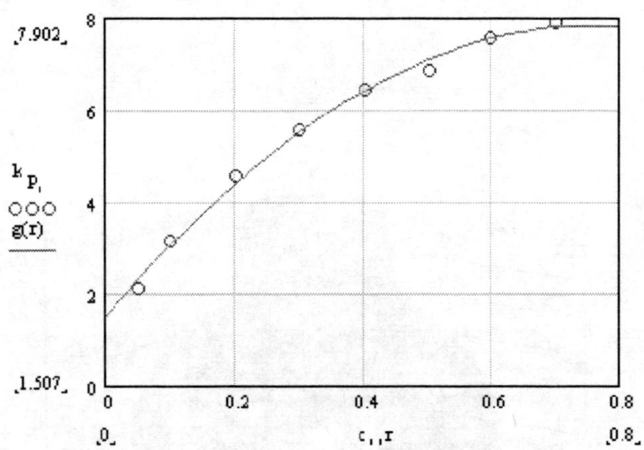

Fig. 5. Effect of dent extension upon pressure riser

As expected, the increase of radius r_0 leads to an asymptotic decrease of pressure riser coefficient towards unity at high radii. The values of pressure riser k_p exceed 2 at small values of radius.

On the other hand, the increase of the diameter *c* has an opposite effect, the coefficient of pressure rising increasing asymptotically towards a value just lower than 8.

CONCLUSIONS

The work herein reported can be summarized by following conclusions:

- A realistic geometry of the dent is analyzed, this having a circular cross-section and a parabolic shoulder profile.
- The integral equation of contact pressure is derived by applying the half-space superposition principle to the solution of Erdogan for a surface ring of tractions.
- This equation is digitized and a system of algebraic linear equations is obtained. Because the inner and outer radii of contact area are not known in advance, a numeric recurrent program written in Mathcad solves this system.
- The numerical solutions show that a high pressure riser acts in the neighborhood of dent shoulder which depends on dent geometry.
- The effects of two parameters of dent geometry, namely the curvature radius of dent shoulder and the radial extension of the dent upon pressure rising effect is numerically analyzed and graphically evidenced. These results are in agreement with general principles of stress concentration.

REFERENCES

[1] Chiu, Y.P., and Liu, J.Y., 1970, "An Analytical Study of the Stress Concentration Around a Furrow Shaped Surface Defect in Rolling Contact," Journal of Lubrication Technology, Transaction of the ASME, **92**, pp. 258-263.

[2] Lubrecht, A.A., Dwyer-Joyce, R.S., and Ioannides, E., 1992, "Analysis of the Influence of Indentations on Contact Life," *Proceeding of 18th Leeds-Lyon Symposium on Tribology*, Elsevier.

[3] Gao, N., Dwyer-Joyce, R.S., and Beynon, J.H., 1999, "Effects of Surface Defects on Rolling Contact Fatigue of 60/40 Brass," Wear, pp. 225-229.

[4] Diaconescu, E.N., and Cerlincă, D.F., 2001, "The Pressure Distribution in a Hertz Line Contact Affected by a Longitudinal Furrow," Second World Tribology Congress, September 2-7, Vienna, 6p.

[5] Lubrecht, A.A., and Ioannides, E., 1991, "A Fast Solution of the Dry Contact Problem and the Associated Sub-Surface Stress Field, Using Multilevel Techniques," Journal of Tribology, Transaction of the ASME, **113**, pp. 128-133, 1991.

[6] Hills, D.A., Nowell, D., and Sackfield, A., 1993, *Mechanics of Elastic Contacts*, Oxford Press Ltd.

Proceedings of
2003 STLE/ASME Joint International Tribology Conference
Ponte Vedra Beach, Florida USA, October 26-29, 2003

2003-TRIB-0275

A CORRELATION BETWEEN PRESSURE DISTRIBUTION AND BOUNDING SURFACES IN ELASTIC CONTACTS

Emanuel DIACONESCU/University of Suceava, ROMANIA

ABSTRACT

Many authors attempted to establish a direct correlation between the equation of the surface of an equivalent rigid punch and the pressure distribution arising when this punch is pressed against a corresponding elastic half-space. The results are encouraging, but they have a limited applicability and the involved calculations are still complicated. This paper advances a simple correlation based on a new interpretation of integral condition of deformation of a contact and on the dependence of pressure or pressure gradient on singularities in surface gradient or curvature. Based on this correlation, a method to find the pressure distribution for contacts bound by surfaces described by up to second order polynomials is advanced. The new method is applied in several classical contact problems and a very good agreement is found with existing solutions.

Key words: elastic contact, pressure distribution, bounding surface, singularities, discontinuities.

INTRODUCTION

In contact elasticity, the pressure distribution and surface displacements are connected by the integral deformation condition. The solution of this integral equation is seldom obtainable analytically, the Hertz contact being one of the few cases when the problem can be solved. Numerical solutions can nowadays be obtained for nearly all practical situations, but theoretical results are more general and easier to apply. Therefore, numerous attempts were made to relate in a simpler fashion the surface geometry and pressure distribution. Shtaerman [1] showed that in the case of an axisymmetric punch, bounded by a surface of equation $z = c\,r^{2n}$, the pressure distribution is the product between an even order polynomial of degree 2n-2 in r and $\sqrt{1-r^2/a^2}$, r being the current radius and a the outer radius of contact area. For an elliptical contact domain, Shtaerman [1] showed that a pressure of equation:

$$p(x,y) = \left[\sum_{m=0}^{n} a_m\left(1 - x^2/a^2 - y^2/b^2\right)^m\right] \Big/ \sqrt{1 - x^2/a^2 - y^2/b^2}$$

gives rise to a surface potential, and therefore to a surface normal displacement, expressed by an even polynomial of order 2n. Galin [2] proofed that if a punch of front surface given by a polynomial of degree n is pressed against an elastic half-space over a domain bounded by an ellipse of semi-axes a and b, the contact pressure can be written as the ratio between another polynomial of the same degree n and the square root $\sqrt{1-x^2/a^2-y^2/b^2}$. This statement is known as Galin's theorem. Barber [3] advanced a new theorem to determine the contact area in the indentation of an elastic half-space and to assess the contact pressure. Gladwell [4] supplied an alternative proof of Galin's theorem and proposed explicit relationships between the two involved polynomials for a transversely isotropic half-space.

This paper advances a new simple method to derive the pressure distribution when the punch surface depends on a single parameter, its equation is expressed by a polynomial of degree $n \le 2$ and the shape of boundary of contact area is expected to be elliptical.

NOMENCLATURE

- A: contact area;
- B, C: coefficients of second order polynomial surfaces;
- E_1, E_2: Young moduli of involved materials;
- Q: normal load applied to a finite contact;
- R: equivalent radius of a contact between cylinders;
- a, b: half-axes of contact ellipse;
- e, β: eccentricity and aspect ratio of contact ellipse;
- p(x,y), p(ρ), p: contact pressure;
- p_0: central value of contact pressure;
- q: normal line load applied to a line contact;
- r_r: round-off radius of circular edges of a cylindrical roller;
- t: contact parameter;
- x, y and z: co-ordinate axes, with z normal to the contact;
- w(x,y) or w: normal surface displacement;
- δ: normal approach between contacting bodies;

$\eta = [(1-\nu_1^2)/E_1] + [(1-\nu_2^2)/E_2]$: contact elastic constant;

- ν_1, ν_2: Poisson's ratios of involved materials;
- ρ: local polar radius or contact parameter;
- ρ_c - distance of current point in contact area to contour.

SURFACE GEOMETRIC DISCONTINUITIES AND CONTACT PRESSURE

Boussinesq [5] was the first to report pressure singularities in the points of contact contour when a flat ended rigid punch is pressed against an elastic half-space. Later on, similar conclusions reached Shtaerman [1], Galin [2], Solomon [6], Gladwell [7], Johnson [8], Mossakovski et al [9], Fabrikant [10], Hill et al [11] and Glovnea [12]. Infinite pressure identified at slope discontinuities inside the contact, Shtaerman [1], Gladwell [7], Johnson [8] and Love, quoted in [8], for blunt cones and/or blunt wedges.

In the case of an infinite long line contact, Hall and Savage [13] proofed that at an end point of the contact zone, "*singularities in pressure and pressure gradient arise as a consequence of discontinuities in* dw/dx *and* d^2w/dx^2, *respectively*". *Both, pressure and pressure gradient exhibit algebraic (square root) singularities.* In his PhD thesis, quoted in [13], Hall showed that, *in an inner point* x_m *of a line contact, both pressure and pressure gradient record logarithmic singularities of the kind* $\ln|x-x_m|$ *if, respectively,* dw/dx *and* d^2w/dx^2 *are discontinuous at* x_m.

It is obvious that most investigations on the correlation between contact pressure and geometric discontinuities of bounding surfaces are carried out on line contacts. Consequently, new investigations are required in the field of finite contact area, typical for point and surface contacts.

The integral condition of deformation for the equivalent contact rigid punch–elastic half-space has the known form:

$$w(x,y)=\frac{\eta}{\pi}\iint_A \frac{p(u,v)}{\sqrt{(x-u)^2+(y-v)^2}}\,dudv=\delta-z(x,y), \quad (1)$$

where $z(x,y)=z_1(x,y)+z_2(x,y)$ is the equation of the equivalent punch surface.

In line contacts, when $z=z(x)$, this equation is expressed in terms of derivatives with respect to abscissa *x* as follows:

$$\frac{\partial w}{\partial x}=2\frac{\eta}{\pi}\int_{-a}^{a}\frac{p(u)}{x-u}\,du=\frac{\partial z}{\partial x}. \quad (2)$$

It is a simple matter to show that the module of displacement gradient and its Laplacian take the forms:

$$|\mathrm{grad}w|=\frac{\eta}{\pi}\iint_A \frac{p(u,v)}{(x-u)^2+(y-v)^2}\,dudv; \quad (3)$$

$$\Delta w=\frac{\eta}{\pi}\iint_A \frac{p(u,v)}{\left((x-u)^2+(y-v)^2\right)^{3/2}}\,dudv. \quad (4)$$

By using the substitution:

$$u=x+\rho\cos\theta; \qquad v=y+\rho\sin\theta;$$
$$\rho\in[-\rho_1,\rho_2]; \qquad \theta\in[0,\pi], \quad (5)$$

Eqs. (3) and (4) take the following new forms:

$$|\mathrm{grad}\,w|=\frac{\eta}{\pi}\int_0^{\pi}\int_{-\rho_1}^{\rho_2}\frac{p(x+\rho\cos\theta,\,y+\rho\sin\theta)}{\rho}\,d\rho d\theta; \quad (6)$$

$$\Delta w=\frac{\eta}{\pi}\int_0^{\pi}\int_{-\rho_1}^{\rho_2}\frac{p(x+\rho\cos\theta,\,y+\rho\sin\theta)}{\rho^2}\,d\rho d\theta, \quad (7)$$

where ρ_1 and ρ_2 are the distances from a current point $M(x,y)$ in contact area to each of the contour points $C_{1,2}\ (x+\rho_{1,2}\cos\theta,\ y+\rho_{1,2}\sin\theta)$.

After performing the integration by parts, Eqs. 6 and 7 can be further written as:

$$|\mathrm{grad}w|=\frac{\eta}{\pi}\int_0^{\pi}\left[p_{c2}\ln\rho_2-p_{c1}\ln\rho_1-\int_{-\rho_1}^{\rho_2}\ln\rho\frac{\partial p}{\partial\rho}d\rho\right]d\theta; \quad (8)$$

$$\Delta w=\frac{\eta}{\pi}\int_0^{\pi}\left[-\frac{p_{c2}}{\rho_2}-\frac{p_{c1}}{\rho_1}+\int_{-\rho_1}^{\rho_2}\frac{1}{\rho}\cdot\frac{\partial p}{\partial\rho}d\rho\right]d\theta, \quad (9)$$

p_{c1} and p_{c2} being the values of contact pressure in contour points C_1 and C_2.

According to Eqs. 8 and 9, if the contour pressure is non-zero, |gradw| and Δw possess respectively, logarithmic, ln(0), and simple algebraic, (1/0), singularity, when the point (x,y) reaches the contour, that is when $\rho_1=0$ or $\rho_2=0$. In most situations, the pressure vanishes in the contour points and the behavior of |gradw| and Δw depends on pressure or pressure gradient as analyzed below.

If the pressure distribution has an algebraic square root singularity in contour points, $p=p^*\rho_c\big/\sqrt{\rho_c^2-\rho^2}$, p^* being a characteristic pressure, such as central pressure or maximum Hertz pressure, the Eq. 6 can be written as follows:

$$|\mathrm{grad}\,w|=\frac{\eta}{\pi}\int_0^{\pi}\int_{-\rho_1}^{\rho_2}\frac{p^*\rho_c}{\rho\sqrt{\rho_c^2-\rho^2}}\,d\rho d\theta=$$
$$=\frac{\eta p^*}{2\pi}\int_0^{\pi}\left|\ln\frac{\rho_c-\sqrt{\rho_c^2-\rho_2^2}}{\rho_c+\sqrt{\rho_c^2-\rho_2^2}}-\ln\frac{\rho_c-\sqrt{\rho_c^2-\rho_1^2}}{\rho_c+\sqrt{\rho_c^2-\rho_1^2}}\right|d\theta. \quad (10)$$

The point (x,y) falls on contact area contour when either ρ_1 or ρ_2 is zero. Assuming that $\rho_1=0$, which implies $\rho_c=\rho_2$, Eq. 10 leads to a logarithmic singularity in |gradw|. By virtue of the theorem of uniqueness of solution in elastostatics, it can be concluded that *a singularity in the slope of equivalent surface at the contact boundary generates a square root algebraic singularity in pressure*. It is also obvious that the logarithmic singularity in slope corresponds to a simple singularity in Δw. Therefore, *in contact contour points a simple singularity in* Δw *produces a square root algebraic singularity in pressure.*

If the pressure gradient has in contour points a square root algebraic singularity, of kind $\partial p/\partial\rho=p^*\big/\sqrt{\rho_c^2-\rho^2}$, Eq. 9 becomes:

$$\Delta w=\frac{\eta}{\pi}\int_0^{\pi}\left[p^*\int_{-\rho_1}^{\rho_2}\frac{1}{\rho\sqrt{\rho_c^2-\rho^2}}\,d\rho\right]d\theta=$$
$$=\frac{\eta p^*}{2\pi}\int_0^{\pi}\frac{1}{\rho_c}\left[\ln\frac{\rho_c-\sqrt{\rho_c^2-\rho_2^2}}{\rho_c+\sqrt{\rho_c^2-\rho_2^2}}-\ln\frac{\rho_c-\sqrt{\rho_c^2-\rho_1^2}}{\rho_c+\sqrt{\rho_c^2-\rho_1^2}}\right]d\theta. \quad (11)$$

Under the same assumptions $\rho_1=0$, $\rho_2=\rho_c$, Eq. 11 indicates that in the contour points the second derivative of normal displacement possesses a logarithmic singularity that corresponds to the square root algebraic singularity in pressure gradient. It follows that, *on the contour of contact area, in the points of logarithmic singularity in* Δw *, the pressure gradient has a square root algebraic singularity.*

An interesting situation occurs when the pressure gradient is singular in an inner point M(x,y) of contact domain. To find the behavior of |gradw| and Δw in such a point, a circle centered on (x,y), having an infinitely small radius, ε, and an area A_ε is considered. The pressure is extended by continuity of its gradient over entire domain A and therefore the singularity affects only the area A_ε. In this situation, Eq. 4 of Δw takes the following expression, in which p_a indicates the pressure with continuous pressure gradient over A and p_e represents the excess pressure directly resulting from the singularity:

$$\Delta w = \frac{\eta}{\pi}\iint_A \frac{p_a(u,v)}{\left((x-u)^2+(y-v)^2\right)^{3/2}}\,du\,dv + \\ +\frac{\eta}{\pi}\iint_{A_\varepsilon} \frac{p_e(u,v)}{\left((x-u)^2+(y-v)^2\right)^{3/2}}\,du\,dv. \tag{12}$$

The first integral, on A, is continuous, but the second, on A_ε, may be improper. Obviously, the excess pressure vanishes on the circle of radius ε, $p_e(\varepsilon)=0$. The application of co-ordinate transform expressed by Eqs. 5 for area A_ε, leads to following equation:

$$\Delta w_\varepsilon = \frac{\eta}{\pi}\int_0^\pi\int_{-\varepsilon}^{\varepsilon}\frac{p_e(\rho)}{\rho^2}\,d\rho\,d\theta = \frac{\eta}{\pi}\left[\int_0^\pi\int_{-\varepsilon}^{\varepsilon}\frac{1}{\rho}\frac{\partial p_e}{\partial \rho}\,d\rho\;d\theta\right]. \tag{13}$$

If the pressure gradient has a simple algebraic singularity in the point M(x,y), of the form $\partial p_e/\partial \rho = -p^*/\rho$, where the sign minus occurs because the pressure decreases with radius ρ, Eq. 12 becomes:

$$\Delta w_\varepsilon = \frac{\eta}{\pi}\left[-\int_0^\pi\int_{-\varepsilon}^{\varepsilon}\frac{p^*}{\rho^2}\,d\rho\;d\theta\right] = \frac{2\eta p^*}{\varepsilon}. \tag{14}$$

Equation 13 shows that , as $\varepsilon \to 0$, Δw_ε, and therefore, Δw, possesses a simple algebraic singularity. Therefore, *in any inner contact point in which Δw has a simple singularity,* (1/0), *the pressure gradient becomes simply singular,* (1/0) *and the pressure possesses a logarithmic singularity.*

CONTACT PRESSURE FOR NONSINGULAR SURFACES

According to Galin's theorem, an equivalent punch bounded by a front surface expressed by a polynomial P_m with constant coefficients:

$$P_m(x,y) = \sum_{i=0}^{m}\left(a_{xi}x^i + a_{yi}y^i\right), \tag{15}$$

pressed against an elastic half-space on an elliptical area A, generates a pressure distribution given by following equation:

$$p(\rho) = \frac{P_m^*(x,y)}{\sqrt{1-\frac{x^2}{a^2}-\frac{y^2}{b^2}}} = \frac{\sum_{i=0}^{m}\left(a_i^*x^i + b_i^*y^i\right)}{\sqrt{1-\frac{x^2}{a^2}-\frac{y^2}{b^2}}}. \tag{16}$$

This general result can be simplified for certain surfaces. For instance, if the cross-sections of the punch are bounded by curves that can be expressed in terms of a single parameter, $t=\text{constant}$, and the surface is symmetrical with respect to the co-ordinate planes, Eq. 15 takes the form:

$$P_{2n}(t) = \sum_{i=0}^{n} a_i\, t^{2i} \tag{17}$$

and, consequently, the pressure distribution becomes:

$$p(t) = p_0\left(1-t^2\right)^{-\frac{1}{2}}P_{2n}^*(t) = p_0\left(1-t^2\right)^{-\frac{1}{2}}\sum_{i=0}^{n} b_i t^{2i}. \tag{18}$$

In the case the punch cross-sections are elliptical, the parameter t is defined by following transform:

$$x = at\cos\alpha\,;\quad y = bt\sin\alpha\,;\quad t^2 = \frac{x^2}{a^2}+\frac{y^2}{b^2}. \tag{19}$$

Because the contact pressure vanishes on elliptical contour, at $t=\pm1$, t^2-1 divides $P_{2n}^*(t)$. This means $P_{2n}^*(\pm1)=0$ and therefore $\sum_{i=0}^{n} b_i = 0$. The condition $p(0)=p_0$, yields $b_0=1$, and the pressure distribution takes the form:

$$p(t) = p_0\sqrt{1-t^2}\,P_{2n-2}(t) = p_H(t)\sum_{i=0}^{n-1} c_i t^{2i}, \tag{20}$$

where $c_0 = 1$ and $c_i = -\sum_{j=i+1}^{n} b_j$.

This pressure equation is the product of a Hertz like pressure, $p_H = p_0\sqrt{1-t^2}$, and an even polynomial of degree 2n-2, as mentioned in the introduction for circular contacts. The Hertz type pressure extends over entire contact area.

By using a reverse method of solution, this author [14] proofed that if the pressure distribution is described by Eq. 20, with t expressed in terms of x and y, the punch surface is described by following even polynomial of order 2n:

$$z(x,y) = \delta - \eta p_0 b\sum_{i=0}^{n} c_i \sum_{k=0}^{i}\frac{(2k+1)!!}{2^{k+1}k!} I_k, \tag{21}$$

where δ is the normal approach and:

$$I_k = \int_{-\frac{\pi}{2}}^{\frac{\pi}{2}} \frac{\left[1-\left(\frac{x}{a}\cos\alpha - \beta\frac{y}{b}\sin\alpha\right)^2\left(1-e^2\sin^2\alpha\right)^{-1}\right]^{k+1} d\alpha}{\sqrt{1-e^2\sin^2\alpha}}. \tag{22}$$

Equations 17 to 22 define a correlation between punch surface and pressure distribution for elliptical domains.

If one takes into account that on elliptical contour the parameter t has the value $t_c = 1$, he can write the pressure p_H under following form:

$$p_H(t) = p_0\left(1-\frac{t^2}{t_c^2}\right)\left(1-\frac{t^2}{t_c^2}\right)^{-\frac{1}{2}}. \tag{23}$$

Equation 23 allows to point out an interesting interpretation of pressure p_H. Although both factors entering this equation depend on the same variable and can be simplified, they have different geometric significances and deserve to be written apart. In order to find the significance of the nominator, it is remembered that p_H is the pressure for a punch surface:

$$z_H(x,y) = Bx^2 + Cy^2, \tag{24}$$

B and C being chosen to yield the same contact area as the whole pressure.

If expressions 19 of x and y are replaced in Eq. 24, it results:

$$z_H(t) = t^2 \left(B a^2 \cos^2 \alpha + C b^2 \sin^2 \alpha\right). \tag{25}$$

The ordinate corresponding to contact ellipse is:

$$z_{Hc}(t_c) = t_c^2 \left(B a^2 \cos^2 \alpha + C b^2 \sin^2 \alpha\right). \tag{26}$$

The ratio of these ordinates:

$$\frac{z_H(t)}{z_{Hc}(t_c)} = \frac{t^2}{t_c^2}. \tag{27}$$

shows that in the nominator of Eq. 23 enters the ratio between the ordinate z_H in a current point of contact area and the ordinate of a contour point, having the same angle α, whereas the denominator remains unchanged.

In polar coordinates, at a polar angle θ, the contour polar radius is $r_c(\theta) = b\,t_c \big/ \sqrt{1 - e^2 \cos^2 \theta}$, whereas the polar radius of an ellipse of parameter t is $r(\theta) = bt \big/ \sqrt{1 - e^2 \cos^2 \theta}$. It is obvious that the ratio t/t_c is equal to $r(\theta)/r_c(\theta)$, the correlation between α and θ being: $a \tan(\theta) = b \tan(\alpha)$.

As a result, Eq. 23 takes one of following final forms:

$$p_H(t) = p_0 \left(1 - \frac{z_H(t)}{z_{Hc}(t_c)}\right)\left(1 - \frac{t^2}{t_c^2}\right)^{-\frac{1}{2}}; \tag{28}$$

$$p_H(r,\theta) = p_0 \left(1 - \frac{z_H(r,\theta)}{z_{Hc}(r,\theta)}\right)\left(1 - \frac{r^2(\theta)}{r_c^2(\theta)}\right)^{-\frac{1}{2}}. \tag{29}$$

These simple equations offer a quick method to assess the Hertz like pressure if the bounding surface is known and the contact contour is elliptical, even degenerate into straight lines, or it is made of segments of ellipse and straight lines.

A phenomenological justification of this result is suggested below, based on observation that two factors interact to build the contact pressure, as illustrated in Fig. 1.

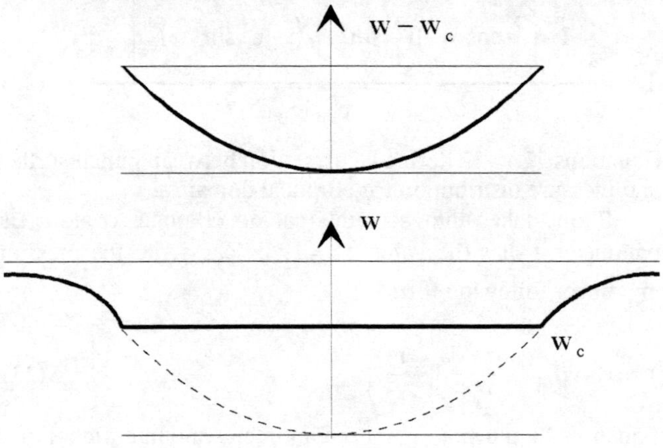

Fig. 1. Decomposition of deformed contact surface

The first factor is the elastic deformation effectively involved on z direction on contact area. Because the displacement w has a value w_c at the boundary of contact area, it is assumed that only the difference w-w_c, assessed at the same polar angle, generates an initial pressure proportional to the displacement difference:

$$p_i(t) = K\left(w - w_c\right). \tag{30}$$

Equation 1 yields the following expression for this difference:

$$w - w_c = z(x_c, y_c) - z(x, y) = z_c - z. \tag{31}$$

As a result of Eq. 31, the pressure p_i becomes:

$$p_i(t) = K\left[z_c(t_c) - z(t)\right]. \tag{32}$$

The second factor that affects the pressure generation is the pressure riser generated by the slope discontinuity and by the simple singularity in Δw which occurs in contour points, as evidenced in Fig. 1. As explained in the discussions following Eq. 10, in this situation the pressure possesses a square root algebraic singularity at the contact contour and, in polar coordinates, it is expressed as $p_s = p^* r_c \big/ \sqrt{r_c^2 - r^2}$. The pressure rising factor is defined by following ratio: $k_p = p_s / p^* = r_c \big/ \sqrt{r_c^2 - r^2} = t_c \big/ \sqrt{t_c^2 - t^2}$, as $r/r_c = t/t_c$.

The whole contact pressure is the product of $p_i(t)$ and k_p:

$$p_H(t) = p_i(t)\,k_p = K\,t_c \frac{z_{Hc} - z_H}{\sqrt{t_c^2 - t^2}}, \tag{33}$$

or, in polar coordinates:

$$p(r,\theta) = K\,r_c(\theta) \frac{z_{Hc} - z_H}{\sqrt{r_c^2(\theta) - r^2(\theta)}}. \tag{34}$$

In the centre of the contact $z = 0$ and Eqs. 30 and 33 indicate that $p_H(0) = p_i(0) = K z_c$ and, consequently,:

$$K z_c = p_0. \tag{35}$$

As a result, relations 32 and 28 become identical.

Once the Hertz like pressure p_H is found by this simple method, the final pressure distribution is given by Eq. 20.

When the punch surface is expressed by a second order polynomial, by a combination of such polynomials or by a constant, $z(t) = z_H(t)$ and $p(t) = p_H(t)$. These results simplify considerably the determination of the solution.

CONTACT PRESSURE FOR SINGULAR SURFACES

Often, the bounding surface of the equivalent punch possesses a slope discontinuity or a singularity in Δw, which may occur at the boundary or inside contact area. According to the discussion concluding Eq. 14, in inner contact points of such a discontinuity the pressure gradient becomes simply singular. Therefore, as opposed to the previous case when a correlation was established between surface ordinate and contact pressure, when a singularity occurs inside contact the pressure gradient is to be related to the surface gradient. Consequently, Eq. 32 is replaced by:

$$\frac{\partial p_i}{\partial t} = -K \frac{\partial z}{\partial t}, \tag{36}$$

where the minus sign takes into account that the pressure decreases with t, while z increases.

Two risers of pressure gradient occur in this case, one of kind t_c/t, in the inner points of slope discontinuity, and the second, of form $t_c \big/ \sqrt{t_c^2 - t^2}$, in contour points. The global rising factor for pressure gradient can be defined as follows:

$$k_g = \frac{t_c^2}{t\sqrt{t_c^2 - t^2}}. \tag{37}$$

The pressure gradient is the product of k_g and $\partial p_i / \partial t$:

$$\frac{\partial p}{\partial t} = k_g \frac{\partial p_i}{\partial t} = -K \frac{t_c^2}{t\sqrt{t_c^2 - t^2}} \frac{\partial z}{\partial t}. \tag{38}$$

The pressure distribution is found by integrating Eq. 36:

$$p(t) = -K\, t_c^2 \int \frac{1}{t\sqrt{t_c^2 - t^2}} \frac{\partial z}{\partial t} dt + D, \tag{39}$$

where D is a constant of integration with dimensions of pressure. If use if made of equality $t/t_c = r(\theta)/r_c(\theta)$, Eq. 36 takes the following form:

$$p(r,\theta) = -K\, r_c^2(\theta) \int \frac{1}{r(\theta)\sqrt{r_c^2(\theta) - r^2(\theta)}} \frac{\partial z}{\partial r} dr + D. \tag{40}$$

APPLICATIONS

The formulae 28, 29, 39 or 40 can be employed to obtain a rapid assessment of pressure distribution in contacts modeled by a rigid punch bounded either by a nonsingular surface of equation expressed by a constant or a second order polynomial, or by a surface possessing slope discontinuity inside contact area or at its boundary. This method is applied below in several solved cases to check its validity or in new cases of interest.

1. Hertz point contact

The expression 24 of punch surface and Eq. 28 yield:

$$p(t) = p_0\sqrt{1 - t^2}. \tag{41}$$

Rewritten in terms of x and y, Eq. 41 becomes the well-known Hertz pressure relation:

$$p(x, y) = p_0\sqrt{1 - \frac{x^2}{a^2} - \frac{y^2}{b^2}}. \tag{42}$$

The force balance equation on z direction:

$$Q = p_0 \iint_A \sqrt{1 - \frac{x^2}{a^2} - \frac{y^2}{b^2}} dxdy = \frac{2}{3}\pi a b p_0, \tag{43}$$

yields the expression of maximum Hertz pressure:

$$p_0 = \frac{3}{2}\frac{Q}{\pi a b}. \tag{44}$$

The identification of coefficients of polynomials entering both members of Eq. 21 leads to:

$$\delta = \eta p_0 b K(e);$$
$$B a^2 = \eta p_0 b D(e); \tag{45}$$
$$C b = \eta p_0 [K(e) - D(e)].$$

In Eqs. 45, K(e) and E(e) represent, respectively, the complete elliptic integrals of first and second kind whereas D(e) is defined in terms K(e) and E(e) by $D(e) = [K(e) - E(e)]/e^2$.

The known relation of contact curvature:

$$k = 2(B + C), \tag{46}$$

and Eqs. 44 and 45 yield the following well known expressions of contact ellipse semi-axes, normal approach and maximum Hertz pressure:

$$a = \sqrt[3]{\frac{3}{2}\frac{\eta Q D(e)}{\pi B}} = \sqrt[3]{\frac{2}{\pi}\left(1 + \frac{C}{B}\right)D(e)}\sqrt[3]{\frac{3}{2}\frac{\eta Q}{k}} = n_a\sqrt[3]{\frac{3}{2}\frac{\eta Q}{k}};$$

$$b = \sqrt[3]{\frac{3}{2}\frac{\beta \eta Q}{\pi C}[K(e) - D(e)]} =$$

$$= \sqrt[3]{\frac{2}{\pi}\left(1 + \frac{B}{C}\right)[K(e) - D(e)]\sqrt{1 - e^2}}\sqrt[3]{\frac{3}{2}\frac{\eta Q}{k}} = n_b\sqrt[3]{\frac{3}{2}\frac{\eta Q}{k}}; \tag{47}$$

$$\delta = \frac{3}{2}\frac{\eta Q}{\pi a}K(e) = \frac{1}{2}K(e)\sqrt[3]{\frac{4}{\pi^2}\frac{B}{B + C}\frac{1}{D(e)}}\sqrt[3]{\frac{9}{4}\eta^2 Q^2 k} =$$

$$= \frac{n_\delta}{2}\sqrt[3]{\frac{9}{4}\eta^2 Q^2 k};$$

$$p_0 = \frac{1}{\pi n_a n_b}\sqrt[3]{\frac{3}{2}\frac{k^2 Q}{\eta^2}} = \frac{n_p}{\pi}\sqrt[3]{\frac{3}{2}\frac{k^2 Q}{\eta^2}},$$

where the eccentricity of contact ellipse yields from the condition $n_b = \beta n_a = \sqrt{1 - e^2}\, n_a$, and is found to be the solution of following transcendental equation:

$$\left(1 - e^2\right)\frac{D(e)}{K(e) - D(e)} = \frac{B}{C}. \tag{48}$$

It is obvious that above were found all results of Hertz theory by means of a simple method herein advanced.

2. Hertz line contact

If the curvature radius in the plane xz becomes infinite, as is the case of the contact between two cylinders with parallel axes, the ellipse degenerates into two parallel straight lines, $x = \pm a$, a being the half-width of contact strip. The eccentricity of contact ellipse becomes $e = 1$. For B = 1/2R, the surface equation is $z(x) = x^2/2R$. The abscissa y becomes the parameter t of the lines z = constant and, according to Eq. 28, the contact pressure is:

$$p(x) = p_0\sqrt{1 - \frac{x^2}{a^2}}. \tag{49}$$

The balance equation for a unit contact length yields:

$$q = \int_{-a}^{a} p(x)dx = \frac{\pi}{2} a\, p_0. \tag{50}$$

If Eq. 2 is applied for this contact, the following supplementary relation is found:

$$a = 2 R \eta p_0, \tag{51}$$

which implies $K = 1/\eta a$.

The parameters of the contact are the solutions of the system made of Eqs. 50 and 51:

$$p_0 = \sqrt{\frac{q}{\pi \eta R}}; \tag{52}$$

$$a = 2\sqrt{\frac{1}{\rho} q \eta R}. \tag{53}$$

Equations 49, 52 and 53 express well known results in Hertz theory of line contacts.

3. Flat ended punch

In this case, the ordinate z of the punch surface is zero inside contact area and finite in the contour points. The role of surface parameter is played by the polar radius r of a current point; the contour value of r, let it be r_c, must be considered at the same polar angle θ as r. Accordingly, the pressure distribution is given by following known equation:

$$p(r,\theta) = p_0 \frac{r_c(\theta)}{\sqrt{r_c^2(\theta) - r^2(\theta)}}. \quad (54)$$

The pressure is at minimum in the contact center and tends to infinity in contour points, thus evidencing a strong edge effect. The force balance equation of the contact, $Q = \iint_A p(r)\,dr d\theta$, shows that the central pressure is half of the average pressure:

$$p_0 = \frac{Q}{2A} = \frac{1}{2} p_m. \quad (55)$$

4. Circular flat ended punch with rounded-off edge

A simple solution to reduce the edge effect of previous punch is to round off the edge that bounds the front surface. If the punch has a circular cross-section, this is achieved by a quarter of a torus of transverse radius r_r, leaving the flat region bounded by a circle of radius r_0. After expansion in series, the punch profile is described by following equations:

$$z(r) = \begin{cases} 0 & \text{if } r \le r_0; \\ \frac{(r - r_0)^2}{2\,r_r} & \text{if } r \ge r_0. \end{cases} \quad (56)$$

If the outer radius of contact area is a, according to Eq. 29, the pressure distribution is given by:

$$p(r) = \begin{cases} p_0 \frac{a}{\sqrt{a^2 - r^2}} & \text{if } r \le r_0; \\ p_0 \frac{a(a + r - 2r_0)}{(a - r_0)^2} \sqrt{\frac{a - r}{a + r}} & \text{otherwise.} \end{cases} \quad (57)$$

The force balance equation $Q = 2\pi \int_0^a p(r)\, r\, dr$, leads to following relation between central pressure and outer radius:

$$Q = 2\pi a^2 p_0 \left[1 + \frac{\pi a r_0}{2(a - r_0)^2} - \frac{2a^2 + r_0^2}{3(a - r_0)^2} \sqrt{1 - \frac{r_0^2}{a^2}} - \frac{a r_0}{(a - r_0)^2} \operatorname{asin}\left(\frac{r_0}{a}\right) \right]. \quad (58)$$

Equation 56 has as solution the outer radius of contact area, a, when the applied load is known.

Plots of radial pressure profile predicted by Eq. 57 for three values of ratio a/r_0, namely 1.05, 1.1 and 1.2 are shown in Fig. 2, for the same normal load applied to the punch. The pressure is presented in the dimensionless form used by Shtaerman [1] as ratio to $Q/\pi r_0^2$. A comparison of these results with those predicted by Shtaerman [1] via a different, but considerably lengthier procedure, reveals an excellent agreement.

5. Cylindrical roller with rounded-off edges

When a finite length cylindrical roller of radius R and length 2ℓ contacts a half-plane along a generatrix, very strong end effects occur. The simplest way to attenuate these effects, is the round-off of the lateral circular edges by a circular torus of cross radius r_r. If ℓ is the half-length of cylindrical zone, after expansion in power series, the surface equation is:

$$z(x,y) = \begin{cases} \frac{x^2}{2R} & \text{if } |y| \le \ell; \\ \frac{x^2}{2R} + \frac{(y - \ell)^2}{2\,r_r} & \text{otherwise.} \end{cases} \quad (59)$$

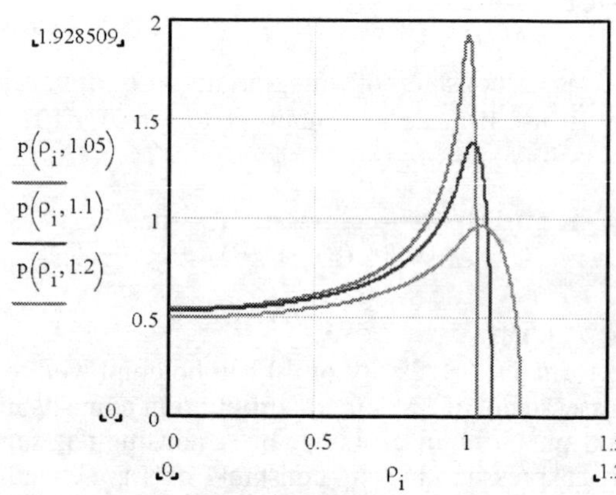

Fig. 2. Pressure distributions for $a/r_0 = 1.05$, 1.1 and 1.2

The contact area has a dog-bone shape, not known in advance. Therefore, the problem cannot be fully solved, but some useful features of pressure distribution can be assessed.

For instance, in a transverse plane, parallel to zx, the surface profile is parabolic and, consequently, Eq. 28 leads to a semi-elliptic pressure profile. This very simple result proofs that in numerical analysis of contact of profiled rollers the transverse profile of pressure is correctly assumed as being semi-elliptical.

The axial pressure profile can be predicted as well because the axial contact area profile is known, being made of a straight line ending in parabola segments. Under a normal load, the contact extends from ±ℓ to ±a, with a>ℓ. In the axial plane, the role of t is played by y and that of t_c by a. The involved ordinates are $z = z(0, y)$ and $z_c = z(0, a)$, as follows:

$$\frac{z}{z_c} = \begin{cases} 0 & \text{if } |y| \le \ell; \\ \frac{(y - \ell)^2}{(a - \ell)^2} & \text{otherwise.} \end{cases} \quad (60)$$

According to Eq. 28, the profile of maximum pressure along the contact, $p_0(y)$, is given by following equations:

$$p_0(y) = \begin{cases} p_0 \frac{a}{\sqrt{a^2 - y^2}} & \text{if } |y| \le \ell; \\ p_0 \frac{a(a + y - 2\ell)}{(a - \ell)^2} \sqrt{\frac{a - y}{a + y}} & \text{otherwise.} \end{cases} \quad (61)$$

As an illustration, the pressure profiles predicted by Eqs. 61 are shown in Fig. 3 for three values of a, namely 1.05ℓ, 1.1ℓ and 1.2ℓ.

Once the variation of maximum pressure along the contact is known, the determination of width of the contact at abscissa y by numerical methods becomes much simpler and quicker.

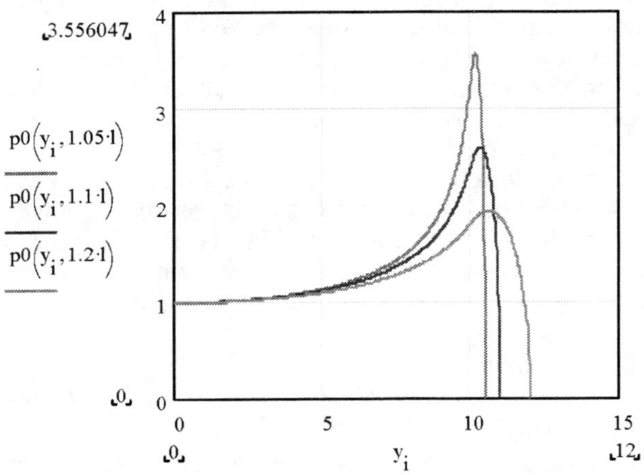

Fig. 3. Maximum pressure for $a/\ell = 1.05$, 1.1 and 1.2

6. Wedge contact

It is assumed that a two-dimensional symmetrical wedge, of semi-angle α, bounded by flat surfaces, indents a half-space. The axis x is chosen perpendicular to the apex line. A discontinuity in slope and a singularity in Δw occur at $x = 0$. The semi-width of contact strip is a. Consequently, the pressure distribution is given by relation 39, in which t is replaced by x and t_c by a.

The derivative $\partial z/\partial t$ becomes:

$$\frac{\partial z}{\partial x} = \operatorname{sign} x \cdot \cot\alpha . \tag{62}$$

Equation 39 leads to following expression of contact pressure:

$$p(x) = -K \cot\alpha \, a^2 \int \frac{\operatorname{sign} x}{x\sqrt{a^2 - x^2}} dx + D , \tag{63}$$

which yields the solutions:

$$\begin{aligned} p(x) - D &= Ka \cot\alpha \operatorname{atanh}\frac{\sqrt{a^2 - x^2}}{a} = \\ &= \frac{1}{2} Ka \cot\alpha \ln\frac{a + \sqrt{a^2 - x^2}}{a - \sqrt{a^2 - x^2}} = Ka \cot\alpha \operatorname{acosh}\frac{a}{x}. \end{aligned} \tag{64}$$

Because the pressure vanishes at the contact limits, the constant D is zero and the pressure distribution becomes:

$$\begin{aligned} p(x) &= Ka \cot\alpha \operatorname{atanh}\frac{\sqrt{a^2 - x^2}}{a} = \\ &= \frac{1}{2} Ka \cot\alpha \ln\frac{a + \sqrt{a^2 - x^2}}{a - \sqrt{a^2 - x^2}} = Ka \cot\alpha \operatorname{acosh}\frac{a}{x}. \end{aligned} \tag{65}$$

If one of these expressions of pressure, for instance the first, is substituted in Eq. 2, it yields:

$$2\frac{\eta K a}{\pi} \cot\alpha \int_{-a}^{a} \frac{\operatorname{atanh}\frac{\sqrt{a^2 - u^2}}{a}}{x - u} du = \operatorname{sign} x \cot\alpha . \tag{66}$$

The integral entering Eq. 66 does not seem to possess an explicit primitive. Nevertheless, its value can be found by numerical integration. This way, it is found that this integral is equal to $(\pi^2/2)\cdot\operatorname{sign}(x)$. As a result, the following expression is found for K:

$$K = \frac{1}{\pi \eta a} . \tag{67}$$

With this value of K, the pressure distribution becomes:

$$\begin{aligned} p(x) &= \frac{1}{\pi\eta} \cot\alpha \operatorname{atanh}\frac{\sqrt{a^2 - x^2}}{a} = \\ &= \frac{1}{2}\frac{1}{\pi\eta} \cot\alpha \ln\frac{a + \sqrt{a^2 - x^2}}{a - \sqrt{a^2 - x^2}} = \frac{1}{\pi\eta} \cot\alpha \operatorname{acosh}\frac{a}{x}, \end{aligned} \tag{68}$$

which is identical with the results of Shtaerman [1] and Johnson [8].

The force balance equation for a unit length of contact, loaded with a line force, q, leads to the following expression of contact half-width:

$$a = \frac{\eta q}{\cot\alpha} . \tag{69}$$

Alternatively, the pressure equations can be written as follows:

$$\begin{aligned} p(x) &= \frac{q}{\pi a} \operatorname{atanh}\frac{\sqrt{a^2 - x^2}}{a} = \\ &= \frac{1}{2}\frac{q}{\pi a} \ln\frac{a + \sqrt{a^2 - x^2}}{a - \sqrt{a^2 - x^2}} = \frac{q}{\pi a} \operatorname{acosh}\frac{a}{x}. \end{aligned} \tag{70}$$

7. Contact of circular cones

Similar results can be obtained for the contact between a circular cone of apex semi-angle α and a half-space, because in a radial plane the same slope discontinuity and singularity in Δw occur as in the previous case. Consequently, the pressure distribution is obtained from Eq. 65 by replacing the abscissa x with the radius r and denoting by a the outer radius of contact circle:

$$\begin{aligned} p(r) &= Ka \cot\alpha \operatorname{atanh}\frac{\sqrt{a^2 - r^2}}{a} = \\ &= \frac{1}{2} Ka \cot\alpha \ln\frac{a + \sqrt{a^2 - r^2}}{a - \sqrt{a^2 - r^2}} = Ka \cot\alpha \operatorname{acosh}\frac{a}{r}. \end{aligned} \tag{71}$$

To find the value of K, the derivative of Eq. 1 with respect to the radius r must be evaluated. To this end, Eq. 1 is rewritten first in polar coordinates, as follows:

$$w(r) = \frac{\eta}{\pi} \int_0^{2\pi} \int_0^{a} \frac{\rho \, p(\rho) \, d\rho d\theta}{\sqrt{(r - \rho\cos\theta)^2 + \rho^2 \sin^2\theta}} = \delta - r\cot\alpha . \tag{72}$$

The derivative of this equation is:

$$\frac{\eta}{\pi} \int_0^{2\pi} \int_0^{a} \frac{\rho (r - \rho\cos\theta) \, p(\rho) \, d\rho d\theta}{\left[(r - \rho\cos\theta)^2 + \rho^2 \sin^2\theta\right]^{3/2}} = -\cot\alpha . \tag{73}$$

By using for pressure one of expressions 71 and performing numerically the integration, the following value of K is found:

$$K = \frac{1}{2\eta a} . \tag{74}$$

As a result, the pressure distribution equation is:

$$\begin{aligned} p(r) &= \frac{1}{2\eta} \cot\alpha \operatorname{atanh}\frac{\sqrt{a^2 - r^2}}{a} = \\ &= \frac{1}{4\eta} \cot\alpha \ln\frac{a + \sqrt{a^2 - r^2}}{a - \sqrt{a^2 - r^2}} = \frac{1}{2\eta} \cot\alpha \operatorname{acosh}\frac{a}{r}. \end{aligned} \tag{75}$$

The force balance equation on z direction takes the form $Q = 2\pi\int_0^a p(r)\, r\, dr$ and leads to the following expression of a:

$$a = \sqrt{\frac{2}{\pi} Q\, \eta \tan\alpha}\ . \tag{76}$$

Alternatively, the pressure distribution can be written as:

$$p(r) = \frac{Q}{\pi a^2}\, \mathrm{atanh}\frac{\sqrt{a^2 - r^2}}{a} = \frac{Q}{\pi a^2}\, \mathrm{acosh}\frac{a}{r} =$$
$$= \frac{1}{2}\frac{Q}{\pi a^2} \ln\frac{a + \sqrt{a^2 - r^2}}{a - \sqrt{a^2 - r^2}} = -\frac{Q}{\pi a^2} \ln\left(\frac{a}{r} - \sqrt{\frac{a^2}{r^2} - 1}\right). \tag{77}$$

The last form of pressure in Eq. 77 is due to Shtaerman [1] and the second was derived by Love, as mentioned in [8].

CONCLUSIONS

The results reported herein can be summarized as follows:

Several reciprocal correlations between the surface singularities and the features of pressure distribution were proofed in the case of finite dimensions contacts, as follows:

- A logarithmic singularity in surface gradient (slope) at contact boundary generates a square root algebraic singularity in pressure.
- A simple singularity in surface Laplacian (curvature) at contact boundary generates a square root algebraic singularity in pressure.
- A logarithmic singularity in surface Laplacian at contact boundary generates a square root algebraic singularity in pressure gradient.
- A slope discontinuity and hence a simple singularity in Laplacian in an inner point generates a simple singularity in pressure gradient and a logarithmic singularity in pressure.

For an elliptical contact area, Eqs. 20 to 27 establish a correlation between pressure distribution and the punch surface, when the latter is expressed by an even order polynomial of a degree 2n.

If the punch surface is expressed by a second order even polynomial, the pressure distribution can be expressed directly in terms of surface equation and either the surface parameter or the polar radius in the contact domain.

A simple phenomenological interpretation of this result is based on two concurrent factors: i. the proportionality between contact pressure and the difference of surface displacements $w - w_c$; ii. the pressure riser effect generated by the singularity in surface Laplacian arising in contour points.

For simple surfaces possessing singularities inside contact area, a direct proportionality between surface gradient and pressure gradient as well as the pressure riser effect in contour points lead to an integral expression of pressure distribution.

The application of the new method in several cases of known solutions, such as point Hertz contact, line Hertz contact, flat ended punch, circular flat ended punch with round-off edge, wedge contact and the contact of circular cones yields correct results.

An analytical expression of the axial profile of maximum pressure for a quasi-linear Hertz contact, between a cylindrical roller with round-off edges and an elastic half-space, was found. This reduces significantly the volume of calculations implied by a numerical full solution of the problem.

REFERENCES

[1] Shtaerman, I., 1949, *Contact Problems in the Theory of Elasticity*, (in Russian), Gostehizdat, Moscow.

[2] Galin, L.A., 1953, *Contact Problems in the Theory of Elasticity*, (in Russian), Gostehizdat, Moscow.

[3] Barber, J.R., 1974, "Determining the Contact Area in Elastic Indentation Problems," Jnl. of Strain Analysis, **9**, pp. 230-232.

[4] Gladwell, G.M.L., 1978, "Polynomial Solutions for an Ellipse on an Anisotropic Elastic Half-Space," Journal. Mech. Appl. Math., vol. XXXI, Part.2, 251-260.

[5] Boussinesq, J., 1969, *Application des potentiels á l'etude de l'equilibre et du mouvement des solides élastiques*, Reed. A. Blanchard, Paris.

[6] Solomon, L., 1968, *Elasticité linéaire*, Masson, Paris.

[7] Gladwell, G.M., 1980, *Contact Problems in the Classical Theory of Elasticity*, Sijthoh & Noordhoff.

[8] Johnson, K.L., 1985, *Contact Mechanics*, Cambridge University Press.

[9] Mossakovski, V.I., Kacialovskaia, H.E., and Golikova, S.S., *Contact Problems in the Mathematical Theory of Elasticity*, (in Russian), Naukova Dumka, Kiev.

[10] Fabrikant, V.I., 1989, *Applications of Potential Theory in Mechanics*, Kluwer Acad. Publ.

[11] Hill, D.A., Nowell, D., and Sackfield, A., 1993, *Mechanics of Elastic Contacts*, Oxford, Butterworth Heinemann Ltd.

[12] Glovnea, M.L., 1999, "Effect of Surface Geometric Discontinuities upon Elastic Contacts," (in Romanian), Ph.D. Thesis, University of Suceava.

[13] Hall, R., and Savage, M.D., 1988, "Two-Dimensional Elastohydro-dynamic Lubrication, Part 1: The Associated Dry Contact Problem," Proc. Inst. Mech. Engr., **202**, pp. 347-360.

[14] Diaconescu, E.N., 2003, "Hertz Theory Revisited," Private communication, Opening Plenary Session of *10th Romanian Tribology Conference*, Galatzi, 25-27 Sept. 2003.

Proceedings of
2003 STLE/ASME Joint International Tribology Conference
Ponte Vedra Beach, Florida USA, October 26-29, 2003

2003-TRIB-0276

ULTRA-THIN FILM ULTRASONIC CHARACTERIZATION

Antanas Daugela / Hysitron Inc., USA

Vytautas Blechertas / Kaunas University of Technology, Lithuania

Oden L. Warren / Hysitron Inc., USA

Hiroshi Kutomi/ Gifu University, Japan

Thomas J. Wyrobek / Hysitron Inc. , USA

ABSTRACT

Ultrasonically induced nanoscale fatigue and friction reduction phenomena are researched for nanometer order thick protective overcoats. The newly developed method described here is a synergy of the nanometer scale quantitative nanoindentation/scratch technique and ultrasonic excitation. A standing ultrasonic wave is generated at the sample holder during the quasi-static nanoindentation/scratch test at an excitation frequency of several hundreds of kilohertz. Due to the fact that high strain rates are being generated at the surface of the sample, nanofatigue phenomenon followed by delamination and overcoat chipping can be observed. Statistical investigation of fatigue inducing parameters, such as critical quasistatic load and amplitude of oscillations, leads to a means of comparative sample characterization.

Keywords: nanoindentation/scratch, ultrasonic excitation, nano fatigue, friction reduction

INTRODUCTION

Due to the rapid development of overcoats for data storage and semiconducter industries, traditional quasistatic nanoindentation/scratch techniques based on *in-situ* evaluation of hardness and Young's modulus are reaching their technical limits. Therefore, the more physical properties and phenomena that can be characterized simultaneously, the better the opportunity for achieving important insights for advanced materials development.

Acoustic methods have been used successfully for microscale contact non-destructive evaluation, but until very recently, were not available for nanometer scale measurements. Non-contact sub-micrometer resolution ultrasonic microscopy and Scanning Probe Microscopy (SPM) operated at higher acoustic modes have great potential to be useful qualitative tools but face data interpretation difficulties. Nanomechanical test instruments offer nanometer scale quantitative characterization. These tools coupled with *in-situ* SPM-type imaging [1] and simultaneous acoustic response monitoring open new instrumentation horizons for micro/nano fracture mechanics [2]. Both active and passive acoustic methods can help to characterize a large variety of data storage substrates and coatings. Acoustic Emission (AE) monitoring during nanoindentation/scratch is one way of obtaining additional information at the local contact. AE events induced by martensitic phase transformation were monitored during nanoindentation, as reported by Daugela et al [3]. Plasticity yield point induced AE events on W(100) have been presented by several researchers [3, 4]. AE events induced by the twinning phenomenon on various orientations of Sapphire were revealed in the recent paper by Tymiak et al [5]. Nanoscratch induced AE events on magneto-optical disk media were reported recently [6]. Due to the uniqueness of the AE monitored indentation technique, this technique can be considered as a measure of the non-homogeneity of the sample.

Active acoustic methods are attractive because of simple implementation and have been used for years in various non-destructive testing devices. Here, a contact resonance phenomenon between the measuring tool tip and the sample is employed in order to obtain quasi-static, dynamic or acoustic response and correlate it with material properties [7]. A nanoindentation assisted fracture toughness measurement by means of external sample excitation at 500Hz was reported by Beake et al [8]. Preliminary results on nanoindentation induced nanofatigue and friction reduction measurements during nanoscratch were presented by Kutomi et al [9].

In the present study, ultrasonically excited and load controlled nanoindentation induced fatigue results are presented together with friction reduction phenomenon observed during nanoscratch testing. Further discussions and recommendations on test methodology are provided.

EXPERIMENTAL SETUP AND PROCEDURES

The experimental setup shown in Fig. 1 consists of the three-plate capacitor quasi-static nanoindentation/scratch tester (Hysitron, Inc.) with the *in–situ* SPM imaging option. The noise floors for lateral and vertical quasi-static displacement and vertical force sensing are 10nm, 0.1nm, and 100nN, respectively. The nanoindentation/scratch tester is interfaced to a SPM scanner (Digital Instruments Co.). A nanoindentation/scratch tester is attached to the SPM scanner base. An originally designed ultrasonic transducer is placed on the SPM scanner. It consists of a piezo actuator with sectioned electrodes for input and output, an acoustic isolator attached to the bottom and a DLC sample to be tested. An ultrasonic transducer is driven at one of the resonance standing modes of 206kHz frequency by a wave generator. A DLC film coated silica sample is placed permanently on the nodal plane of the piezo actuator. The nanoscratch/nanoindentation tester and the SPM scanner were controlled by two dedicated data acquisition systems. The SPM scanner was used for *in-situ* imaging of post indents/scratches. A Berkovich diamond tip with a radius of 120nm was used as an indentation and scratching tool.

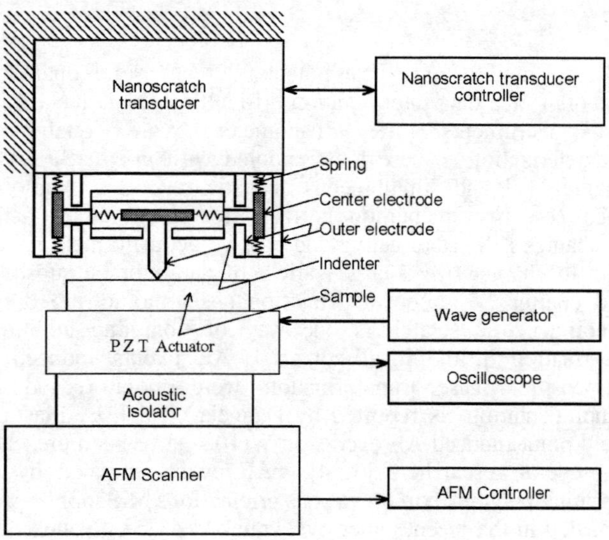

Fig. 1. Experimental setup.

EXPERIMENTAL RESULTS

1. Ultrasonically Excited Nanoindentation

Approximately 400 ultrasonically assisted nanoindentation tests were performed on DLC coated silica samples at estimated oscillation amplitudes of 4 to 20nm and quasistatic loads of 2000 to 8000µN. A typical nanoindentation loading-unloading curve for the ultrasonically excited DLC film sample is shown in Fig. 2 for a load controlled indent of 3200µN. An ultrasonically modified 5s loading, 40s holding and 5s unloading profile was used for the experiment. The large excursion on the holding segment and the lifted up unloading segment suggest that fracture followed by delamination occurred. The fracturing phenomenon is depicted in Fig. 3, where a large displacement discontinuity is observed at 39 seconds. The total number of nanoscale ultrasonic loading cycles delivered till the fracture event was 8034000. A 3µm *in-situ* SPM type image illustrating delamination of the film is shown in Fig. 4 where a crater was formed as a consequence of material removal. Here, the vertical scale is 200nm. The initiation of delamination is depicted in Fig. 5 where buckled film area is adjacent to the indent. In this case, the displacement versus time curve did not show such drastic changes.

Reduced Young's modulus and hardness values derived from unloading curves by the classical nanoindentation approach yielded 127GPa and 38GPa, respectively. Uncertainties in modulus and hardness values were within 3%.

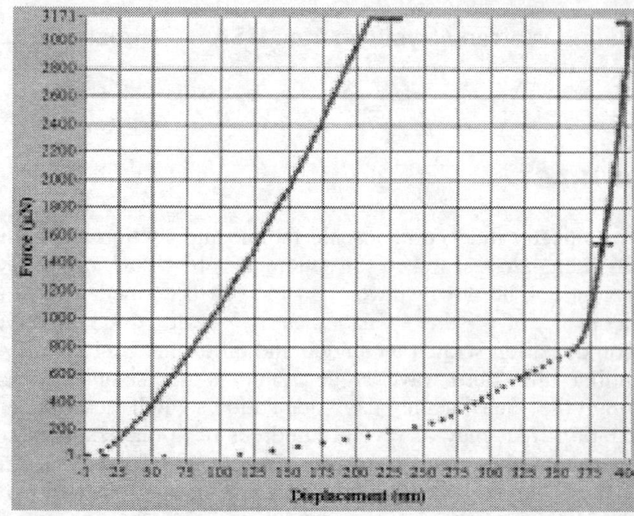

Fig. 2. Loading-unloading curve: force versus displacement.

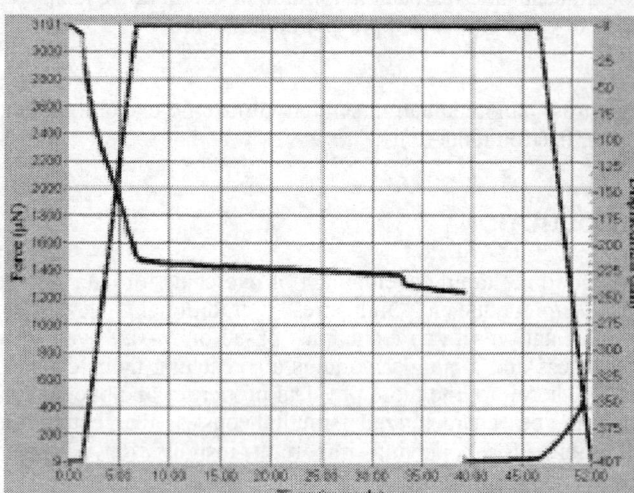

Fig. 3. Loading-unloading curve: force and displacement versus time.

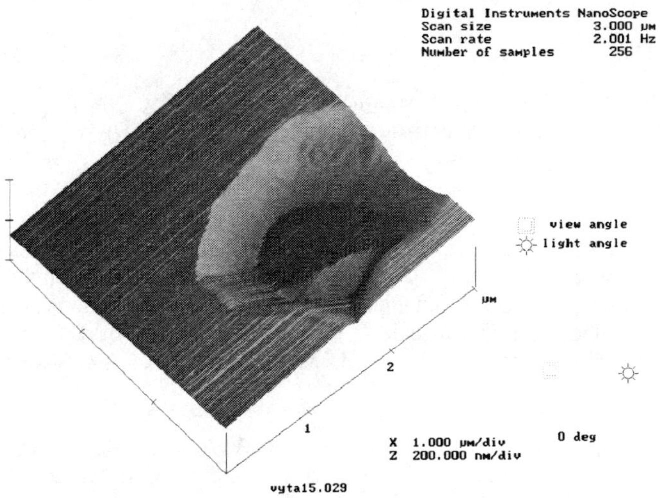

Fig. 4. Delaminated film as a result of ultrasonic sample excitation for 3200µN quasistatic load.

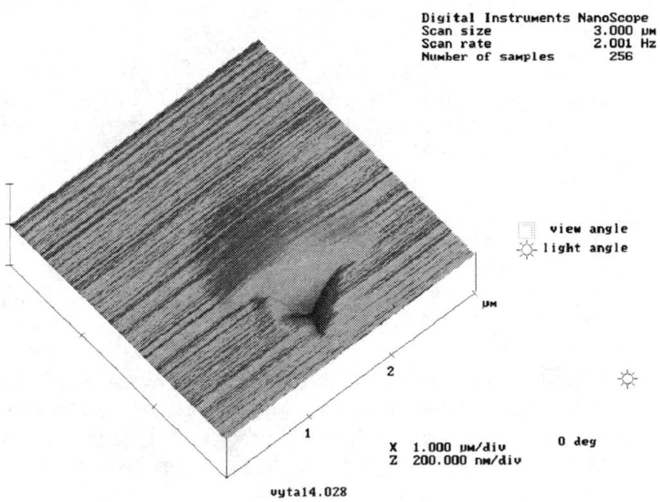

Fig. 5. Formation of delamination due to ultrasonic sample excitation for 2000µN quasistatic load.

The time of onset of nanofatigue failure occurred in three distinct groups, at ca. 20, 30 and 40 seconds, corresponding to $4x10^6$, $6x10^6$ and $8x10^6$ ultrasonic loading cycles.

2. Ultrasonically Excited Nanoscratch

Friction coefficient was investigated as a function of quasi-static scratch loading force and ultrasonic vibration amplitude. A 20nm thick DLC film with R_a = 4nm roughness coated on silica was exited in elastic standing wave mode at 206kHz. Approximately 300 quasi-static normal load nanoscratches of 50, 100, 150, 200, 250µN and 5µm length were placed on the sample vibration nodes.

A typical result is depicted in Fig. 6 where friction coefficient taken as a ratio of lateral and normal forces is plotted versus scratch length. Here, position "#1", corresponds to the case with no ultrasonic excitation. Positions "#2" and "#3" correspond to ultrasonic excitation at normalized amplitudes of 0.175 and 0.25 of the sample vibration maximum amplitude, respectively. As it can be observed, the absolute value of friction coefficient was reduced from 0.18 to 0.06 due to the ultrasonic excitation.

A 10µm SPM type post-scratch scan is shown in Fig. 7. Here, position "#1", corresponds to the case with no ultrasonic excitation. Positions "#2", "#3", "#4" and "#5" correspond to the cases of ultrasonic excitation at normalized amplitudes of 0.175, 0.25, 0.5 and 1 of the sample vibration maximum amplitude, respectively. The measured maximum amplitude of ultrasonic vibration was 6nm.

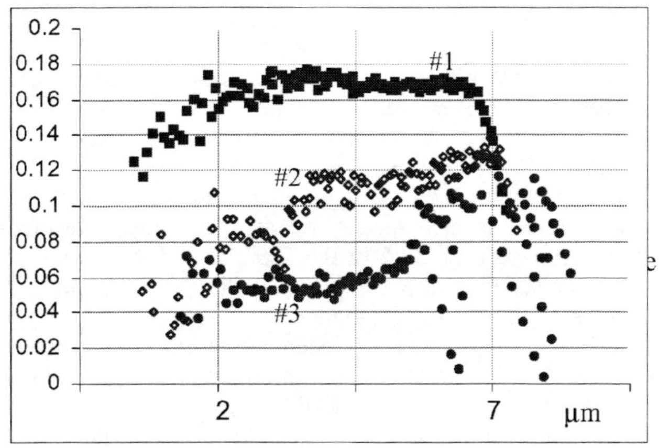

Fig. 6. Friction coefficient versus scratch length for ultrasonically modified and not modified cases.

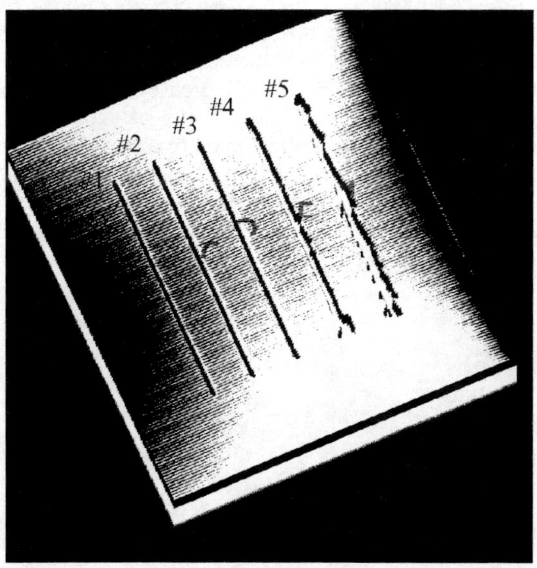

Fig. 7. SPM type 10µm scan of nanoscratches.

The post scratch SPM image was processed by Scion Image Co. software and revealed larger amount of material removed by the larger ultrasonic vibration amplitudes (cases

"#4" and "#5") where DLC chipping and delamination effects were found. The friction coefficient in these cases exceeded values for non-excitation cases by a factor of 5. Cases "#1", "#2" and "#3" yield similar volumes of material removal.

It is observed that friction coefficient measurements on DLC film in ambient environment without ultrasonic excitation does not always yield repeatable data. In contrast, ultrasonically excited friction coefficient measurements were very consistent.

CONCLUSIONS

Two innovative experimental methods are presented for characterization of DLC type overcoat films. These methods are based on the ultrasonically excited nanoindentation/scratch technique and can provide a very fast wear/fatigue test at the local contact. A quasi–static nanoindentation/scratch instrument interfaced with a Scanning Probe Microscope (SPM) and a specially designed ultrasonic transducer were used, where various modes of standing wave resonance vibrations were generated in the DLC coated silica substrate. Indents and scratches were placed within SPM precision at locations of vibration nodal planes.

Ultrasonically-excited indentation-induced interfacial fatigue and delamination was observed for DLC film during the load holding segment and verified by post indentation *in-situ* SPM imaging. The number of cycles and the sample vibration amplitude are two variation parameters for finding the critical number of interfacial fatigue cycles that produce film-interface delamination.

Ultrasonically excited nanoscratch test revealed changes in friction coefficient. A quasi-statically monitored 20 percent reduction in friction coefficient was observed at rather small excitation amplitudes. Increase in sample oscillation amplitude yields a drastic (up to 5 times) increase in friction coefficient. Post-scratch *in-situ* SPM type images correlated friction coefficient changes with ultrasonically induced wear marks and film chipping. A combination of shear and normal stresses modifies the contact pressure under the scratch tool, which changes the lateral force and affects the quasistatically determined friction coefficient.

REFERENCES

1. Lilleodden, E.T., Bonin, W., Nelson, J., Wyrobek, J. T., Gerberich, W.W, 1995, J. Mater. Res., Vol. 10, pp. 2162-2168.

2. Warren, O. L., Daugela, A., Asif, S.A.S., Tymiak, N.I., Wyrobek, T.J., 2003, "Scanning Ultra-Nanoindentation: Measuring and Imaging Quantitative Nanomechanical Properties of Nanophase Materials and Nanostructures at the True Nanoscale," Nanoengineering World Forum Proc., June 23–25, 2003, Marlborough, MA (CD-ROM: www.iec.org).

3. Daugela, A., Kutomi, H., Wyrobek, T.J., 2001, "Nanoindentation Induced Acoustic Emission Monitoring of Native Oxide Fracture and Phase Transformations," Zeitschrift fur Metallkunde, Vol. 92 (9) 1052-1056.

4. Bahr, D.F., Gerberich, W.W., 1998, J. Mater. Res., Vol. 13, pp.1065-1072.

5. Tymiak, N.I., Daugela, A., Wyrobek, T.J., Warren, O.L., 2003, "Highly Localized Acoustic Emission Monitoring of Nanoscale Indentation Contacts," J. Mater. Res., Vol. 18(5), pp.784-796.

6. Daugela, A., Warren, O. L., Wyrobek, J.T., 2003, "Contact Acoustic Characterization at Nanometer Scale," CD-ROM Proc. of 2003 SEM Annual Conference & Expo on Experimental and Applied Mechanics, June 2-4, 2003, Charlotte, NC (ISBN: 0-912053-83-6).

7. Borovsky, B., Krim, J., Syed Asif, S. A., Wahl, K., 2001, "Measuring Nanomechanical Properties of a Dynamic Contact Using an Indenter Probe and Quartz Crystal Microbalance, " J. Appl. Phys., Vol. 90, pp. 6391-6396.

8. Beake, B. D., Garcia, M. J. I., Smith, J. F., 2001, Thin Solid Films, Vol. 398-399, pp. 438-444.

9. Kutomi, H., Daugela, A., Gerberich, W.W., Fujii, H., Wyrobek, T.J., 2003, "Nanoscale Friction Reduction and Fatigue Monitoring Due to Ultrasonic Excitation," Tribology International, Vol. 36(4-6), pp. 255-259.

Proceedings of
2003 STLE/ASME Joint International Tribology Conference
Ponte Vedra Beach, Florida USA, October 26-29, 2003

2003-TRIB-0277

PARTICLE ADHESION AND REMOVAL IN THE SEMICONDUCTOR INDUSTRY

A. A. Busnaina, X. Xiong, NSF Center for Microcontamination Control, Northeastern University, Boston, MA 02115, www.cmc.neu.edu

J. G. Park, Dept. of Metallurgy and Materials Eng., Hanyang University, Ansan, 425-791, Korea

INTRODUCTION

The Adhesion and removal of small particles to surfaces is very important to many industries such as semiconductor manufacturing, imaging, aerospace, etc. In most cases, particles represents undesirable contaminants that lower product yield and cause defects. These particulate contaminants may be generated by the process or the product environment. The industry sets threshold specification for acceptable contamination for each generation of products.

For example, the International Technology Roadmap for Semiconductors sets the level of acceptable particulate defects at 125nm (at a concentration of 0.3-0.75/cm^2) in 1997. That is projected to decrease to the nanoscale at 25nm with even a smaller concentration at 0.01-0.15/cm^2 in 2011[1]. This is a requirement for particle removal (surface cleaning) that specifies acceptable contamination levels. These requirements are considered as one of the most challenging cleaning applications in semiconductor manufacturing. The path to particle removal starts with understanding the adhesion force and how it changes with time and other process or environmental conditions. The adhesion force of a particle on a surface is influenced by many factors, such as aging time, humidity, temperature, particle composition, particle size, etc. The adhesion force can cause deformation of the particle and/or the substrate generating stress between materials resulting in a finite contact area. Depending on the material properties of the particle and the substrate, the adhesion force can cause either elastic or plastic deformation. In this study, the effect of time and humidity effect on the adhesion of submicron PSL particles on silicon substrates were investigated using field emission scanning electron microscopy.

When it comes to the removal of particles whether we are considering nano or micro scale particles, there is a need to deliver the removal force (mostly hydrodynamic drag) to the particle. In the past decade, megasonic cleaning has been widely used as a non-contact cleaning technique. This technique shows quite a promise of being a good candidate for the removal of nanoscale particles. In megasonic cleaning, wafers are immersed in a cleaning liquid medium to which sonic energy is applied. High intensity sound waves generate pressure fluctuations and acoustic streaming which generates a drag force capable of detaching particles from surfaces. Busnaina *et al* [2-3] studied megasonic particle removal and the effect of acoustic streaming on the cleaning process especially in post-CMP applications.

1. PARTICLE ADHESION

Van der Waals attractive force is the predominant adhesion force for small uncharged particles (smaller than 50μm) [4,5]

$$F_{vdw} = \frac{AR}{6z_0^{\ 2}} \tag{1}$$

where, A is Hamaker constant, R the radius of the spherical particle, and z_0 the separation distance between the particle and the substrate. For smooth surfaces, the separation distance is 4Å [4,5]. When we have soft particles on hard substrates or vice versa, adhesion-induced-deformation occurs. This deformation results in the increase of the contact area and consequently the adhesion force. Bowling [5] gave the total adhesion force including the component due to the deformation as:

$$F_a = \frac{AR}{6z_0^2}\left(1+\frac{a^2}{Rz_0}\right) \quad (2)$$

where, a is contact radius between particle and substrate.

In the absence of any applied load and depending of the material property, we could have an elastic or plastic deformation. A well known model for predicting the contact area between a particle and a substrate as a result of elastic deformation is the JKR [6] model and is given by:

$$a^3 = \frac{6w_A \pi R^2}{K} \quad (3)$$

Where a is the contact radius, R is the particle radius, w_A is the work of adhesion and is related to the surface energies γ_1 and γ_2 of the two materials and their interfacial energy γ_{12} by:

$$w_A = \gamma_1 + \gamma_2 - \gamma_{12} \quad (4)$$

$$K = \frac{4}{3\left[\frac{1-\upsilon_1^2}{E_1}+\frac{1-\upsilon_2^2}{E_2}\right]} \quad (5)$$

E_1 , E_2, υ_1, and υ_2 are the Young's modulus and Poisson's ratio of two contact materials.

For plastic deformation, the Maugis Pollock model (MP) [7] model is given by:

$$a = \left(\frac{2w_A}{3Y}\right)^{1/2} R^{1/2} \quad (6)$$

where Y is the yield stress of the material.

1.1 Experimental

Polystyrene latex (PSL) spherical particles having diameter of 400nm and 600nm, obtained from Duke Scientific, Inc. were used in this study. Prior to deposition of the particles, the Si substrate was cleaned with 1:1:20 SC1 solution (1 ammonium hydroxide:1 hydrogen peroxide:20 DI water) to remove particles and organic contamination. The Si substrates were then rinsed with DI water. A few drops of highly concentrated PSL suspension were diluted with isopropyl alcohol (IPA) to form a dilute solution. The particles were deposited onto the substrate by atomizing the particle suspension using a nebulizer. After the deposition, the samples were kept in two desiccators at 40%RH and 95%RH respectively, and aged from one to seven days. The contact radii between the PSL particles and the silicon substrate were measured using a Hitachi S4500 FESEM. The samples were mounted onto cross-sectional stubs to allow the contact radius to be observed at an 88° tilt angle.

1.2 Results

Figure1 shows a sample of aged particle of polystyrene having nominal diameter of 400nm and 600nm, respectively on polished silicon wafer [8]. The contact area between particles and substrate resulting from adhesion-induced deformation can be observed. Figures 2 and 3 show the dependence of particle deformation for 400nm and 600nm particle at 40%RH and 95%RH aging conditions The figures show the contact diameter/particle diameter as a function of time at 40%RH and 95%RH respectively. The contact diameter between the particles and the silicon oxide substrate was found to increase with time. For the same aging time, the 95% relative humidity conditions show significant increase in particle deformation compared to the 40%RH conditions. At 95%RH condensed water vapor between particle and substrate induces a capillary force that increases the total adhesion force thereby increasing the contact diameter. The deformation is shown to be a function of size in figure 2 when it was induced by the van der Waals force only. However, at high humidity, the capillary force dominates van der Waals [9] and the deformation was not affected by size. The 95% relative humidity actually accelerated the increase of the total adhesion force. Maugis and Pollock's plastic deformation model [7] predicted that the adhesion force is five time smaller than the two day 40% relative humidity results.

Following the analysis by Tabor [4] and Bowling [5], The average pressure exerted on by the particle on the substrate is:

$$P = \frac{2\omega_A}{z_0} \quad (7)$$

where z_0 is the separation distance for a van der Waals bonded solid.($\approx 4\mathring{A}$). Substituting the experimentally obtained value for the work of adhesion for PSL particle [10,11], the pressure p is found to be 1.6×10^9 Pa, which is comparable to the Young's modulus of the polystyrene (2.55×10^9 Pa [5]) and far in excess of its yield stress (10.8×10^6 Pa [5]). This shows that plastic deformation occurred.

(a)

(b)

Figure 1. PSL particle on silicon wafer (a) 400nm (b) 600nm

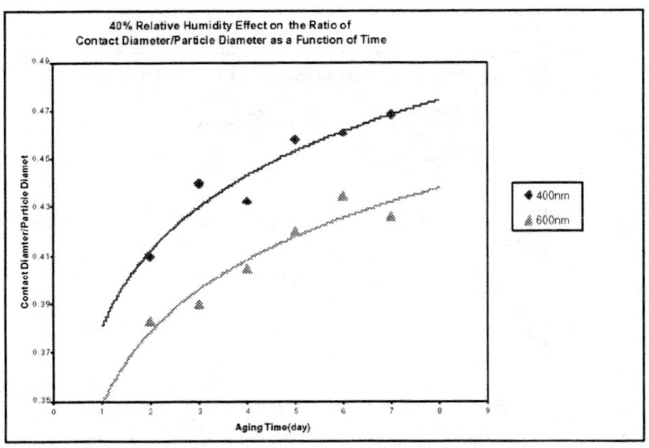

Figure 2. The effect 40%RH on the deformation of 400nm and 600nm PSL spherical particles as a function of time

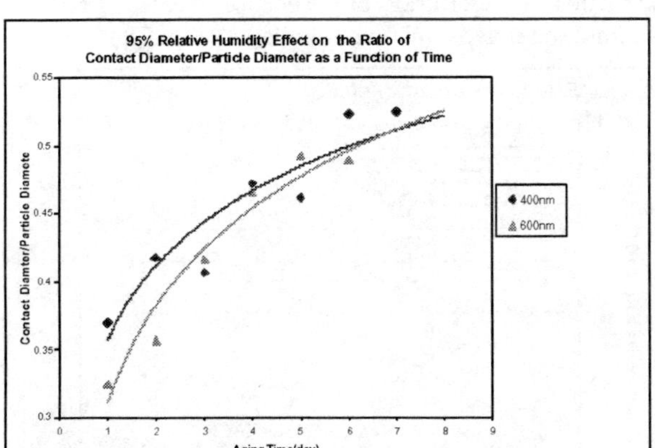

Figure 3. The effect of 95%RH on the deformation of 400nm and 600nm PSL spherical particles as a function of time

2. PARTICLE REMOVAL

2.1 PARTICLE REMOVAL MECHANSIM

Particle can be removed through the following mechanisms; lifting, sliding and rolling. In megasonic non-contact cleaning, the rolling mechanism is dominant. However, before we discuss the mechanism, we need to consider important parameters that are responsible of increasing (or decreasing) the removal force as well as electrostatic forces that may enhance or hinder removal.

a. Boundary layer thickness

One of the most important aspects of megasonic cleaning is the thickness of the acoustic boundary layer, which is very small compared to a typical hydrodynamic boundary layer at the same velocity. The acoustic boundary layer thickness is a function of the acoustic frequency ω ($\omega = 2\pi f$) and the viscosity of the cleaning liquid ν,

$$\delta_{ac} = (2\nu/\omega)^{0.5} \tag{8}$$

The hydrodynamic (turbulent) boundary layer thickness is given by

$$\delta_H = 0.16(\nu/Ux)^{\frac{1}{7}} \cdot x \tag{9}$$

where U is the fluid velocity and x is the distance from the leading edge of the wafer.

The turbulent hydrodynamic boundary layer thickness is larger than 2500 micron (at the center of an 8" wafer) in water with u = 5 m/s. In water, for frequency higher than 850k Hz, the acoustic boundary layer thickness is 0.6 micron. The thinner acoustic boundary layer exposes submicron and nano-scale particles on the surface to larger velocities and therefore increases the drag force and the particle removal efficiency.

b. Acoustic streaming velocity

Acoustic streaming velocity at center of the tank is defined as [12]

$$u = \frac{8\pi^2}{3\rho_l c^4} \cdot If^2\left(\frac{h^2}{4} - z_1^2\right) \tag{10}$$

where ρ_l is the density of the cleaning liquid, c the sound speed in liquid, I the intensity of the megasonic wave, f the frequency of the megasonic wave, h the distance between the walls of the tank, z_1 the distance between the wall and the edge of the sound beam.

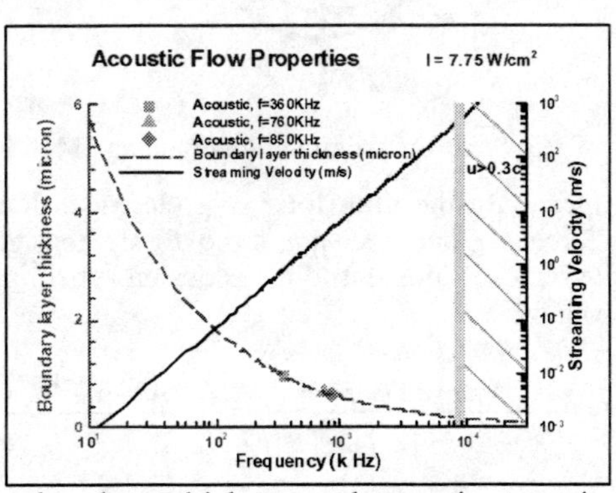

Fig.4 Boundary layer thickness and acoustic streaming velocity *vs.* acoustic frequency.

The acoustic streaming velocity is proportional to the intensity and square of frequency. As frequency increases, acoustic streaming velocity also increases and the boundary layer thickness decreases rapidly (Figure 4) with both effects increasing particle removal.

c. Drag Force

The drag force on a spherical particle in a Newtonian fluid can be expressed by:

$$F_d = \frac{\pi}{8} C_D \rho_l d_p^2 u^2 \tag{11}$$

where C_D is drag coefficient, ρ_l the density of the cleaning liquid, d_p the diameter of the particle and u the streaming velocity.

d. Electrical double layer force

According to HHF model, the electrical double layer force interacting between a sphere and a plate with constant potential can be expressed as [13]:

$$F_{el}^{\Psi} = \pi\varepsilon_r\varepsilon_0 R\left(\Psi_{01}^2+\Psi_{02}^2\right)\frac{\kappa e^{-\kappa H}}{1-e^{-2\kappa H}}\left[\frac{2\Psi_{01}\Psi_{02}}{\Psi_{01}^2+\Psi_{02}^2}-e^{-\kappa H}\right] \tag{12}$$

where Ψ_{01} is the zeta potential of the particle of radius R, Ψ_{02} the zeta potential of the substrate, ε_r the dielectric constant of the medium, ε_0 the dielectric permittivity of a vacuum and κ the Debye-Huckel parameter of the electrolyte solution. Using a high pH cleaning solution, such as SC1 or ammonia, a strong repulsion occurs between the particle and the substrate reducing the total adhesion force and enhancing the removal of these particles.

e. Rolling Removal Mechanism

Particle rolling removal mechanism is shown in Figure5.

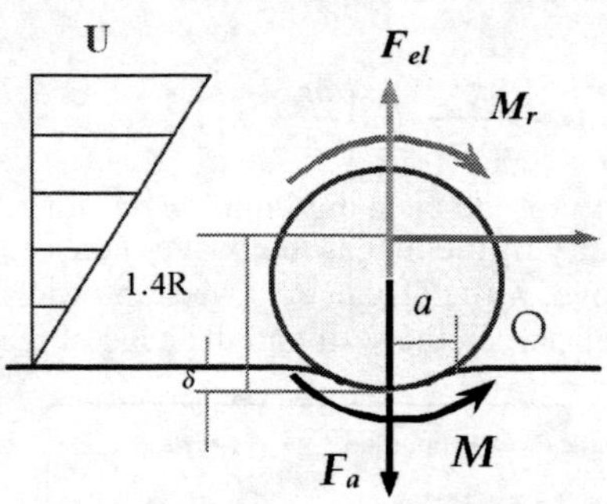

Fig.5 Rolling removal mechanism [14]

In this paper, only the drag force F_d, electrical double layer repulsive force F_{el}, and adhesion force F_a are considered. The ratio of the removal moment to the adhesion resisting moment, RM, is given by:

$$RM = \frac{\text{Removal moment}}{\text{Adhesion resisting moment}} = \frac{F_d(1.4R-\delta)+F_{el}\cdot a}{F_a\cdot a} \tag{13}$$

where R is the particle radius and a the contact radius. When the removal moment overcomes the adhesion resisting moment, namely, when RM>1, the particle is removed by rolling. The force required to remove a particle by rolling is less than that required by sliding or lifting.

2.2 RESULTS AND DISCUSSION

a. The effect of particle size on the adhesion and removal forces

Figure 6 shows that adhesion and removal forces increase as the particle size increases. The drag force increases much faster than the adhesion force. The drag force induced by acoustic streaming is two orders larger than that induced by hydrodynamic flow at the same velocity. When d>100nm, acoustic flow drag force dominates; when 30nm<d<100nm, drag force and electrical double layer force are on the same order of magnitude; and when d<30nm, electrical double layer force dominates.

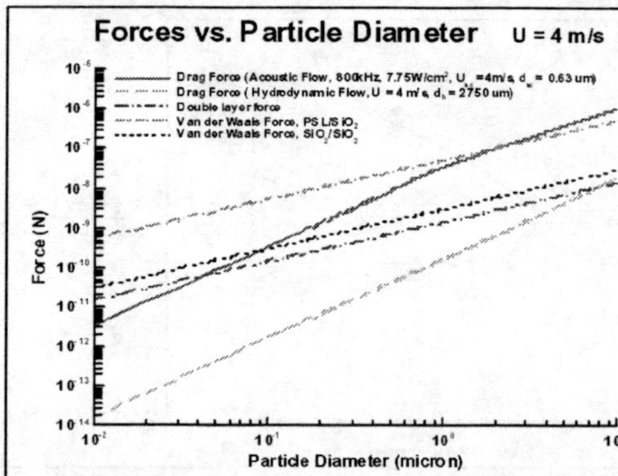

Fig.6 Adhesion and removal forces *vs*. particle size.

b. The effect of acoustic wave frequency

For acoustic streaming flow with constant intensity, higher frequency leads to larger drag force because of larger streaming velocity and thinner acoustic boundary layer (Fig. 7).

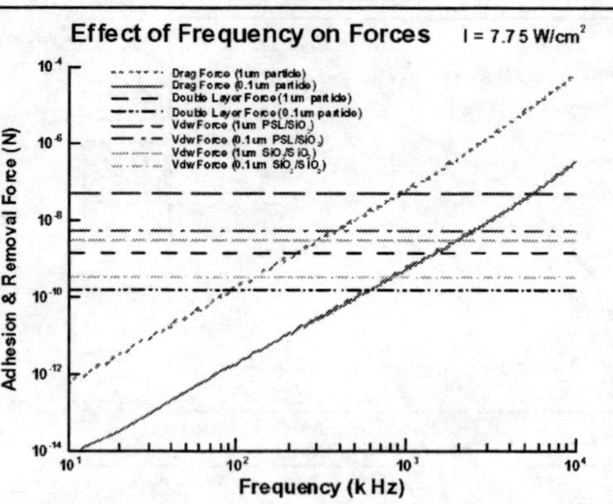

Fig. 7 Removal Forces *vs*. Frequency

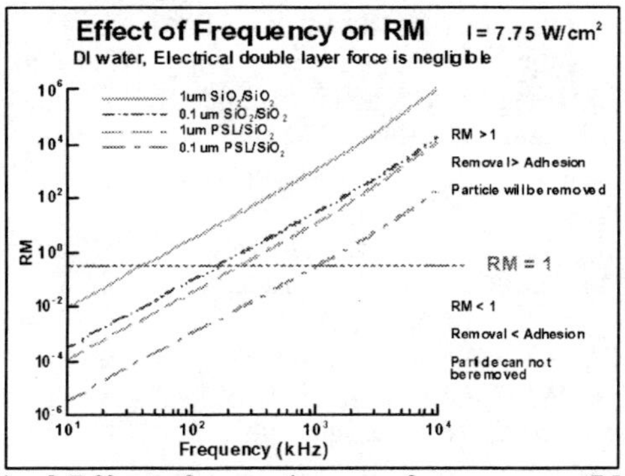

Fig. 8 Effect of acoustic wave frequency on RM

Figure 8 shows that higher acoustic frequency leads to larger removal/adhesion moment ratio. Higher frequency is needed to remove smaller particles. The removal of 100 nm silica and PSL particles are achievable by using DI water and acoustic streaming with frequencies larger than 270 kHz and 1.6MHz, respectively.

c. The effect of cleaning solution

To physically remove silica particles down to 10nm, megasonic flow with frequency larger than 1.3MHz is necessary when using DI water only (Fig. 9a). The experimental data in Fig. 10 shows that complete removal of silica particles down to 100 nm is achieved by using a single wafer megasonic cleaning process and DI water only (1t 760 kHz).

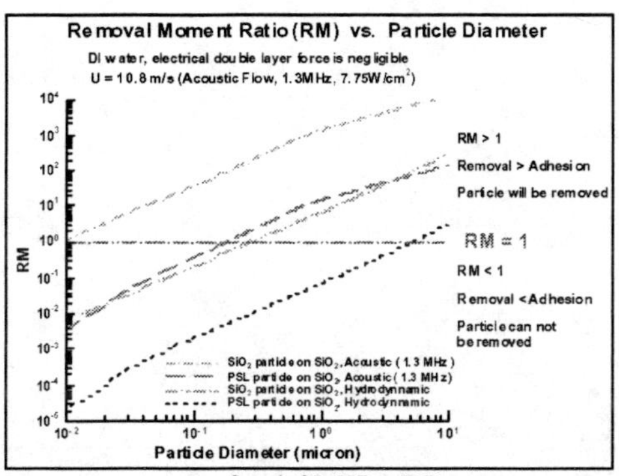

a. f = 1.3MHz, DI water

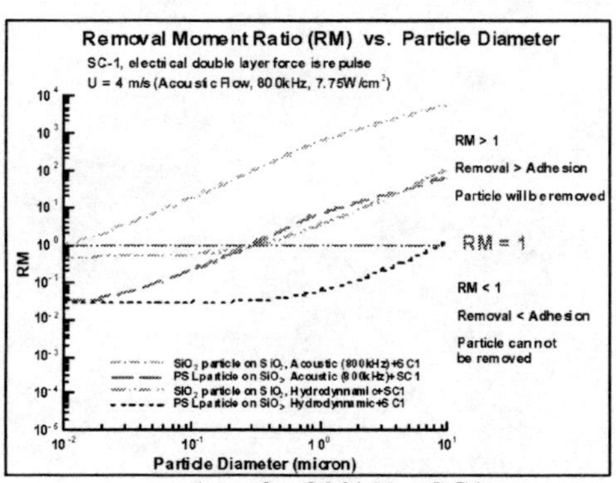

b. f = 800kHz, SC1

Fig.9 Effect of particle size on RM

In addition to the physical megasonic effect in removing particles, the use of chemistry has shown a big improvement in cleaning efficiency. Using SC1 (Fig. 9b), typical megasonic flow (7.75W/cm^2, 800kHz, 4m/s streaming velocity) can remove all silica particles in the shown range (10nm ~ 10micron) and PSL particles larger than 310 nm. Hydrodynamic flow without megasonics (at 4m/s) can remove silica particles larger than 0.34 micron and PSL particles larger than 9.07 micron.

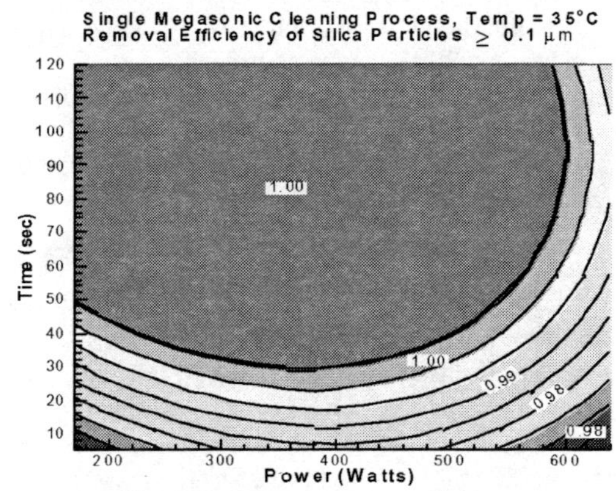

Fig. 10 Single wafer megasonic cleaning

3. CONCLUSIONS

The contact diameter between the particle and the substrate was found to increase with time within one to seven days. The increase at 95%RH is larger than that at 40%RH. The smaller the particle, the larger the ratio of contact diameter/particle diameter. The resulting plastic deformation is caused by the fact that the adhesion stress is sufficiently large to exceed the yield stress of PSL particles. The deformation was a function of size when it was caused by the van der Waals force. However, at high humidity, the capillary force dominated and the deformation was not affected by size.

High frequency megasonic cleaning is a promising technique for nano-scale particles removal. Using DI water, the removal of nano-size particles down to 10 nm can be best accomplished using acoustic streaming with frequency above 1.3 MHz. Megasonic induced acoustic streaming is essential to the removal of submicron and nano-size particles. As the frequency increases, the acoustic boundary layer thickness decreases and the streaming velocity increases thereby increasing the drag force and consequently the removal moment. By utilizing the electrical double layer force as a repulsive force (by using basic chemistry,) the removal of 10nm silica particles can be accomplished using megasonic cleaning above 800 kHz. Softer particles (such as PSL) are more difficult to remove than hard particles (such as silica), because of adhesion-induced deformation, need a higher frequency. The experimental results show that a complete removal of silica particles down to 100 nm is achievable using a single wafer non-contact cleaning process with DI water only.

4. ACKNOWLEDGMENTS

The authors wish to express their gratitude for the support of the NSF center for Microcontamination control and support under NSF grant for U.S.-Korea Cooperative Research.

5. REFERENCES

1. The International Technology Roadmap for Semiconductors, 2001, (www.sematech.org).
2. Busnaina, A. A. and Elsawy, T. M., *Journal of Electronic Materials*, 27(10), 1095, 1998.
3. Busnaina, A. A. and Gale, G. W, *Journal of Particulate Science and Technology*, 17(3), 1999.
4. Tabor, D., J. Colloid Interface Sci. 58, 2 (1977).
5. Bowling, R. A., in: Particles on Surfaces I: Detection, adhesion, and Removal, K.L. Mittal (Ed.), Plenum Press, New York (1988).
6. Johnson, K., Kendall, K., and Roberts, A. D., R. Soc. London A, 324, 301 (1971).
7. Maugis, D. and Pollock, H.M., Acta Metall..32,1323 (1984).
8. Feng, Jiang-wei , Busnaina, A., Steel, E. B., and Small, J. A., "Time and Humidity Effect on the Adhesion Induced Deformation of Submicron PSL Particles," *Proceedings*, 24th Annual Meeting of The Adhesion Society, Williamsburg, VA Feb. 25-28, 2001, pp. 41-43.
9. Tang, J., and Busnaina, "The Effect of Time and Humidity on Particle Adhesion and Removal," *J. of Adhesion*, 74, 411 (2000).
10. Rimai, D.S., Moore, R.S., Bowen, R.C., Smith, V.K. and Woodgate, P.E., J. Mater. Res., Vol.8, No.3, 662, (1993).
11. Krishnan, S., Busnaina, A. A., Rimai, D. S. and DeMejo, D. P., "The Adhesion-induced Deformation and the Removal of Submicrometer Particles," *Journal of Adhesion Science and Technology*, vol. 8, No. 11, 1994, pp. 1357-1370.
12. Nyborg, W. L., "Acoustic streaming", in Physical acoustics, Vol. IIB, edited by W. P. Mason, Academic Press, New York, 1965.
13. Visser, J., 13, 169, 1995.
14. Soltani, M. and Ahmadi, G., *J. of Adhesion Sci. Technol*., 8(7), 763, 1994.

Proceedings of
2003 STLE/ASME Joint International Tribology Conference
Ponte Vedra Beach, Florida USA, October 26-29, 2003

2003-TRIB-0278

A NEW EXPERIMENTAL TECHNIQUE TO MEASURE CONTACT PRESSURE

Emanuel N. Diaconescu/
University of Suceava, Romania

Marilena L. Glovnea/
University of Suceava, Romania

Ovidiu Petrosel/University of Suceava, Romania

ABSTRACT

A new technique to measure the pressure in a real contact is proposed. One of contacting surfaces is covered, prior to contact establishment, by a special gel. The contact closing removes the excess gel and, during a certain time interval, the contact pressure transforms the entrapped substance in an amorphous solid. In each point, the refractive index of this solid depends on the pressure acting during transformation. After contact opening, the reflectivity of this coating depends on the former contact pressure and it is mapped by aid of a laser profilometer and becomes an indicator of contact pressure. Theoretical considerations show that the gel must possess certain optical parameters for the method to work. Several experimental reported results offer a clear image upon pressure distribution in Hertz point contacts, end effects in finite length line contacts, pressure distribution between rough surfaces and indicate the onset of plastic deformation.
Keywords: contact pressure, refractive index, reflectivity, adsorption, gel coating.

INTRODUCTION

The contact mechanics is essentially a theoretical science, supplying various analytical results concerning the elements of the contact and the stress states inside contacting bodies. Nevertheless, because the theoretical approach uses frequently simplifying assumptions upon the elastic behavior of the materials, the geometry of bounding surfaces and the tangential interaction between these surfaces, an experimental verification of theoretical results is often needed.

Glovnea and Diaconescu [1] and Glovnea [2] analyzed the existing experimental investigations into the contact mechanics and suggested that these may fall into two classes, namely the measurement of displacements or strains in the contact region or in its neighborhood and the evaluation of either contact pressure or stresses.

The first class comprises the following procedures:

- accurate measurement of normal approach between the contacting bodies;
- investigation of form and dimensions of contact area;
- evaluation of deformation state in the central planes of the contact by moiré fringes;
- strain gauge measurement of the strain in the points of half space boundary;
- evaluation of deformed surface of one of the contacting bodies by interferometry or by laser profilometry.

Two basic methods fall in the second class, as follows:

- direct measurement of contact pressure by special pressure sensitive gauges or by Raman spectrometry;
- evaluation of contact pressure by measuring the reflectivity of contacting surfaces prior and after the contact;
- investigation of contact stresses by photo-stress techniques or by holography.

From these methods, the most advanced results offer those that supply point to point information upon some contact features, such as the deformed surface of one or both of contacting bodies or measurement of contact pressure and contact stresses. A precise method to find the deformed surface of a metallic equivalent punch pressed against a thick sapphire window as well as the actual contact area was recently advanced by Diaconescu and Glovnea [3] by aid of laser profilometry. By using these experimental results as input data for normal displacement, numerical calculations yield the contact pressure responsible for these deformations.

Yamaguchi, Uchida and Abraha [4] advanced an interesting method of contact pressure evaluation, based on measurement of intensity of a reflected laser beam by the same surface, prior and after the contact. They found that after contact, the intensity of reflected light increases in the former points of contact area and becomes proportional to contact pressure. Eventually, the etching of the surface improves the sensibility of the method.

The idea of Yamaguchi, Uchida and Abraha to investigate optically the surface after the contact, led the authors of present paper to develop a new technique for evaluation of contact pressure in real contacts. This consists in measuring the reflectivity of a thin coating formed on one of the contacting surfaces as a result of transformation of a gel into an amorphous solid at contact pressure. As the refractive index of the coating depends on the phase change pressure, the measured reflectivity is a useful indication of contact pressure.

NOMENCLATURE

- M: molecular mass;
- N: Avogadro's Number;
- R: molecular refraction;
- c: velocity of light in vacuum;
- i: imaginary unit;
- k: wave number;
- n: real part of refractive index;
- $\tilde{n}$ complex refractive index;
- v: velocity of light in a given medium;
- α: absorption coefficient;
- β: polarizability;
- ε: dielectric constant or permittivity;
- μ: magnetic permeability;
- ν: frequency of monochromatic light;
- ρ: gel density;
- σ: the intermolecular distance when the potential is zero;
- φ: minimum value of interaction potential, at equilibrium;
- χ: extinction coefficient.

THE REFRACTIVE INDEX OF GEL COATING

As shown in the introduction, the new method consists in covering one of contacting surfaces with a special gel, prior to contact establishment. After a well-defined time, the contact is closed, the normal load is applied and the system is maintained in this state another adequate time interval. Although the majority of the gel is removed when the contact is established, a minute amount of substance remains caught inside contact. Under the action of contact pressure, the entrapped gel suffers a phase transformation. Because the rate of pressure increase to the nominal value is quite high, the available time for molecular rearrangement in a low viscosity state is short, a few seconds. The viscosity of the gel, already high at contact closing, increases immensely at contact pressure and impedes the molecular rearrangement. Consequently, the solid state resulting from transformation is amorphous and, therefore, isotropic. Finally, the contact is opened and a very thin coating of solidified gel is found on previously covered surface. This is an optical medium, characterized by a refractive index.

Primarily, the absolute refractive index of a medium is defined as the ratio between the velocities the electromagnetic wave possesses in vacuum and in that medium:

$$n = \frac{c}{v} = \sqrt{\varepsilon\mu}\,. \qquad (1)$$

The optical media can be transparent or absorbing.

A transparent medium has zero conductivity and its magnetic relative permeability differs from unity by a negligible amount. Consequently, for such media the refractive index is:

$$n = \sqrt{\varepsilon}\,. \qquad (2)$$

Let **r** be the position vector of a point in a transparent medium and **s** a unit vector in a fixed direction. Any solution of one of Maxwell equations of the form **V**=**V**(**r**·**s**,t) represents a *plane wave* that propagates in the direction of **s**. As shown by Born and Wolf [5], **V** is a function of **r**·**s**-vt only, that is it can be written under the form **V**=**V**(**r**·**s**-vt).

No medium, except a vacuum, is perfectly transparent. All material media show strong absorption, at least in some regions of the electromagnetic spectrum. An absorbing medium has a finite conductivity and, consequently, a finite current density. Nevertheless, in such media the volume charge density is zero. The permittivity is constant but is complex because a phase-shift occurs between the component field vectors. Similarly, the conductivity is also complex. As a result, the refractive index is complex and is symbolized by $\tilde{n}$, the dielectric constant by $\tilde{\varepsilon}$ and the conductivity by $\tilde{\sigma}$. According to Lambert's law, known also as Bouguer's law, when a plane wave of monochromatic light enters an absorbing medium on a direction z perpendicular to a plane interface, the light intensity decreases as a result of absorption. The amplitude A(z) of one of the field vectors inside the absorbing medium depends on its incident amplitude A_0 and on depth z as follows:

$$A(z) = A_0 \exp(-\alpha z)\,. \qquad (3)$$

In an exponential form, the wave equation becomes:

$$V = A_0 \exp[-\alpha z + i(\omega t - kz)]\,, \qquad (4)$$

where k is the wave number ($k = \omega / v = n\omega / c$) and $\omega = 2\pi\nu$. The use of the extinction coefficient $\chi = \alpha/k$ allows writing the wave equation under following form:

$$V = A_0 \exp i\omega\left(t - \frac{\tilde{n}}{c} z\right). \qquad (5)$$

Because in this case **r**·**s**=z, Eq. (5) is formally similar to the expression **V**=**V**(**r**·**s**-vt) for a wave in a transparent medium and, for this reason, $\tilde{n}$ is called *complex refractive index*. According to Ditchburn [6], this index is given by equation:

$$\tilde{n} = n(1 - i\chi) = n - n\chi i\,, \qquad (6)$$

where n is its real part.

Optically, the gel coating belongs to the class of absorbing media and therefore its refractive index is complex. The real part of this index depends on pressure. The starting point for evaluation of this dependence is the law of Clausius-Mosotti, which correlates the dielectric constant of non-polar substances with pressure by the relation:

$$\frac{\varepsilon - 1}{\varepsilon + 2}\frac{M}{\rho} = \frac{4\pi}{3}N\beta\,. \qquad (7)$$

The right hand member of Eq. (7) is called the molecular polarization and it is approximately equal to the real volume occupied by molecules in a mole of substance. Its magnitude is independent of pressure and temperature and changes slightly between liquid and solid states. The polarization of a molecular mixture is cumulative, as the molecular volume is. For substances having the same magnetic permeability as the vacuum, expression 2 is substituted for the dielectric constant in Eq. (7) and Lorentz-Lorentz formula is obtained:

$$\frac{n^2-1}{n^2+2}\frac{M}{\rho}=\frac{4\pi}{3}N\beta=\text{constant}. \quad (8)$$

If Eq. (8) is applied at two pressures, namely a reference pressure p_r for which the refractive index n_r is known and a current pressure p at which refractive index n is needed, the following relation is obtained:

$$\frac{n_r^2-1}{n_r^2+2}\frac{1}{\rho_r}=\frac{n^2-1}{n^2+2}\frac{1}{\rho}. \quad (9)$$

Equation (9) leads to following expression of n:

$$n=\sqrt{\frac{2+n_r^2+2\left(n_r^2-1\right)\overline{\rho}}{2+n_r^2-\left(n_r^2-1\right)\overline{\rho}}}, \quad (10)$$

where $\overline{\rho}$ is dimensionless density, defined as the ratio of density at pressure p to that at reference pressure.

This dimensionless density can be expressed as a function of pressure if the molecular interaction of specified substance is known. Because the gel is a non-polar molecular substance, the interaction between its structural units can be described quite accurately by Lennard-Jones 6/12 potential. As shown by Diaconescu [7], the dimensionless density for such an intermolecular potential can be found as follows. The equilibrium dimensionless intermolecular distance at reference pressure, chosen equal to atmospheric pressure, is:

$$r_0=\sqrt[6]{2\frac{H}{G}}\cdot\sigma, \quad (11)$$

where G=14.45395361538747 and H=12.13197182474035.

The equilibrium distance r at pressure p is the root of following dimensionless equation:

$$12\frac{G^2}{H}\left[\left(\frac{r_0}{r}\right)^{15}-\left(\frac{r_0}{r}\right)^{9}\right]=3\sqrt{2}\left(\frac{r_0}{\sigma}\right)^{3}\overline{p}, \quad (12)$$

where:

$$\overline{p}=\frac{\sigma^3}{\varphi}p \quad (13)$$

is the dimensionless pressure.

The dimensionless density at pressure p is now given by:

$$\overline{\rho}=\left(\frac{r_0}{r}\right)^{3}. \quad (14)$$

If the refractive index n_r at atmospheric pressure is known and the dimensionless pressure is calculated by Eqs. (11) to (14), the refractive index at a pressure p is found from Eq. (10).

The refractive index of solidified gel coating is generated during solidification and depends on contact pressure acting in each point. Therefore, after contact opening, the coating memorizes the contact pressure by its refractive index. As this index is variable across the contact area, the coating reflectivity is a function of former contact pressure.

REFLECTIVITY OF A THREE-LAYERED MEDIUM

The reflectivity of surface with gel coating is scanned and measured by aid of a laser profilometer. The laser beam falls from air on the solidified gel coating. A part of incident light energy returns into air by reflection. The remaining energy traverses the optically absorbing coating. The absorption extinguishes part of incoming energy and the remaining suffers a phenomenon of reflection-refraction by the interface between gel coating and the metal. Again, a part of incident energy is reflected by metallic surface and the metal absorbs the rest. The former traverses the coating, where it is partially absorbed, and returns into air by another process of reflection-refraction at first interface. At exit it sums up to the light reflected primarily on the interface air-gel. This sum is measured and then, divided by initially incident energy, defines the reflectivity of the system. This is a typical problem of reflection-refraction of light by a three-layered optical medium with two optical interfaces, namely air-gel coating and gel coating-metal.

At each traversal of an interface, the laser beam suffers a process of reflection-refraction, as illustrated in Fig. 1, where θ_i and θ_t are, respectively, the incidence and the refraction angles.

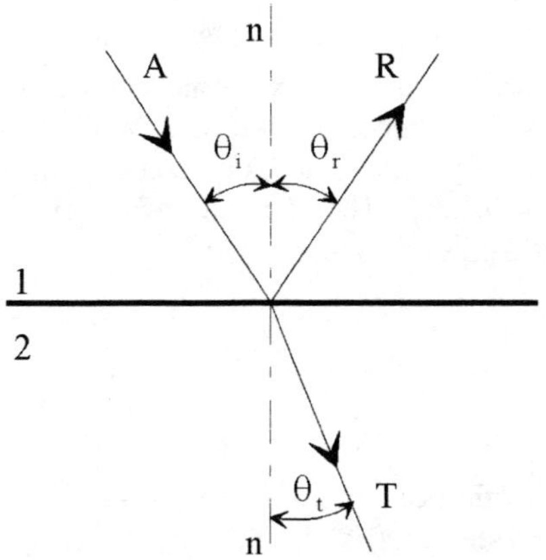

Fig. 1. Reflection-refraction of light by an optical interface

As stated by Born and Wolf [5], when a plane light wave falls onto a boundary between two homogenous media with different optical properties, it is split into two waves, namely a *transmitted wave*, proceeding into the second medium and a *reflected wave*, propagated back into the first medium. This process is often called reflection-refraction of light by the second medium and depends essentially on the optical properties of adjacent media. Let *A*, *R* and *T* be the amplitudes of anyone of the field vectors, magnetic or electric, in incident, reflected and transmitted fields, respectively. Each of these is resolved into components parallel and perpendicular to the plane of incidence denoted, respectively, by subscript $\parallel$ and $\perp$. When adjacent media are transparent, Snell's law governs the process of reflection-refraction and the components of reflected and transmitted waves are expressed in terms of incident wave by Fresnel formulae. According to Born and Wolf [5], the reflectivity R and transmissivity T at an optical interface are defined, respectively, as the ratios between the power densities of reflected and transmitted waves and the power density in the primary wave. Alternatively, [8], these parameters can be defined as the ratios of corresponding Poynting vectors fluxes. In agreement with the law of energy conservation, the following relation holds:

$$R+T=1. \quad (15)$$

Normally, these factors have different values for the two wave components, parallel and perpendicular to the incidence plane.

Ditchburn [6] and Moisil [8] show that the formal similarity of Eqs. 5 and 6 with those for transparent bodies allows writing formally a law similar to Snell's law for an interface separating two absorbing media:

$$\sin\tilde{\theta}_t = \frac{\tilde{n}_1}{\tilde{n}_2}\sin\theta_i\,, \tag{16}$$

where θ_t is complex, except for normal incidence when $\theta_i=\tilde{\theta}_t=0$. As a result, the refraction of light by absorbing media can be modeled by following complex Fresnel formulae, [8]:

$$\tilde{T}_{\parallel} = \frac{2\,\tilde{n}_1\cos\theta_i}{\tilde{n}_2\cos\theta_i + \tilde{n}_1\cos\tilde{\theta}_t}A_{\parallel}\,;\ \tilde{T}_{\perp} = \frac{2\,\tilde{n}_1\cos\theta_i}{\tilde{n}_1\cos\theta_i + \tilde{n}_2\cos\tilde{\theta}_t}A_{\perp}\,; \tag{17}$$

$$\tilde{R}_{\parallel} = \frac{\tilde{n}_2\cos\theta_i - \tilde{n}_1\cos\tilde{\theta}_t}{\tilde{n}_2\cos\theta_i + \tilde{n}_1\cos\tilde{\theta}_t}A_{\parallel}\,;\tilde{R}_{\perp} = \frac{\tilde{n}_1\cos\theta_i - \tilde{n}_2\cos\tilde{\theta}_t}{\tilde{n}_1\cos\theta_i + \tilde{n}_2\cos\tilde{\theta}_t}A_{\perp}\,.$$

The factors $\tilde{r}_{\parallel,\perp}$ and $\tilde{t}_{\parallel,\perp}$ multiplying the amplitudes of incident field vectors are called complex Fresnel coefficients.

For normal incidence $\theta_i=\tilde{\theta}_t=0$ and the distinction between the components parallel and perpendicular to the plane of incidence disappears. Therefore, the expressions of complex Fresnel coefficients become:

$$\tilde{r} = \frac{\tilde{n}_2 - \tilde{n}_1}{\tilde{n}_2 + \tilde{n}_1}\,; \tag{18}$$

$$\tilde{t} = \frac{2\,\tilde{n}_1}{\tilde{n}_2 + \tilde{n}_1}\,, \tag{19}$$

When the direction of light propagation reverses, the modules of complex Fresnel coefficients do not change.

By using these coefficients and indexing the involved optical interfaces by i,j in order of light propagation (i,j=12, 23, 21) and taking into account the absorption in the gel layer according to Eq. 3, the following formulae can be written for components of light wave at various interface transversal:

- the light vector reflected back into air by interface air-gel:

$$\tilde{R}_{12} = \frac{\tilde{n}_2 - n_1}{\tilde{n}_2 + n_1}A\,; \tag{20}$$

- the light vector transmitted into the solidified gel layer:

$$\tilde{T}_{12} = \frac{2\,n_1}{\tilde{n}_2 + n_1}A\,; \tag{21}$$

- the field vector reaching the optical interface gel-metal:

$$\tilde{I}_{23} = \tilde{T}_{12}\exp(-\alpha\ d)\,. \tag{22}$$

- the light reflected by gel-metal optical interface:

$$\tilde{R}_{23} = \frac{\tilde{n}_3 - \tilde{n}_2}{\tilde{n}_3 + \tilde{n}_2}\tilde{I}_{23}\,; \tag{23}$$

- the field vector incident on gel layer-air interface:

$$\tilde{I}_{21} = \tilde{R}_{23}\exp(-\alpha\ d)\,; \tag{24}$$

- the light vector transmitted from the gel layer back into air:

$$\tilde{T}_{21} = \frac{2\,n_1}{\tilde{n}_2 + n_1}\tilde{I}_{23} = \frac{4\,n_1^2(\tilde{n}_3 - \tilde{n}_2)}{(\tilde{n}_2 + n_1)^2(\tilde{n}_3 + \tilde{n}_2)}\exp(-2\alpha d)A\,, \tag{25}$$

where d is the thickness of coating in the considered point. The two waves reflected into air, $\tilde{T}_{21}$ and $\tilde{R}_{12}$ are coherent, but out of phase by a certain angle ψ, which, for a given light wavelength, depends on coating thickness and actual refractive index of amorphous state, not known priory, and metal. The amplitude of light reflected back in the air by the system is given by following equation:

$$R^2 = T_{21}{}^2 + R_{12}{}^2 + 2T_{21}R_{12}\cos\psi\,, \tag{26}$$

R, T_{21} and R_{12} being the amplitudes of corresponding vectors.

The light intensity is proportional to the refractive index of the medium and to the square of light wave amplitude. Therefore, if Eq. (26) is divided by A^2, following correlation between global reflectivity R of the system and its components R_1 and R_2 due to, respectively, $\tilde{T}_{21}$ and $\tilde{R}_{12}$ is found:

$$\mathrm{R}^2 = \mathrm{R}_1^2 + \mathrm{R}_2^2 + 2\sqrt{\mathrm{R}_1\mathrm{R}_2}\cos\psi\,. \tag{27}$$

Because the angle ψ varies randomly, $\cos\psi$ may be replaced by its averaged value 0.5 and relation (27) takes the form:

$$\mathrm{R}^2 = \mathrm{R}_1^2 + \mathrm{R}_2^2 + \sqrt{\mathrm{R}_1\mathrm{R}_2}\,. \tag{28}$$

Each component reflectivity can be calculated as follows:

$$\mathrm{R}_1 = \frac{\tilde{T}_{21}\cdot\tilde{T}_{21}^*}{A^2}\,;\qquad \mathrm{R}_2 = \frac{\tilde{R}_{12}\cdot\tilde{R}_{12}^*}{A^2}\,, \tag{27}$$

where the superscript star denotes the use of complex conjugate refractive indexes in the expressions of $\tilde{T}_{21}$ and $\tilde{R}_{12}$.

A similar result can be arrived at by writing successively the reflectivity and transmissivity at each interface along the light path and by combining adequately the results.

For a unity value of air refractive index, $n_1=1$, Eqs. (20), (25) and (27) lead to following final expressions of R_1 and R_2:

$$\mathrm{R}_1 = \frac{16\,n_2\left[(n_3 - n_2)^2 + (n_3\chi_3 - n_2\chi_2)^2\right]\exp(-4\,\alpha\,d)}{\left[(1+n_2)^2 + n_2^2\chi_2^2\right]^2\left[(n_3 + n_2)^2 + (n_3\chi_3 + n_2\chi_2)^2\right]}\,; \tag{28}$$

$$\mathrm{R}_2 = \frac{(n_2 - 1)^2 + n_2^2\chi_2^2}{(n_2 + 1)^2 + n_2^2\chi_2^2}\,. \tag{29}$$

In order to evaluate the reflectivity of the surface covered by solidified gel, knowledge on gel film thickness d is needed. A typical profile of free surface of a gel film deposited on a flat surface subjected to contact by a ball is shown in Fig. 2. This is parabolic and the thickness is given by the equation:

$$d(r) = d_0 + c\,r^2\,, \tag{30}$$

where, in this set of experimental conditions, $d_0 = 4\cdot10^{-8}$m and $c=5\cdot10^{-7}m^{-1}$.

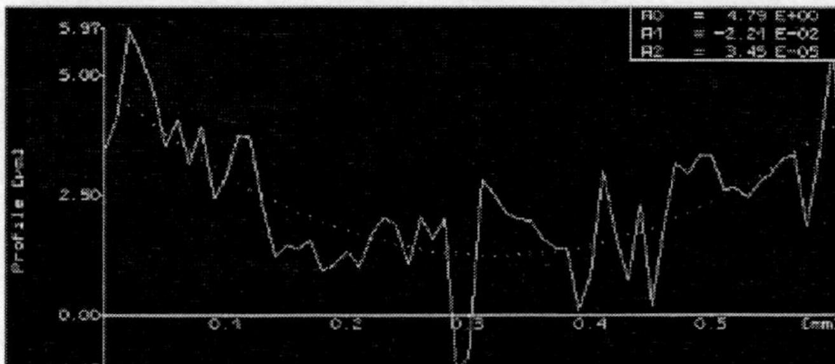

Fig. 2 Radial profile of free surface of solidified gel film

Typical curves of variation of computed global reflectivity with radial distance in a circular Hertz contact subjected to various loads are shown in Fig. 3. The dimensionless Hertz radius at maximum load, which corresponds roughly to 5 GPa maximum Hertz pressure, is chosen to be equal to 1. The reflectivity is plotted for dimensionless Hertz radii of .1, .2, .4, .6, .8 and 1 using following parameters: $n_1 = 1$, $n_{2r} = 1.4$, $n_3 = 2.41$, $\chi_3 = 1.38$, $d_0 = 10^{-8}\,\text{m}$, $c = 1.67 \cdot 10^{-7}\,\text{m}^{-1}$. A special attention was focused upon the extinction coefficient of solidified gel. If this is assumed constant with respect to solidification pressure, the phase shift between the vector fields of light remains constant, which is not the case. Unfortunately, the literature does not offer information upon this subject. Consequently, in order to find similar profiles of calculated and experimentally found reflectivity and on basis of examination of experimental specimens, the following dependence of extinction coefficient on pressure was chosen:

$$\chi_2(p) = \chi_{20}\left[1 - e\left(\frac{p}{p_{00}}\right)^2\right], \qquad (31)$$

where $\chi_{20} = 0.12$ is the gel extinction coefficient at atmospheric pressure, p_{00} is a significant pressure chosen equal to 5 GPa and e is a proportionality constant equal to 0.8.

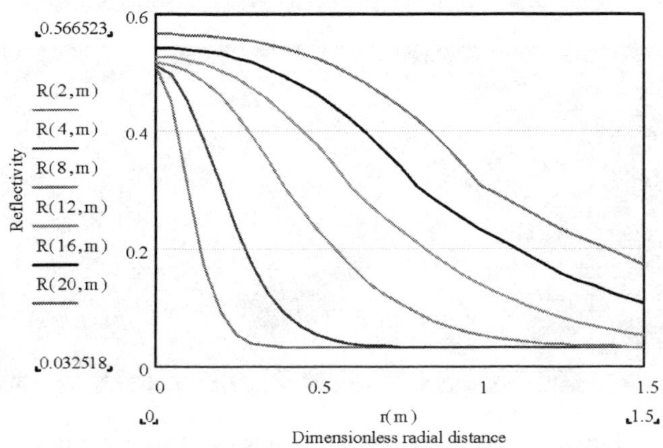

Fig. 3. Typical dependence of reflectivity on radial distance

In Fig. 3, maximum Hertz pressure changes tenfold, which implies a variation of thousand fold in applied load, a fairly large range. It can be seen that the reflectivity profile extends along radial direction about three times more than Hertz radius. Except at the highest load, where a bend in reflectivity occurs at contact boundary, it is quite difficult to guess the contact limit. The numerical analysis shows that the values of some of involved parameters need be placed in certain ranges to obtain good results. For instance, in order to assure a central reflectivity monotonously and consistently increasing with maximum Hertz pressure, χ_{20} must fall between [0.05, 0.2] if d_0 is 10^{-8} m and between [0.02, 0.5] when d_0 is $3 \cdot 10^{-8}$ m. The initial value of refractive index of the gel does not seem to play an important role. The optimum range for other parameters can be found by aid of Eqs. 28 and 29.

EXPERIMENTAL RESULTS

Some preliminary results showing the ability of this method to investigate the contact pressure are included further down.

Circular Hertz Contact

In order to check the new method, experimental investigations were performed for a Hertz contact between a 16 mm diameter bearing ball and a polished flat plate. Three normal loads were applied which generate calculated maximum Hertz pressures of 1.2, 1.73 and 2.05 GPa. After opening the contact, the flat surface covered with gel is placed on the X-Y table of the profilometer UBM 14 and the reflectivity, as well as gel surface geometry are recorded.

Three-dimensional plots of reflectivity obtained are shown in Figs. 4 a,b,c.

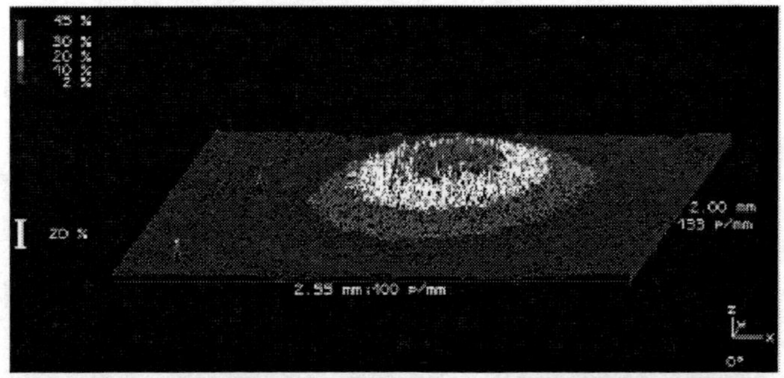

a

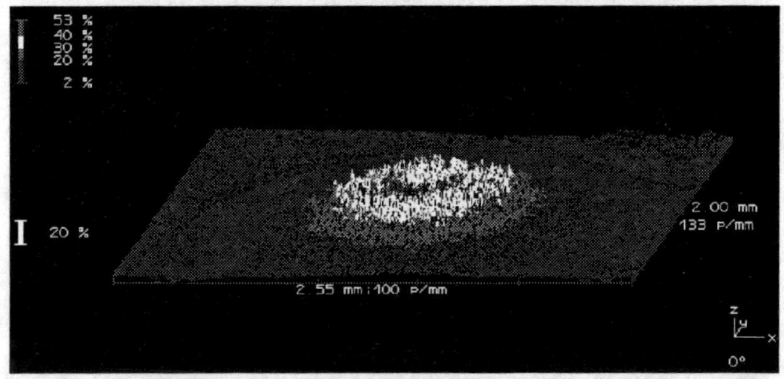

b

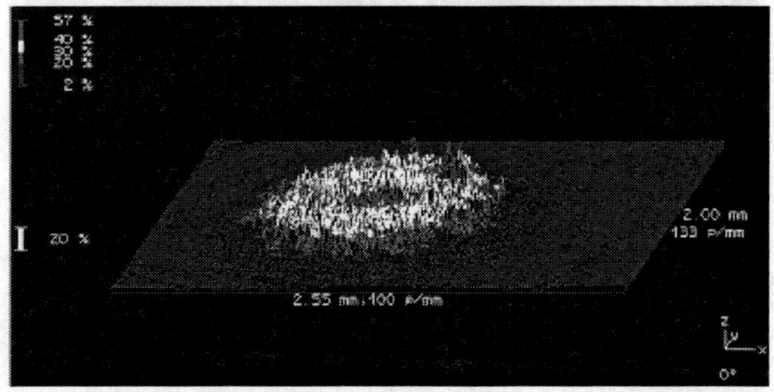

c

Fig. 4. 3-D plot of reflectivity for increasing load

As the load increases, the reflectivity increases slightly and the area of higher reflectivity extends. At higher loads, a dimple occurs in the central part of the plot, probably due to the contact entering in the elasto-plastic range.

The simultaneous surface measured by aid of the laser profilometer is shown in Fig. 5 and it is smooth.

A more detailed plot in the elastic range, using a different gel and a higher resolution is shown in Fig. 6 and a diametric profile of reflectivity is presented in Fig. 7.

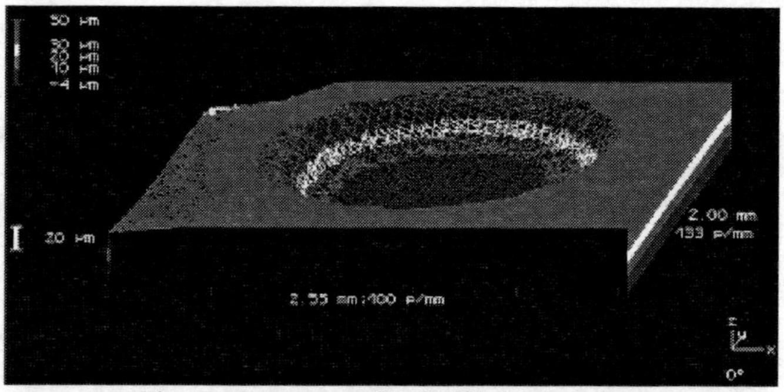

Fig. 5. Geometry of solidified gel surface

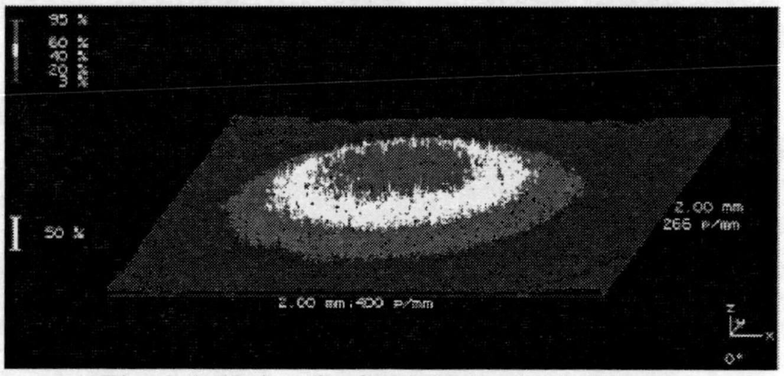

Fig. 6. 3-D plot of reflectivity in elastic domain

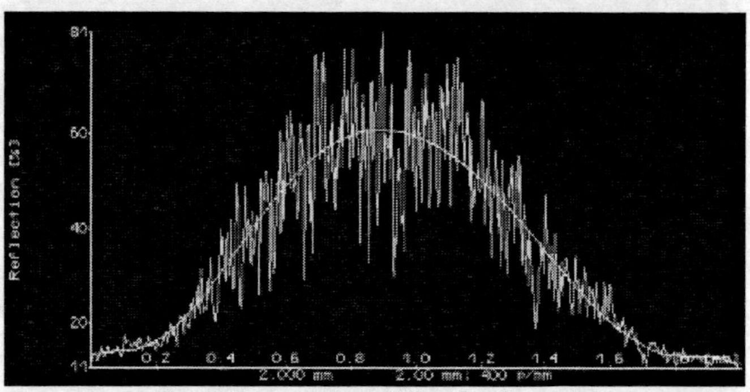

Fig. 7. Diametric profile of reflectivity

The reflectivity profile is not smooth because asperity interaction generates steep peaks and deep valleys with respect to ideal surfaces. Nevertheless, the average line traced as an eighth order polynomial is very similar to the radial reflectivity profiles traced theoretically in Fig. 3. This agreement acts in favor of herein proposed experimental method. At high resolution, the zoom function supply the shape of reflectivity peaks and therefore of asperity pressure. In the valleys, the reflectivity falls to the atmospheric value corresponding to the actual gel film thickness.

Nonsymmetrical Hertz Point Contact

Normally, a Hertz contact is symmetrical with respect to principal curvature planes in the initial contact point. Often, this restriction cannot be satisfied, a typical example being the contact ball-raceway in thrust ball bearings, in which the curvature increases from periphery to bearing center. As a result, the transversal width of contact area narrows in this direction and the contact domain looses the symmetry. To examine such a contact, a ball was pressed against the raceway. Typical experimental results are shown in Figs. 8, a to c.

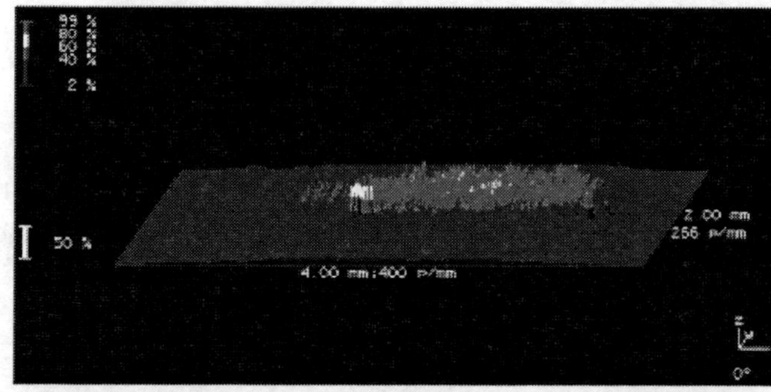

a

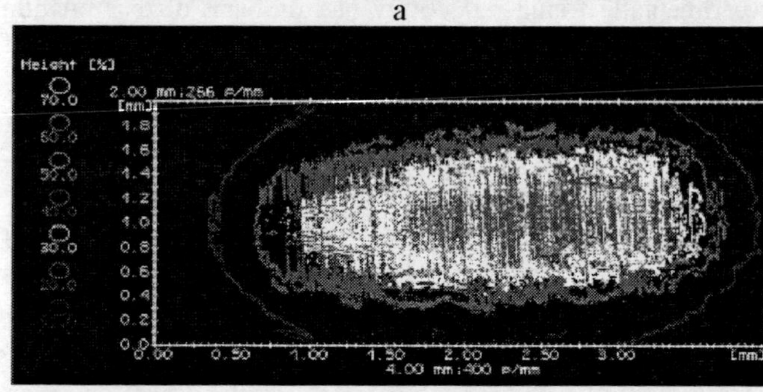

b

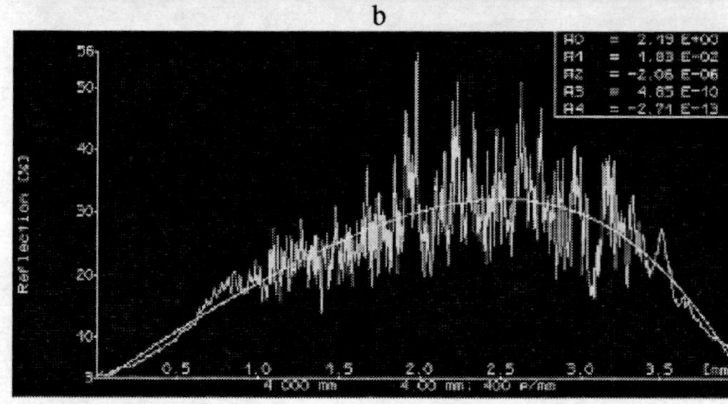

c

Fig. 8. Reflectivity in a nonsymmetrical point contact

The 3-D reflectivity shown in Fig. 8a and the contour lines traced in Fig. 8b indicate that contact area is wider in the right side than in inner points. The effect of surface roughness and of some surface bumps is well evidenced in Fig. 8b. The longitudinal reflectivity profile traced in Fig. 8c shows that maximum reflectivity and, therefore, maximum pressure occurs on the outside of initial contact point.

Finite Length Line Contact

The finite length line contact is yet a subject open to research. Of great importance for these contacts are the solutions of attenuation or of complete removal of pressure and stress risers occurring due to end effects. These methods are essentially theoretical and require supplementary experimental validation. The experimental method herein proposed can be applied to this end. To prove this, the new method was applied to a contact between a cylindrical roller with round-off edges and a polished flat surface, the roller axis being parallel to the surface plane. Typical results obtained at three levels of normal load, Q, namely 42.8, 128.55 and 214.25 N, are shown as 3-D reflectivity plots in Fig. 9, a, b and c, respectively. The load was applied slightly eccentric.

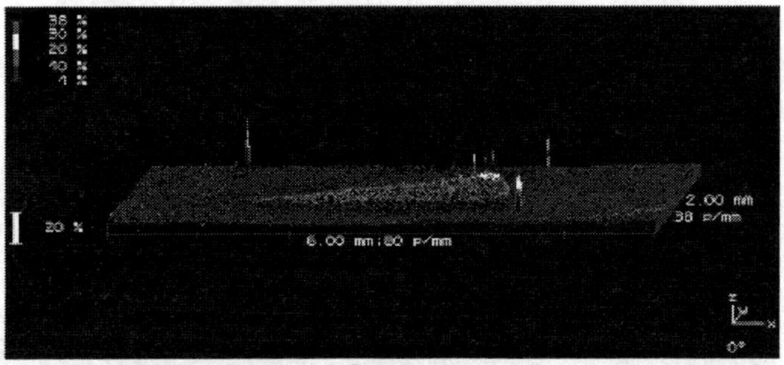

a: Q=42.8 N

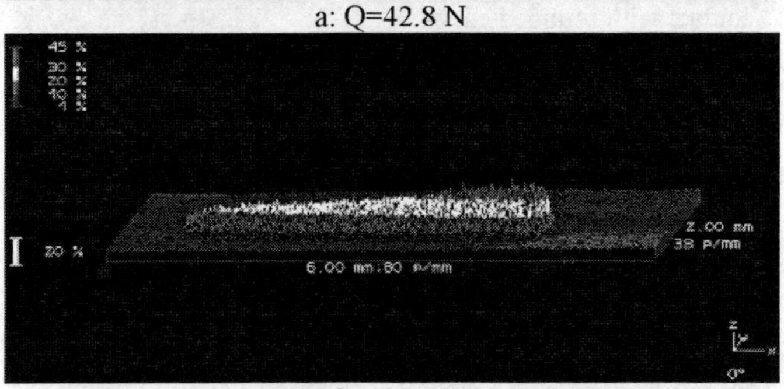

b: Q=128.55 N

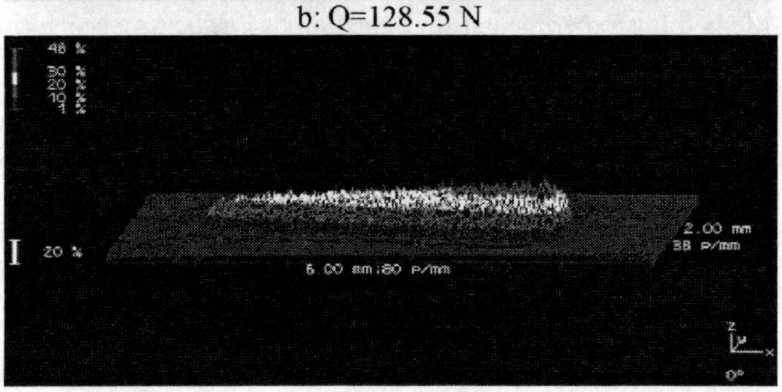

c: Q=214.25 N

Fig. 9. Reflectivity of a line contact, eccentrically loaded

As expected, the reflectivity increase with normal load and the contact area widens. The maximum pressure occurs at the most loaded end, in immediate neighborhood of curvature center.

Detection of Contact Plastic Deformation

When observing the reflectivity of gel layer in a circular Hertz contact, it was noticed that at a certain load level a central dimple occurs in reflectivity, which increases at subsequent load rising. As the flat surface specimen is made of through hardened carbon steel, it occurred the idea that this behavior may be caused by plastic deformation.

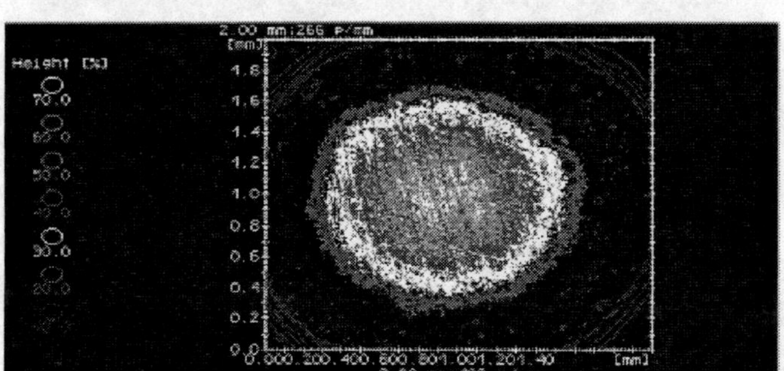

Fig. 10. Contour plot of reflectivity in an elastic contact

To check this assumption, the experimental results were further analyzed. First, the reflectivity of a contact in elastic domain was traced as contour plot in Fig. 10. The reflectivity increases monotonously from border to contact center and reaches a central maximum. Then, the reflectivity of the contact after dimple occurrence was plotted the same way, as shown in Figs. 11, a and b.

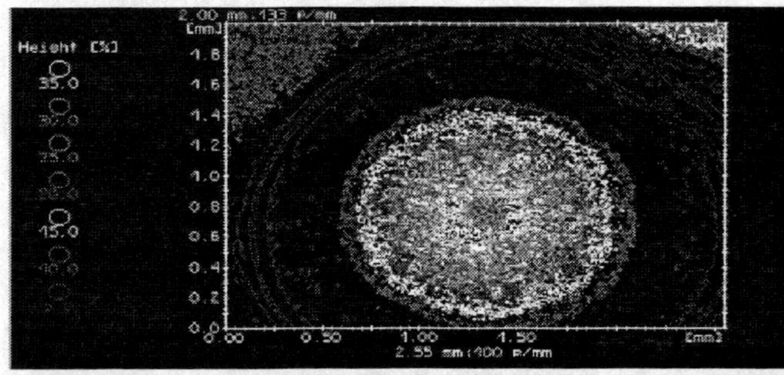

a: Q=128.55 N

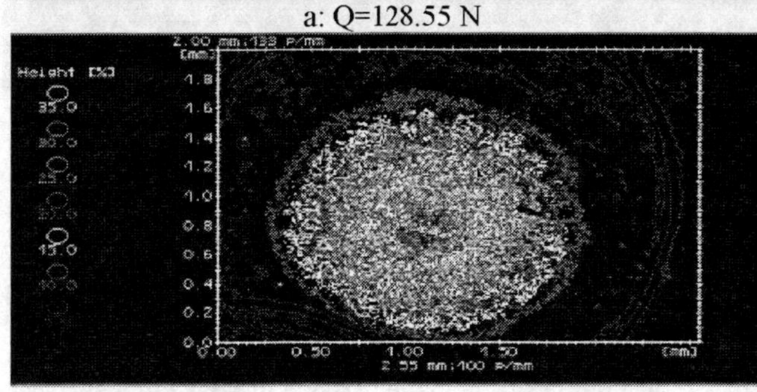

b: Q=214,25 N

Fig. 11. Increase of dimple extension with load

Diametric geometric profiles of gel free surface were then traced, as shown in Fig. 12, a, b and c.

Figure 12 reveals that at a calculated Hertz pressure of 1.2 GPa the gel profile is smooth if surface roughness is disregarded. At a calculated pressure of 1.73 GPa, a small dimple is visible in the contact center, having a diameter of about 0.25 mm and a depth of roughly 7 μm. As the maximum increases just above 2 GPa, the dimple widens to 0.4 mm and its depth reaches about 10 μm.

Further experimental investigations into the subject evidenced that the calculated Hertz pressure, which initiates the reflectivity dimple is 1.5 GPa for the steel of flat specimen. Therefore, this method is a useful tool for detection of onset of plastic deformation in mechanical contacts.

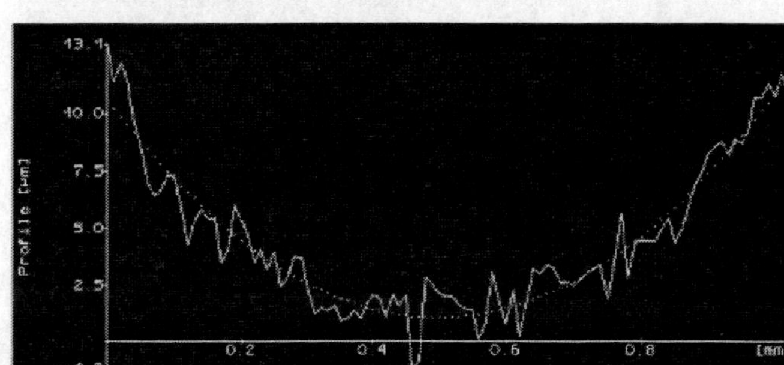

a: gel surface profile at a maximum Hertz pressure of 1.2 GPa

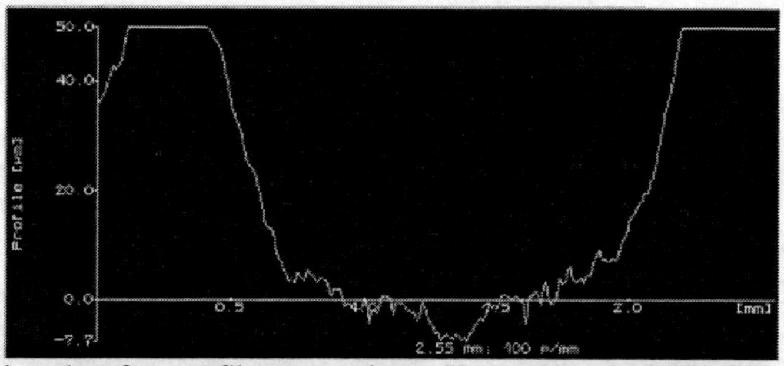

b: gel surface profile at a maximum Hertz pressure of 1.73 GPa

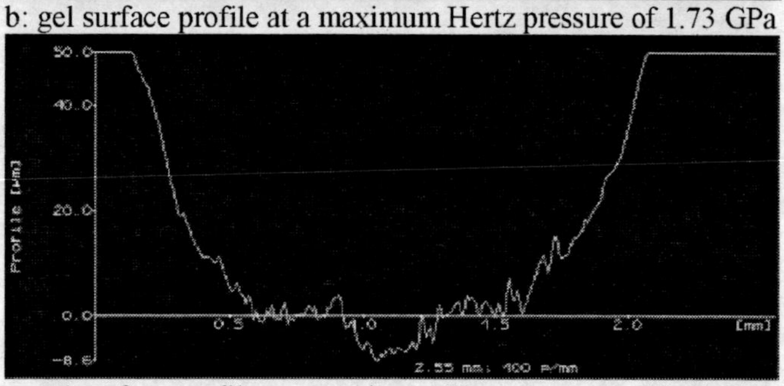

c: gel surface profile at a maximum Hertz pressure of 2.05 GPa

Fig. 12. Gel surface profiles at various normal loads

CONCLUSIONS

The work reported herein can be summarized by the conclusions reviewed below.

- A new optical method for assessment of contact pressure is proposed. This is based on solidification, inside the contact, of a molecular gel film applied on one of contacting surfaces. The refractive index of solidified gel, as well as its extinction coefficient depends on pressure acting during transformation, i.e. on contact pressure. After contact opening, the reflectivity of the surface initially covered with gel is scanned by aid of a laser profilometer. This depends on refractive index, extinction coefficient and local thickness of gel coating.
- A previous theory advanced by one of these authors was applied to predict the dependence of gel density on pressure. This is used in Lorentz-Lorentz formula to find the value of refractive index at various pressure levels.
- A simple formula was derived for the reflectivity of a three-layer medium, as is the case of the media air-surface covered with gel-metal, when two of the layers are optically absorbing. The reflectivity is expressed in terms of refractive index, extinction coefficient and the thickness of solid gel film. The gel thickness was found experimentally to have a parabolic profile and an empirical formula was proposed for the dependence of extinction coefficient on pressure.
- Theoretical curves of reflectivity along radial direction in a circular Hertz contact were traced for given parameters.
- This technique was applied to measure experimentally the reflectivity in a circular Hertz contact and good formal agreement with theoretical predictions was found.
- The new method generates good reflectivity results for both a nonsymmetrical point Hertz contact and a finite length line contact with rounded edges.
- The proposed experimental method seems to detect quite accurately the initiation of contact plastic deformations.
- Finally, this technique promise to become a very useful tool for experimental investigations in contact mechanics.
- Further research work is needed to improve the quality of the gel and the experimental procedure as well as the theoretical understanding of involved phenomena.

REFERENCES

[1] Glovnea, M.L., and Diaconescu, E.N., 1994, "A New Method for Experimental Investigation of Elastic Contacts," (in Romanian), *Symp. on Tradition and Continuity in Railway Research*, Vol.II, Bucharest, pp. 77-82.

[2] Glovnea, M.L., 1999, "The Effect of Geometrical Surface Discontinuities upon the Elastic Contact," (in Romanian), PhD Thesis, University of Suceava.

[3] Diaconescu, E.N., and Glovnea, M.L., 2002, Evaluation of Contact Area by Reflectivity, Proc., *3rd AIMETA International Tribology Conference*, Italy, on CD.

[4] Yamaguchi, K., Uchida, M., and Abraha, P., 1990, "Measurement of Pressure on Contact Surface by Reflection of Light (effect of surface etching)," *Proceedings of the Japan International Tribology Conference*, Nagoya, pp. 1271-1276.

[5] Born, M., and Wolf, E., 1980, *Principles of Optics*, Sixth Edition, Pergamon Press.

[6] Ditchburn, R. W., 1963, *Light*, Vol. II, Blackie & Son, Second Edition.

[7] Diaconescu, E.N., 1993, "Solid-like behaviour of EHD oil film," Scientific Report ERB-CIPA-CT-92-0526.

[8] Moisil G.C., and Curatu E., 1986, *Optics - Theory and Applications,* (in Romanian), Technical Press, Bucharest.

Proceedings of
2003 STLE/ASME Joint International Tribology Conference
Ponte Vedra Beach, Florida USA, October 26-29, 2003

2003-TRIB-0279

NANOINDENTATIONS ON $SrTiO_3$ SUBSTRATES: EFFECTS OF FRACTAL ROUGHNESS ON CONTACT MECHANICS

Renato Buzio/ INFM-UdR Genova, Via Dodecaneso 33, 16146 Genova, Italy
Email: buzio@fisica.unige.it

Karolina Malyska/ Faculty of Mechatronics of the Warsaw University of Technology, ul.Sw.A.Boboli 8, 02-525 Warsaw, Poland

Zygmunt Rymuza/Faculty of Mechatronics of the Warsaw University of Technology, ul.Sw.A.Boboli 8, 02-525 Warsaw, Poland

Corrado Boragno/ INFM-UdR Genova and Dipartimento di Fisica, Via Dodecaneso 33, 16146 Genova, Italy

Francesco Buatier de Mongeot/ INFM-UdR Genova and Dipartimento di Fisica, Via Dodecaneso 33, 16146 Genova, Italy

Ugo Valbusa/ INFM-UdR Genova and Dipartimento di Fisica, Via Dodecaneso 33, 16146 Genova, Italy

ABSTRACT

The non–stationary character of roughness is a widely recognized property of surface morphology and suggests modelling several solid surfaces by fractal geometry. In the field of contact mechanics this demands for novel investigations attempting to clarify the role of multi–scale roughness during physical contact.

Here we propose an experimental investigation of the mechanical response of $SrTiO_3$ substrates probed by a commercial nanoindentor. Load-displacement curves have been acquired respectively on well-defined crystalline surfaces and on mechanically lapped surfaces. We observe the first-loading cycles to be considerably affected by surface roughness whenever the penetration depth is kept below the interface width. The obtained results are analyzed within an elasto-plastic deformation model for fractal surfaces and a picture is developed to describe the deformation process with respect to surface roughness and structural parameters.

INTRODUCTION

Indentation experiments represent a widely employed technique to probe the mechanical properties of surfaces. Usually a hard indenter is pressed into the sample and the indentation depth is recorded for different values of load. Load-displacement curves are thus directly measured and the elastic modulus and hardness can be estimated [1]. Nanoindentors test the mechanical response of bodies by producing deformations in the range of tens or even few nanometers and they allow to investigate thin films and coatings as well as to analyze the role of surface defects in contact stiffness [2,3]. A qualitative difference between nanoindentation and its macro/microscopic counterpart is in the role of surface morphology during the contact process. While surface roughness is negligible with respect to sample deformation in macro/microscopic experiments, this in not the case for nanoindentations on the most commonly encountered surfaces. Engineering surfaces usually show wavy or random nano-roughness produced by machining [4]. Also artificial substrates, like atom- or cluster-assembled thin films, have an intrinsic roughness due to the deposition process and precursors dynamics [5,6]. The advent of the scanning tunnelling microscope and the Atomic Force Microscope (AFM) allowed to measure surface topography with nanometric resolution and fractal geometry has been extensively used to characterize randomness in terms of fractal dimension and saturated surface roughness. Whenever roughness is present, it has to be taken into account for the estimation of nanohardness. Computer simulations treating depth-sensing indentation of Weierstrass-Mandelbrot fractal surfaces have shown that both the mean hardness and scatter are influenced by roughness whenever penetration depth is smaller or comparable to the interface width, since in such case a multi-asperity contact is actually created between the probe and the sample [7]. More recently some of us have conducted experiments on the contact mechanics of self-affine organic thin films by means of custom-designed micrometric flat AFM probes and have observed the strong influence of surface roughness and fractal dimension on the load-displacement curves [8]. Here we present experimental results about the indentation of ceramic substrates by means of a commercial nanoindentor.

EXPERIMENT

Indentations have been performed with a commercial nanoindentor equipped with a diamond Berkovich pyramid probe having curvature radius in the range 100nm-200nm at the very end of the tip [9]. The specimens we have analyzed are commercial single-crystals of $SrTiO_3$, commonly used as substrate for epitaxial growth of oxides [10,11]. One side of each substrate is chemo-mechanically polished by the manufacturer in order to obtain an atomically smooth (001) termination, while the backside is just lapped with a roughness in the micrometer range [12]. In the following we refer to samples 1 and 2 to indicate two $SrTiO_3$ substrates exposing to the indenter respectively the atomically smooth frontside and the micrometric rough backside. Obviously we expect sample 2 to have a distorted, full of defects and not single cristalline surface because of the lapping process. Therefore a comparison of the mechanical response of the two specimens by indentation should reflect differences in roughness as well as in the structure of the surface layer.

First we have characterized the surface morphology of the two samples by AFM. Then load-displacement curves have been acquired for each sample on many different locations and results have been averaged to obtain the mean mechanical response of the surface. We have analysed only *the first loading-cycle* in view of the strong plastic deformations expected under repeated indentations.

RESULTS AND DISCUSSION

Figure 1 shows the AFM morphologies of the samples. Sample 1 displays atomically flat terraces with monoatomic steps of about 3.9Å (Fig.1(a)). Micrometric bumps are present on the contrary on sample 2 (Fig.1(b)); indeed they appear to be structured on nanometric scale, supporting the presence of a multi-scale surface roughness.

Figure 1. AFM morphologies of sample 1 (a) and sample 2 (b). Scan size: 5 x 5μm^2.

In Fig. 2 we report indentation data for sample 1. The figure shows the average load-displacement curve and reveals the elasto-plastic deformation of the surface for loads up to 500μN, with a maximum contact depth of about 40nm and a residual impression depth slightly less than 30nm. Displacement "bursts" of few nanometers, likely due to the onset of plastic deformation [13], were often observed on individual loading curves for P>120μN but they disappeared on the mean curve due to the averaging procedure.

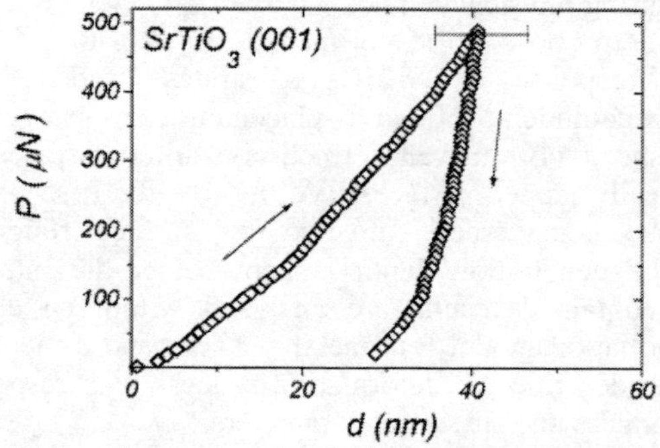

Figure 2. Average load-displacement curve for sample 1.

We analyzed the unloading part of the curve in Fig.2 following Oliver and Pharr method [1] and estimated the reduced Young modulus E_1* to be 360GPa and the hardness H_1=14GPa .

In Fig.3 we report the average load-displacement curve for sample 2. The elasto-plastic response is confirmed by the hysteresis of the cycle, the maximum contact depth being now d~85nm at the compressive load P~500μN. Such observation suggests sample 1 to have a *stiffer surface* than sample 2. To some extent this statement is counterintuitive since we would expect mechanical lapping to considerably strengthen sample 2 with respect to sample 1 [14].

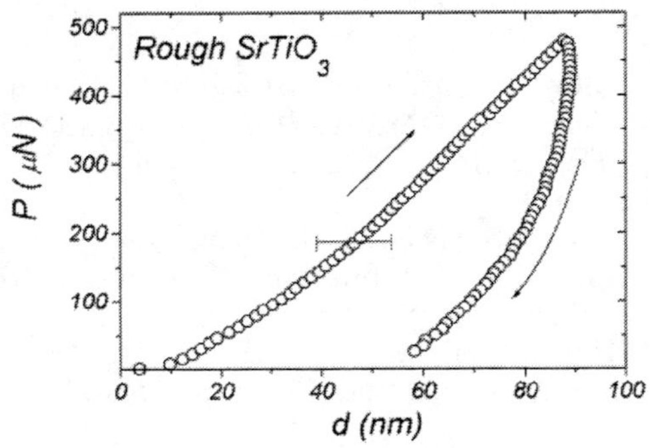

Figure 3. Average load-displacement curve for sample 2.

In order to understand the physical reasons underlying such effect we fitted the mean loading curve for sample 2 with the deformation model proposed by Yan and Komvopoulos [15]. Such theory models sample surface as a 3D Weierstrass-Mandelbrot (WM) function and assumes that it is elasto-plastically deformed by a flat (smooth) rigid counterpart. In our case we assumed the deformed surface to be the sum surface of the tip and the rough sample [7]. First of all, we quantitatively characterized sample morphology by calculating the height-height correlation function G(r) on AFM topographies as described in detail in Ref.6. Then we fitted G(r) with a power-law relationship, as expected for self-affine interfaces[5], thus obtaining the following values for the fractal dimension D_f and the saturated surface roughness σ: D_f =(2.41±0.05) and σ =(400±10)nm. The correlation length was less than 2μm. The estimated values for fractal parameters were used to numerically generate WM surfaces having the same topographical parameters of the analyzed samples. The Berkovich indenter was modelled as an equivalent conical indenter of included angle 140°. The radius of curvature R of the tip was fixed before each simulation and the indentation hardness H and equivalent elastic modulus E* were estimated as fitting parameters. Indeed by varying R in the range 100nm-200nm we observed that the fitted parameters showed only negligible variations. The best fits reported in the following were obtained for R=150nm.

In Fig. 4 we show the comparison of the experimental load-displacement curve for sample 2 and the theoretical model. The red dash line represents a fully-plastic approximation to the experiment. It assumes that sample 2 has a Young modulus comparable or higher than that of sample 1 (E_2^*≥360GPa) but a definitely higher hardness (H_2=50GPa). Deviations of the red curve from data at higher penetration depths (above 70nm) could be easily reduced assuming a decrease of hardness H_2 with deformation volume/depth towards the bulk value. On the contrary the black line represents an elasto-plastic approximation to the experiment: it assumes that sample 2 has a Young modulus E_2^*=200GPa and a hardness H_2= 64GPa.

In view of the good agreement of data with model we suggest the following picture to describe the contact process. On sample 1 the indentation accurs at the tip apex of the indenter (single-asperity contact) and the values of the Young modulus E_1^* and indentation hardness H_1 are due to the crystalline structure of the (001) $SrTiO_3$ surface. On sample 2, because of surface roughness, the contact often occurs over indenter lateral surface more than at the tip apex (multi-asperity contact) while contact area is definitely smaller than expected at similar values of *d* for the smooth (001) surface. This causes sample 2 to appear more compliant than sample1.

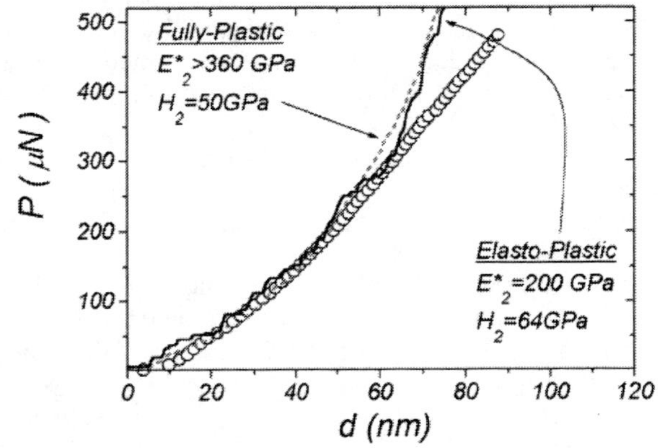

Figure 4. Comparison of experimental loading curve for sample 2 with the theoretical model (see text).

The prediction $H_2 > H_1$ confirms that mechanical polishing has changed the structural properties of the surface layer of sample 2. Defects and policrystalline structure could in principle be responsible also of the predicted reduction of the Young modulus E_2^*=200GPa required by the elasto-plastic approximation. However tha contact behaviour of sample 2 during indentation appears ultimately conditioned by its high surface roughness.

We expect no differences to occur in the mechanical response of our specimens for (micro-)indentations overcoming the interface width (d >>σ) [7].

CONCLUSIONS

In summary, we have investigated the mechanical response of $SrTiO_3$ substrates by means of a commercial nanoindentor. Load-displacement curves have been acquired respectively on well-defined (001) crystalline surfaces and on mechanically lapped surfaces. We have observed that the smooth surfaces appears stiffer than the rougher ones. To clarify this aspect we have analyzed experimental data within an elasto-plastic deformation model for fractal surfaces and we have estimated the Young modulus and indentation hardness. The model suggests rough surfaces to have different structural properties (increased hardness) with respect to the (001) surfaces, likely due to the lapping process, and surface roughness to be responsible for their higher penetration depths. Our findings are in qualitative agreement with computer simulations performed on similar systems.

REFERENCES

[1] Oliver, W.C., Pharr, G.M., 1992 "An improved technique for determining the hardness and elastic modulus using load and displacement sensing indentation experiments", *Journal of Material Research*, **7(6)**, pp.1564-1583.

[2] Li, X., Bhushan, B., 2002, "A review of nanoindentation continuous stiffness measurement technique and its applications", *Materials Characterization*, **48**, pp.11-36.

[3] Rodríguez de la Fuente, O., Zimmerman, J.A., González, M.A., de la Figuera, J., Hamilton, J.C., Woei Wu Pai, Rojo, J.M., 2002, "Dislocation emission around nanoindentations on a (001) fcc metal surface studied by scanning tunneling microscopy and atomistic simulations", *Physical Review Letters*, **88**, pp. 36101-36104.

[4] He, L., Zhu, J., 1997, "The fractal character of processed metal surfaces", *Wear*, **208**, pp.17-24.

[5] Barabasi, L., Stanley, J., 1995, "Fractal Concepts in Surface Growth" (Cambridge University Press, Cambridge, UK).

[6] Buzio, R. Gnecco, E., Boragno, C., Valbusa, U., Barborini, E., Piseri, P., Milani, P., 2000, "Self-affine properties of cluster-assembled carbon thin films", *Surface Science*, **444**, L1-L6.

[7] Bobji, M.S., Biswas, S.K., Pethica, J.B., 1997, "Effect of roughness on the measurement of nanohardness-a computer simulation study", *Applied Physics Letters*, **71**(8), pp.1059-1061.

[8] Buzio, R., Boragno, C., Biscarini, F., Buatier de Mongeot, F. and Valbusa, U., 2003, "The contact mechanics of fractal surfaces", *Nature Materials*, **2**(4), pp. 232-235.

[9] Hysitron Triboindenter, Hysitron, 10025 Valley View Road, Minneapolis, MN 55344, USA.

[10]Crystal-GmbH, Ostendstrasse 2-14, 12459 Berlin, Germany.

[11] Kawasaki, M., *et al.*, 1994, "Atomic control of the SrTiO3 crystal surface", *Science*, **266**, pp. 1540-1542.

[12] Dr. Steffen Sandner (Crystal-GmbH), private communication.

[13] Suresh, S., Nieh, T.G., Choi, B.W., 1999, "Nano-indentation of copper thin films on silicon substrates", *Scripta Materialia*, **41**, pp.951-957.

[14] Bowden, F.P., Tabor, D., 1954, "Friction and Lubrication of Solids" (Oxford University Press, London, UK), pp.325-328.

[15] Yan, W., Komvopoulos, K., 1997, "Contact analysis of elastic-plastic fractal surfaces", *Journal of Applied Physics*, **84**, pp.3617-3624.

Proceedings of
2003 STLE/ASME Joint International Tribology Conference
Ponte Vedra Beach, Florida USA, October 26-29, 2003

2003-TRIB-0280

A NEW METHODOLOGY TO EVALUATE THE ROLLING CONTACT FATIGUE PERFORMANCES OF BEARING STEELS WITH SURFACE DENTS – APPLICATION TO 32CRMOV13 (NITRIDED) AND M50 STEELS.

D. Nélias (LaMCoS – UMR CNRS 5514, INSA Lyon)

C. Jacq (SNECMA Moteurs, Villaroche)

G. Lormand (GEMPPM – UMR CNRS 5510, INSA Lyon)

G. Dudragne (SNR Bearings, Annecy)

A. Vincent (GEMPPM – UMR CNRS 5510, INSA Lyon)

ABSTRACT

A new methodology is proposed to evaluate the rolling contact fatigue (RCF) performances of bearing steels in presence of surface dents. The experimental procedure consists in denting the raceway of the test specimen with a hardness machine using spherical diamond tips of different radius, i.e. 200, 400 and 600 µm, and normal loads ranging from 5 to 50 daN. Analysis of various dent geometries yields to an analytical law with five parameters useful to fit experimental profiles for contact simulation. Besides local residual stresses and plastic strains around the dent have been obtained by finite element simulations of the indentation process. RCF tests performed on a two-disk machine have shown better performances of nitrided 32CrMoV13 steel compared to M50 reference steel. The dominating role of sliding has been highlighted and two areas where damage initiates were identified, while the effects of the normal load and hoop stresses are less marked.

INTRODUCTION

Rolling contact fatigue is a major cause of rolling bearing failure [1]. RCF due to the repeated contact between rolling elements is considerably accelerated when surface defects, such as dents, exist on the surface of the contacting bodies [2]. Dents created by solid particles in suspension in the lubricant cannot be totally removed since perfect filters or seals do not exist. Therefore rolling element bearings should be designed to endure surface dents.

Nitrided steels have been found promising materials for rolling element bearings [3] and seem particularly adapted when raceways present surface defects such as dents. The benefit reached by nitriding the surface of components subjected to high contact pressure arises from the combination of a high surface hardness provided by such surface treatment and a high core toughness. The nitriding treatment hardens the material surface, not only in the area concerned by the surface defect but also up to the Hertzian depth, and introduces compressive residual stresses in the nitrided layer [4] that are known to be beneficial for RCF. For these reasons nitrided 32CrMoV13 steel (AMS 6481) was found to be a good candidate.

To the author's knowledge the literature does not provide quantitative data about the effects of operating conditions on rolling contact fatigue (RCF) and little is known about the damage mechanisms of surface hardened steels in presence of such surface stress raisers. The aim of the present work, therefore, is (i) to evaluate the RCF performances of nitrided 32CrMoV13 steel versus M50 (AMS 6490) – which is a reference steel for aeronautic rolling bearings – when one of the contacting surfaces is artificially dented, (ii) to quantify the influence of the Hertz pressure, slide-to-roll ratio and hoop stresses, and (iii) to identify the areas where damage starts.

The experimental investigation was performed on a high-speed two-disk machine with operating conditions representatives of those encountered for the main-shaft roller bearing of a commercial turbine engine. Geometry and loading conditions have been carefully chosen to reproduce the contact stress profile encountered in an aeronautic rolling bearing and especially with regards to the micro yield shear stress profile determined by Jacq et al [5]. A simple and discriminating procedure is proposed. It consists in denting the raceway of the test specimen with a hardness machine using diamond tips of different radius, i.e. 200, 400 and 600 µm, and normal loads ranging from 5 to 50 daN. Simulations of denting by the finite element method (FEM) have also been performed, in order to determine properly the dent geometry and the residual stress state around the surface defect. Along with the knowledge of the micro-yield stress profile obtained by nano-indentation such data are necessary for 3D cyclic elastic-plastic analysis.

EXPERIMENTAL PROCEDURE

Material. The chemical composition of the 32CrMoV13 steel studied is given in table 1. Before nitriding at 550°C, specimens were austenised at 950°C for 20 min, quenched in oil and then annealed at 630°C for 3 h. The compound layer, which formed at the surface of the material during nitriding, was removed by grinding. The penetration depth of nitrogen obtained after this treatment is about 600 µm.

C	Si	Mn	Ni	Cr	Mo	V
0.334	0.24	0.48	0.10	2.93	0.9	0.32

Table 1. Chemical composition of nitrided 32CrMoV13 steel

Two-disk Machine and operating conditions. Fatigue tests have been performed on a high-speed twin-disk machine [6] whose principle is briefly recalled hereafter. Two spherical rotating disks are brought in contact with an imposed load. The rotating speed of disks is controlled independently, hence enabling various contact conditions, such as pure rolling or controlled sliding. Nitrided 32CrMoV13 steel is surface hardened. Therefore the choice of the loading condition must account not only for the Hertzian stress profile but also for the residual stress versus depth profile inherent to the nitriding process, and that has to be compared to the micro yield shear stress profile. This problem has been stated by Vincent et al. [7], who have described the variation of the RCF endurance limit H1 for a spherical point contact as a function of the contact equivalent radius of curvature in the case of nitrided 32CrMoV13 steel. Subsequently, special care has been brought in the choice of operating conditions so that the stress profile undergone by the tested disk (noted TD in Fig. 1) is close to the one encountered in the aimed application, i.e. a typical aeronautic rolling bearing (ARB) (indicated by ARB in Fig. 1). This consideration has led to the choice of an equivalent radius of 15 mm, obtained with a specimen disk (dented) of 24 mm radius and the opposite disk (smooth) of 40 mm radius. Two loads have been considered, 2.5 and 3 GPa. The radius of the contact zone is then 510 and 610 µm, respectively. Nitrided 32CrMoV13 steel specimen disks have been tested against M50 disks, since in bearing applications, M50 roller are used with nitrided 32CrMoV13 rings. For reference, tests have also been made with M50 specimen disks. The contact is lubricated with a MIL-L-23699 type turbine oil at 80°C. The linear velocity at contact is set to 24 $m.s^{-1}$. To evidence the influence of sliding, tests have been carried out under pure rolling condition, and with a slide to roll ratio of 1.5 %. It should be noted that, in presence of sliding, the slower disk is the nitrided one. Last, the influence of tensile hoop stresses has been investigated, by adding 115 MPa on some nitrided disks by press fitting. Operating conditions corresponding to the set of tests performed are summarized in table 2.

Label	Specimen	Load (GPa)	Sliding	Hoop stress (MPa)
C1	M50	2.5	0	0
C2	M50	3	1.5 %	0
ES1	Nitrided	2.5	0	0
ES2	Nitrided	2.5	0	115
ES3	Nitrided	2.5	1.5 %	0
ES4	Nitrided	2.5	1.5 %	115
ES5	Nitrided	3	0	0
ES6	Nitrided	3	0	115
ES7	Nitrided	3	1.5 %	0
ES8	Nitrided	3	1.5 %	115

Table 2. Fatigue tests operating conditions

Indentation procedure. Artificial controlled dents have been made with spherical diamond indenters of different radius R_p (R_p=200, 400 and 600 µm) at various loads W_i (W_i=5, 10, 20, 30, 50 daN). All dents have been made on the 24 mm radius disk. They are identified by the radius of the indenter and by the applied load. Hence, dent 200-20 is a dent made with R_p = 200 µm under load W_i = 20 daN. The geometrical characteristics of dents thus obtained (diameter ϕ, depth h_p and shoulder height h_b, Fig. 2) are summarized in table 3. Up to 18 dents have been tested simultaneously on each specimen.

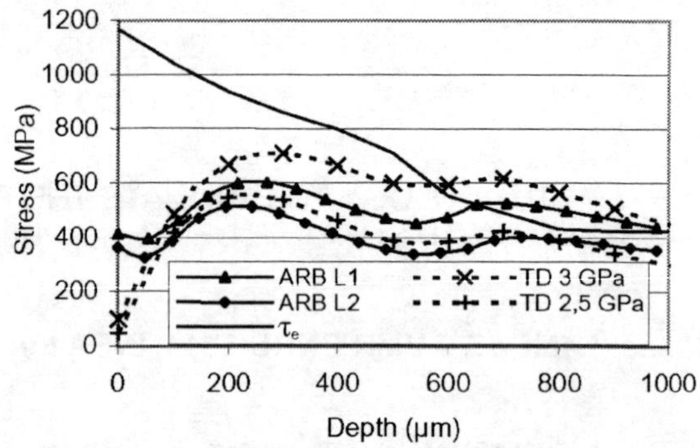

Figure 1. Von Mises and micro yield shear stress profiles.

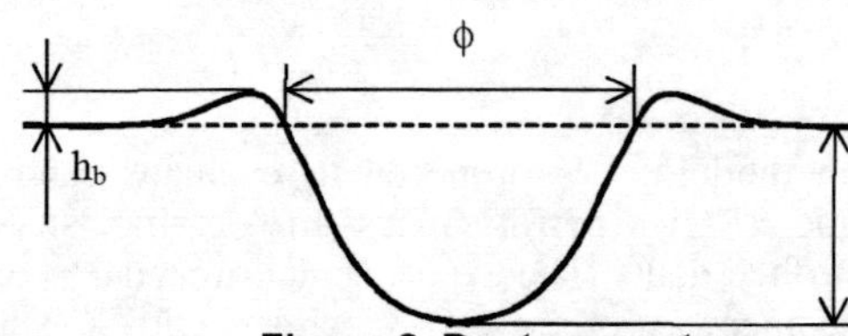

Figure 2. Dent geometry.

32CrMoV13 (nitrided)									
Indentation		Dent			Indentation		Dent		
Radius (µm)	Load (daN)	ϕ (µm)	h_p (µm)	h_b (µm)	Radius (µm)	Load (daN)	ϕ (µm)	h_p (µm)	h_b (µm)
200	5	85	3.5	0.4	400	20	180	6.5	0.6
200	10	120	6	0.7	400	30	210	10	0.8
200	20	170	13	1.4	400	50	290	18	0.9
200	30	200	20	2.7	600	20	195	4	0.3
200	50	260	32	4	600	30	225	6	0.3
400	10	125	2.5	0.6	600	50	290	11	0.7
M50									
Indentation		Dent			Indentation		Dent		
Radius (µm)	Load (daN)	ϕ (µm)	h_p (µm)	h_b (µm)	Radius (µm)	Load (daN)	ϕ (µm)	h_p (µm)	h_b (µm)
200	5	90	3.5	0.3	400	20	190	7.5	0.3
200	10	130	7	0.4	400	30	220	11	0.4
200	20	170	14	1.3	400	50	280	18	0.6
200	30				600	20	190	4.5	0.5
200	50				600	30	230	7	0.6
400	10	140	3	0.2	600	50	300	12	0.7

Table 3. Measured values of the dent geometry parameters.

RESULTS

Damage criterion. During the experiment, it is aimed at determining whether a given dent is harmful or not. A dent is considered as harmful if micro-spalls appear around the dent. To determine a test duration, two tests, C1 and C2, have been run during 40 and $50x10^6$ cycles. On these two tests, all damages visible at the end were initiated before $20x10^6$ cycles. For example in Fig. 3 the dent 200-10 which did not exhibit micro-spalls after 10 millions cycles (Fig. 3a) remained optically identical up to 50 millions cycles. Therefore a dent is considered as harmful for a given operating condition if micro spalls have appeared around the dent at $20x10^6$ fatigue cycles.

It should be emphasized that this strict criteria is focused on the occurrence of localized surface damage, and not on its evolution to generalized micro-pitting or deep spalling.

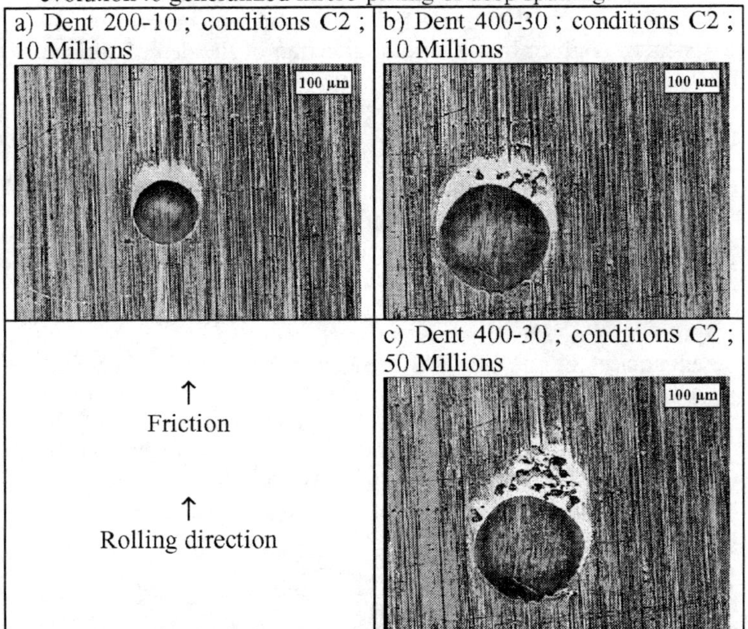

Figure 2. Dent 200-10 and 400-30 at 10x10^6 and 50x10^6 cycles. Operating conditions C2.

Test summary. Results are presented on a table for each operating condition (Tables 4 to 12). Each cell of the table corresponds to the mean behavior of several identical dents (at least three on at least two different disks). Each column of the table corresponds to an indentation load W_i, and each row to an indenter radius R_p. When at least one of the dents has been observed to be responsible for damage, this type of dent is considered as harmful for this operating condition, and the cell is shaded. The cell is blank when no damage has been observed for all identical dents, as well as cells on the same line at the left-hand side (i.e. for lower indentation loads) which in fact have not been tested. A cell with question marks means that the dent has not been tested and the boundary between harmful and harmless dents remains undetermined.

C2	3 daN	5 daN	10 daN	20 daN	30 daN	50 daN
200	???	???				
400						
600	???	???	???			

Table 4. Experimental results for conditions C2

ES1	3 daN	5 daN	10 daN	20 daN	30 daN	50 daN
200						
400						???
600						

Table 5. Experimental results for conditions ES1

ES2	3 daN	5 daN	10 daN	20 daN	30 daN	50 daN
200						
400						???
600						

Table 6. Experimental results for conditions ES2

ES3	3 daN	5 daN	10 daN	20 daN	30 daN	50 daN
200						
400						
600						

Table 7. Experimental results for conditions ES3

ES4	3 daN	5 daN	10 daN	20 daN	30 daN	50 daN
200						
400						
600						

Table 8. Experimental results for conditions ES4

ES5	3 daN	5 daN	10 daN	20 daN	30 daN	50 daN
200						
400						???
600						

Table 9. Experimental results for conditions ES5

ES6	3 daN	5 daN	10 daN	20 daN	30 daN	50 daN
200						
400						???
600						

Table 10. Experimental results for conditions ES6

ES7	3 daN	5 daN	10 daN	20 daN	30 daN	50 daN
200						
400						
600						

Table 11. Experimental results for conditions ES7

ES8	3 daN	5 daN	10 daN	20 daN	30 daN	50 daN
200						
400						
600						

Table 12. Experimental results for conditions ES8

Nitrided 32CrMoV13 vs. M50. Tables 4 and 11 show the results obtained for operating conditions C2 and ES7, i.e. 3GPa, 1.5% sliding, no hoop stress, and dented disk made of M50 for C2 and of nitrided 32CrMoV13 for ES7. The comparison of both tables shows that dents 400-20, 600-20 and 600-30 have been classified as harmful for M50 but not for nitrided 32CrMoV13. The nitrided steel exhibit then improved fatigue strength, compared to M50, as far as dent initiated RCF is concerned.

Effect of sliding. When 1.5% sliding is introduced and results compare to those obtain under pure rolling conditions (ES1-3, ES2-4, ES5-7, ES6-8), 15 cells have changed status. For every operating conditions, the dents 400-30, 400-50 and 600-50 become harmful when sliding occurs. Dent 400-20 becomes also harmful when combined with hoop stress. This result shows that sliding has a major detrimental effect on dent initiated RCF. Therefore an application with sliding, even low, is a severe application as far as dent initiated RCF is concerned.

Effect of hoop stress. When hoop stress of magnitude 115 MPa is introduced (ES1-2, ES3-4, ES5-6, ES7-8), 3 cells have changed status. (dent 400-20 for ES3-4 and dents 400-20 and 600-30 for ES7-8). This result shows that the presence of a hoop stress tends to be detrimental to dent initiated RCF resistance. However, this effect is limited for the given operating conditions. As for the contact load, the hoop stress may be a very important parameter for the growth of nucleated damage to deep spalling.

Effect of the normal load. When the load increases from 2.5 to 3 GPa (ES1-5, ES2-6, ES3-7, ES4-8), a single cell changes its status (dent 600-30 from ES4 to ES8). Except this case, no dent behavior changed while increasing the load. This result shows that this load increase has little influence upon the dent initiated RCF. It should be mentioned that this conclusion is valid only for the load range encountered in aeronautic bearings (2-3 GPa). If the load becomes higher, other phenomena might be activated, and conversely in the very low load range, no damage is likely to occur. This conclusion must be restricted to the nucleation of damage, since the growth stage leading to deep spalling might be more dependent upon the load than the initiation stage, since the growth stage requires a stress high enough at hertzian depth.

Location of surface damage. For each dent tested under pure rolling the damaged area has always been found ahead of the dent along the direction of the load movement. The same conclusion has been found for dents tested under rolling plus sliding conditions, with a damage more pronounced than under pure rolling condition. Both observations are coherent with transient EHL analysis which evidenced a slightly higher peak of pressure ahead of the dent along the load movement direction under pure rolling, and a more significant one under rolling plus sliding when the dent is located on the slower surface [8]. Note that the very detrimental effect of sliding may also be explained by some cavitation phenomenon and subsequently a localized micro-collapse of the lubricant film [8].

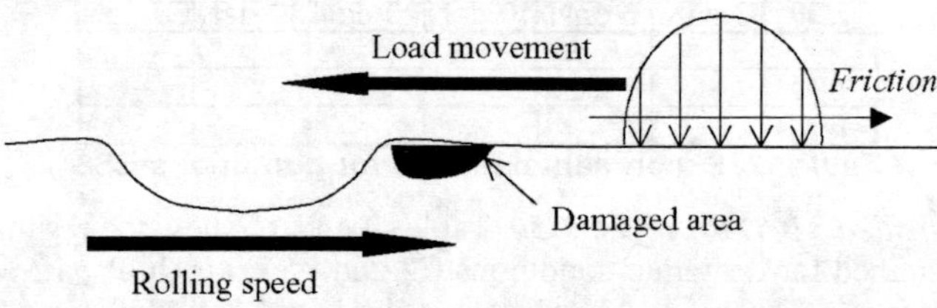

Figure 4. Location of the damaged area for driven surface in presence of sliding.

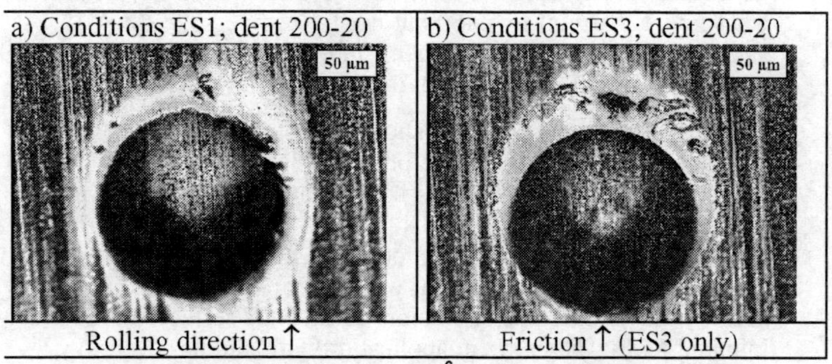

Figure 5. Dent 200-20 at 20x10^6 cycles ; effect of sliding.

A typical damage found in the vicinity of a 200-20 dent is shown in Figs. 5a and b for pure rolling condition and rolling plus sliding conditions, respectively. One may notice that the dent edge is jagged for pure rolling, while it remains smooth and circular for condition with 1.5 % sliding. Moreover, micro-spalls that appear close to the dent edge are always located away from the border.

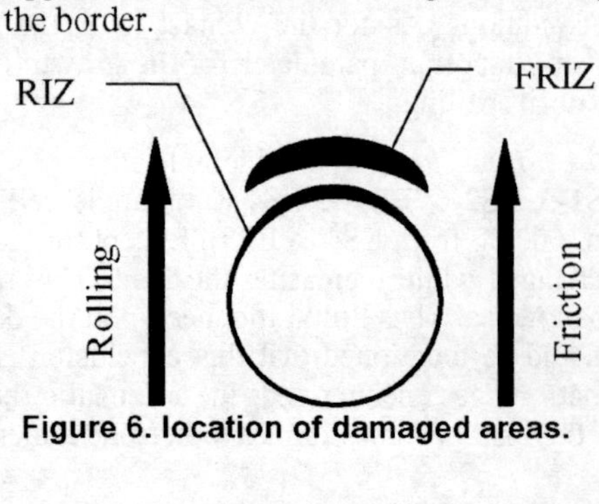

Figure 6. location of damaged areas.

Finally two areas where damage is nucleated have been identified. The first one, called RIZ (Rolling Initiation Zone) in Fig. 6, is activated only under pure rolling conditions and corresponds to the dent edge. In the present experiment, its position is given by the rolling direction of the dented surface. The second area is noted FRIZ (Friction and Rolling Initiation Zone). It is located at about 20µm ahead of the dent along the friction direction.

ANALYSIS

RCF results from repeated over rolling of the dent. The analysis of the fatigue process requires the knowledge of material properties and of loading conditions during initial denting and further over-rolling. These properties are (i) the micro yield shear stress profile, required to define the endurance limit; (ii) the hardening law of the material, necessary to describe the indentation process that involves plasticity ; and (iii) the residual stresses induced by the nitriding treatment. These contribute to stresses undergone by the material during indentation and over-rolling.

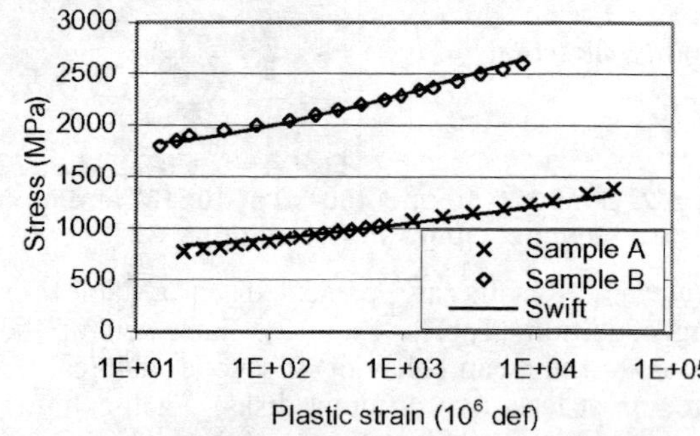

Figure 7. Stress-plastic strain curve for samples A and B

The micro yield shear stress and uniaxial hardening behavior are usually measured by global compressive tests made on material samples. This process gives average values over the section of the sample. Therefore, to be relevant, this test must be made on homogeneous samples. Two types of homogeneous sample (A and B) have been tested. The first type, A, with cylindrical shape, is made of nitrided 32CrMoV13 from which the nitrided layer has been totally removed. It represents then the core material. The second type, B, has been obtained by nitriding homogeneously thin tubes of 32CrMoV13 steel [3]. It represents the behavior of a quasi-uniform nitrided layer. Results obtained from these two tests are plotted in Fig. 7. Two conclusions arise from these tests. First, the plastic behavior can be represented by the Swift law ($\sigma = B(C+10^6\times\varepsilon^p)^n$) [3], where B, C and n are three parameters of the law. This law describes the uniaxial hardening and gives the micro-yield shear stress τ_e ($2\tau_e = B(C+20)^n$). Second, both tests can be described with the same C and n parameters, here equal to 16 and 0.067, respectively. In conclusion the variation of the mechanical behavior in the nitrided layer can be described by the variation of B.

If the dent shape can be simply determined by topographical measurement, local residual stresses and hardening state of the material are not readily available from measurement. It is then necessary to simulate the indentation process to estimate the material state. To attest the quality of such an estimation, a first step will be to compare the real geometry of the dent with the one obtained by simulation. The result will depend on the shape and material of the indenter tip, on the normal load applied during indentation, and also on the elastic-plastic properties of the dented material.

Indenter heads have conical shape with a cone angle of 120°. Their extremity is made of diamond (E = 1140 GPa, ν = 0.07) and is spherical with a radius R_p of 200, 400 or 600 µm. The normal load applied during indentation, W_i, is chosen among 5, 10, 20 or 50 daN. The friction coefficient between the indenter head and the material to dent is unknown during the indentation step. Therefore the first step will be to analyze the effect of this friction coefficient on the dent shape.

Elastic-plastic properties of bearing steels (AISI 52100, M50, nitrided 32CrMoV13) are only known for plastic strain up to 2 %. However, plastic strain may reach up to 40 % during some Rockwell indentations [10]. Therefore an extrapolation of the hardening law used is required, and this should be experimentally validated.

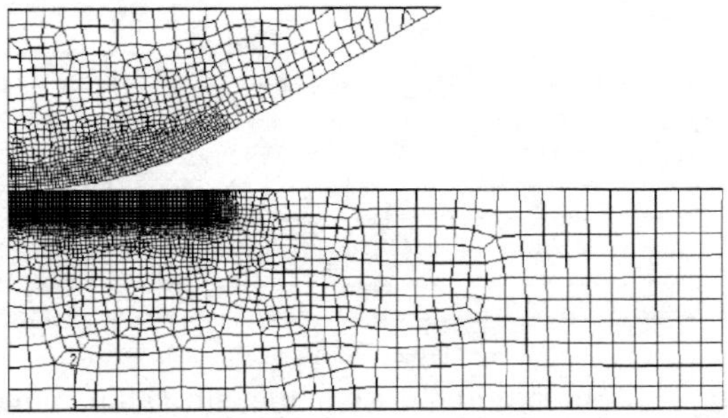

Figure 8. FE mesh used for indentation simulation.

Indentation simulation. The indentation process has been simulated using the ABAQUS commercial FE code. The problem is axisymetric since the punch and the facing body have a common axis of revolution. The mesh used for this simulation is presented in Fig. 8. The mesh size is 2 µm in the contact zone, which is small in comparison to the dent diameter (from 150 to 300µm). The body is meshed by 5050 nodes, and the punch is described by 1000 up to 1700 nodes depending on its radius.

An isotropic plasticity model is used to describe the behavior of the body to dent (M50 or nitrided 32CrMoV13). The punch behavior is assumed purely elastic (diamond). The load is applied by a uniform pressure on the upper surface of the punch, whose nodes are unable to move radially. Boundary conditions are applied through half-infinite elements in order to simulate the half-space body. The problem should be solved with large strains. Therefore the hardening law is expressed in terms of the Cauchy stress versus the logarithmic strain.

First the effect of the friction coefficient between the punch and the specimen is investigated. Results are presented in Fig. 9 for M50 material and the dent 200-20. It can be observed in Fig. 9b that the dent shoulder height decreases significantly from 0.8 to 0.6 µm when the friction coefficient increases up to 0.5, whereas the location of the shoulder top moves radially from 90 to 100 µm. On the contrary, the dent depth and its radius are almost independent of the friction coefficient (Fig. 9a).

In order to evaluate the effect of these geometrical changes, the elastic contact between two spheres of equivalent radius R_{eq} = 15 mm has been solved for dents simulated with the friction coefficient equal to 0, 0.1 and 0.5 (shape given in Fig. 9). Resulting pressure profiles are given in Fig. 10 in a dimensionless form (by P_H, the Hertz pressure). It is observed a marked decrease of the peak of pressure from 2.9 to 2.2 P_H when the friction coefficient increases from 0 to 0.5. The knowledge (and correct simulation) of the real dent shape is therefore very important despite the difficulty for estimating the friction coefficient to use for simulation. Considering that the indentation procedure is a slow process made with some trace of lubricant (boundary lubrication) it will assumed from now a friction coefficient of 0.1.

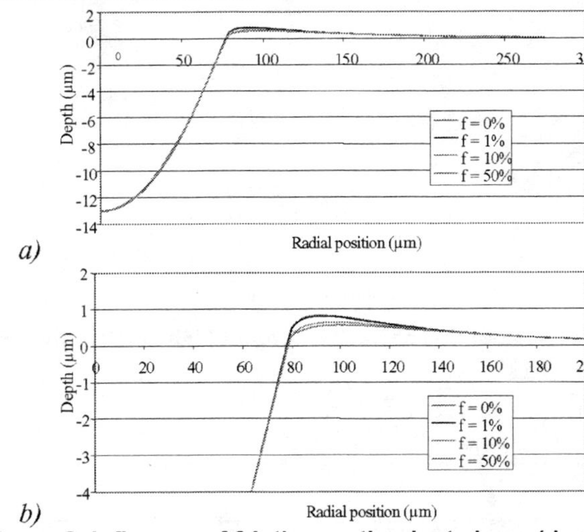

Figure 9. Influence of friction on the dent shape (dent 200-20, M50). a) General view. b) Zoom on the dent shoulder

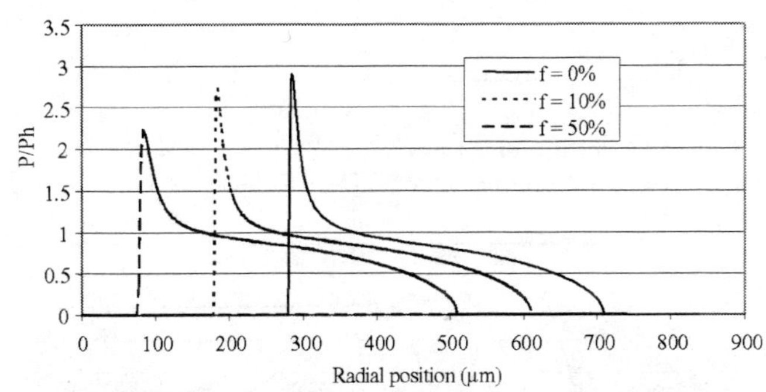

Figure 10. Influence of the friction coefficient during indentation on the contact pressure. (Dent 200-20, M50). For a sake of clarity the profiles f=10% and f=50% are shifted along the radial axis

Extrapolation of the hardening law. As stated above elastic-plastic properties of bearing steels are known from uniaxial compression tests only up to 1 or 2 % of plastic strain. An estimation of the hardening law is proposed below both for M50 and nitrided 32CrMoV13 steels, based on the real geometry of dent 200-20.

M50. The elastic-plastic behavior of M50 could be described up to 2 % of plastic strain by the Swift law, $\sigma_{eq} = B(C + p.10^6)^n$, with B = 1280 MPa, C = 4 and n = 0.095. Three different extrapolations of this hardening law are plotted in Fig. 11:

- The Swift law, originally identified up to 2 % plastic strain, simply extended up to 40 %.

- The modified Swift law, which is the direct transposition of the Swift law in true stress and logarithmic strain. The gap with the Swift law appears only beyond 2 % plastic strain.
- The Voce law, which presents an horizontal asymptote rather than a continuous increase as with the Swift law. This law had been used elsewhere by Cercueil [10] with success for indentation of AISI 52100 steel. The 4 parameters of this law have been chosen to have similar curve as the Swift law in the domain 0 – 2 % plastic strain.

$$\sigma_{eq} = 1444+1996\{1-\exp[-(390.p)^{0.352}]\} \text{ (in MPa)} \quad (1)$$

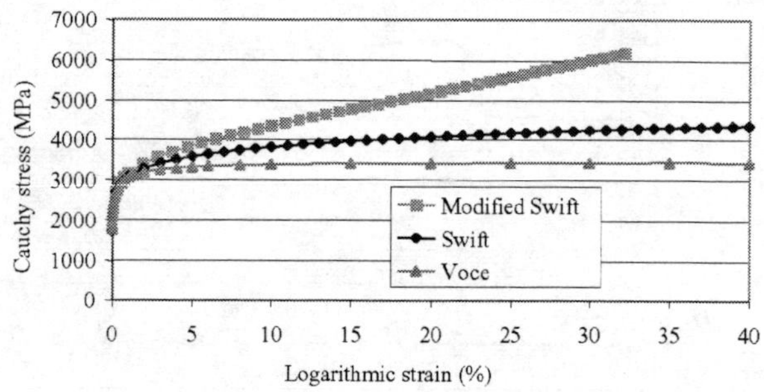

Figure 11. Possible extrapolations of M50 hardening laws.

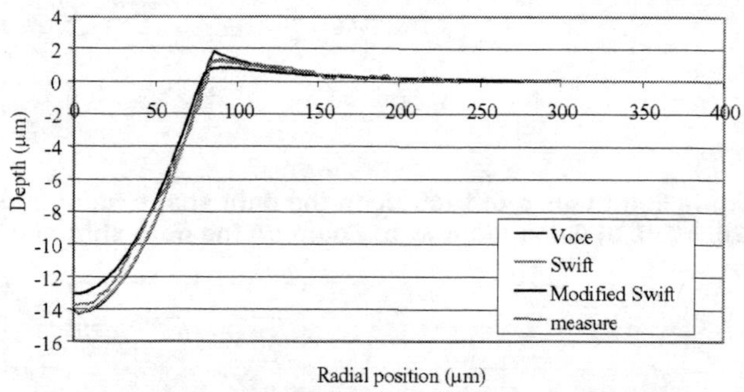

Figure 12. Influence of the extrapolated hardening law on the dent shape for M50. Dent 200-20, cf = 0 %.

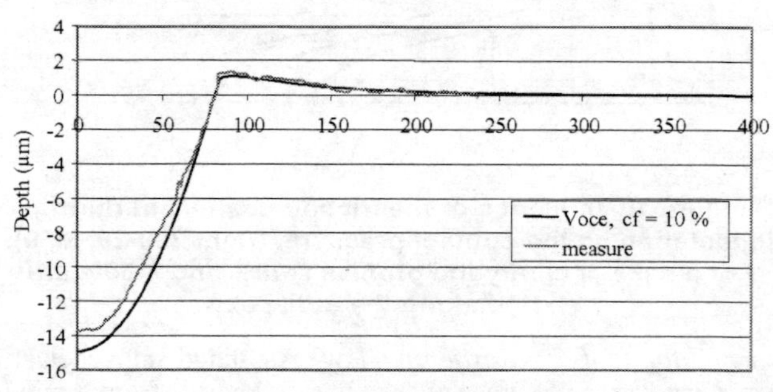

Figure 13. Comparison between measured and simulated 200-20 dent profiles. M50 steel. Voce law ; cf = 10 %.

Figure 12 shows the result of simulations performed without friction for these 3 laws, as well as one real profile measured for the dent 200-20 on M50 steel. The height of the dent shoulder obtained with the modified Swift law is too small compared to the measurement. The one obtained with the Swift law is more important. However, since the friction coefficient has not been introduced in this simulation it corresponds to an overestimation of the shoulder height which again will lead to a significantly smaller shoulder height than the measured value. Finally, the Voce law which results in the highest shoulder in Fig. 12 (cf=0 %) will give a very satisfactory result in presence of friction (10 %), especially in the vicinity of the shoulder as shown in Fig. 13.

Nitrided 32CrMoV13. In the present study, the mechanical properties of the nitrided material can be considered homogeneous at the scale of the influence zone of the dent (less than 150 µm) since it is small in comparison to the nitriding depth (600-800 µm). In a similar manner as for M50 steel the Voce law gave:

$$\sigma_{eq} = 1930+1380\{1-\exp[-(455.p)^{0.490}]\} \text{ (in MPa)} \quad (2)$$

A good agreement is found in Fig. 14 between simulation and measurement for dent 200-20 on nitrided 32CrMoV13 steel, in particular around the dent shoulder, which is of prime interest for the contact problem. It should be mentioned that the initial residual stress due to the nitriding treatment has been considered (i.e. –400 MPa in the stressed zone).

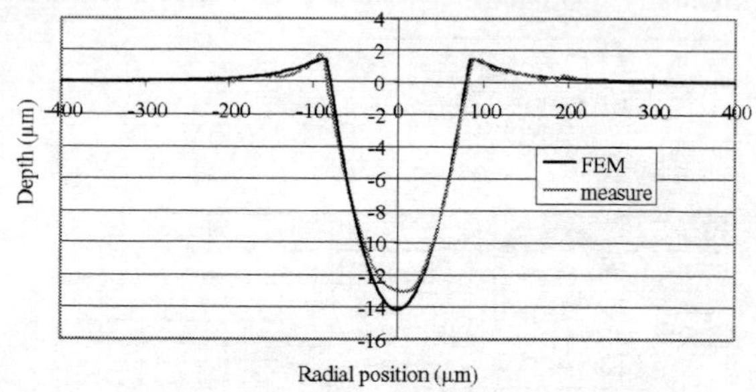

Figure 14. Comparison between measured and simulated 200-20 dent profiles. 32CrMoV13 steel. Voce law ; cf = 10 %.

It can be observed in Fig. 15 that a good correlation with dents produced by punches of radius 400 and 600 µm is also found with Eqs. (1) and (2), for both M50 and nitrided 32CrMoV13 steels, respectively.

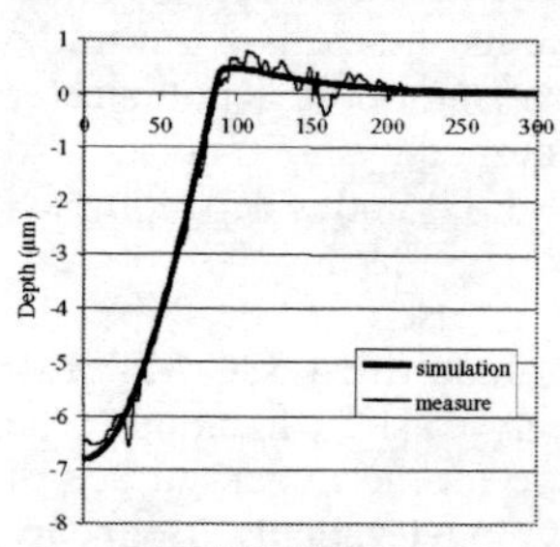

a) 32CrMoV13. 400-20

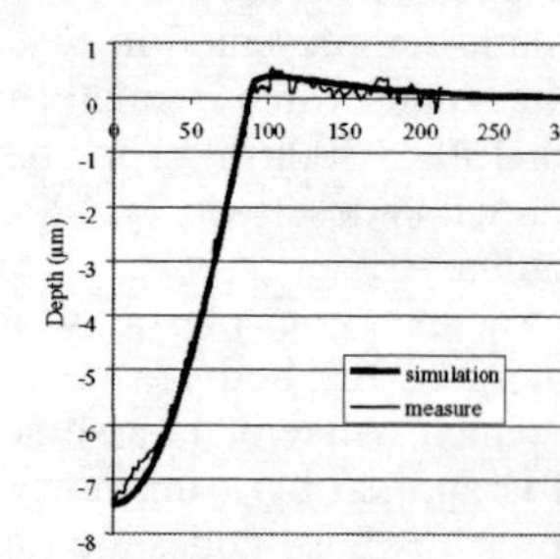

b) M50. 400-20

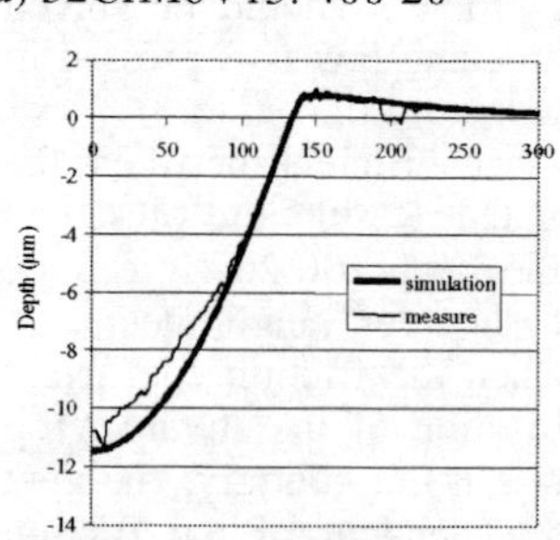

c) 32CrMoV13. 600-50

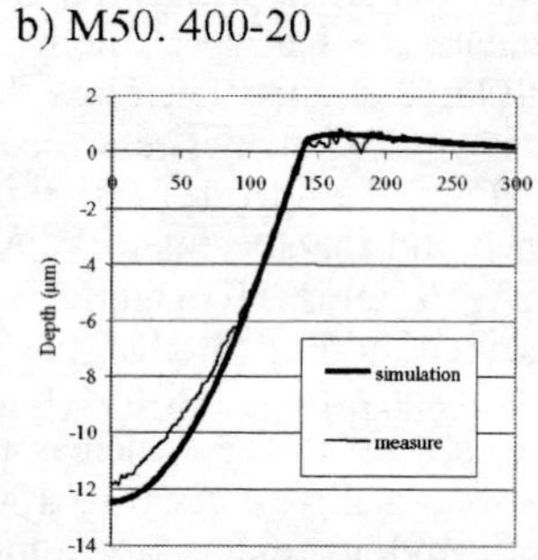

d) M50. 600-50

Figure 15. Comparison of measured and simulated dent profiles for dents 400-20 and 600-50 on 32CrMoV13 and M50.

Residual plastic strain. The cumulated plastic strain (in %) introduced during indentation is given in Fig. 16. The depth (in μm) corresponds to the distance from the deformed surface. The horizontal (radial) axis is normalized by the dent radius. Results are presented for the dents 200-10, 200-20, 400-20, and 600-50 realized on 32CrMoV13. The maximum equivalent plastic strain is found for dent 200-20 (20%). For the 3 other dents, it is lower and ranges from 8 to 13 %. The depth where plastic strain occurs is lower than 150 μm in most cases (and always <250 μm). The maximum strain is found in the vicinity of x/R=1, close to the surface. Beyond this abscissa, i.e. when x/R>1, the plastic strain drops steeply.

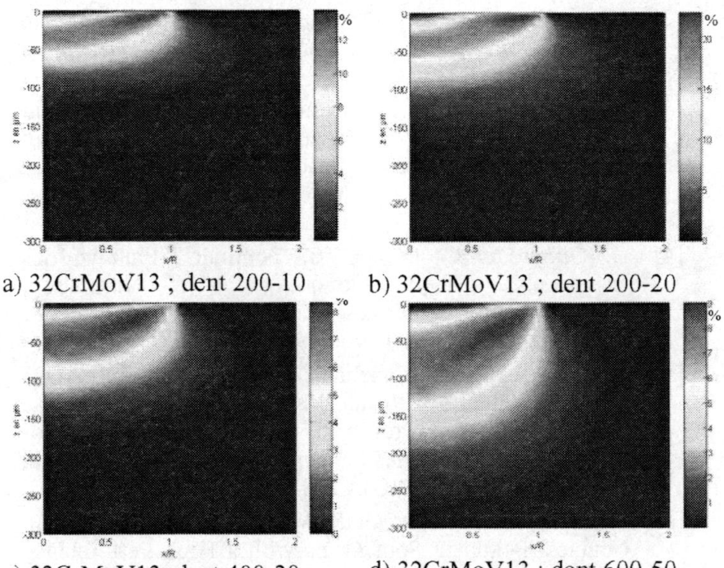

a) 32CrMoV13 ; dent 200-10 b) 32CrMoV13 ; dent 200-20

c) 32CrMoV13; dent 400-20 d) 32CrMoV13 ; dent 600-50

Figure 16. Cumulated plastic strain introduced during indentation on nitrided 32CrMoV13 steel.

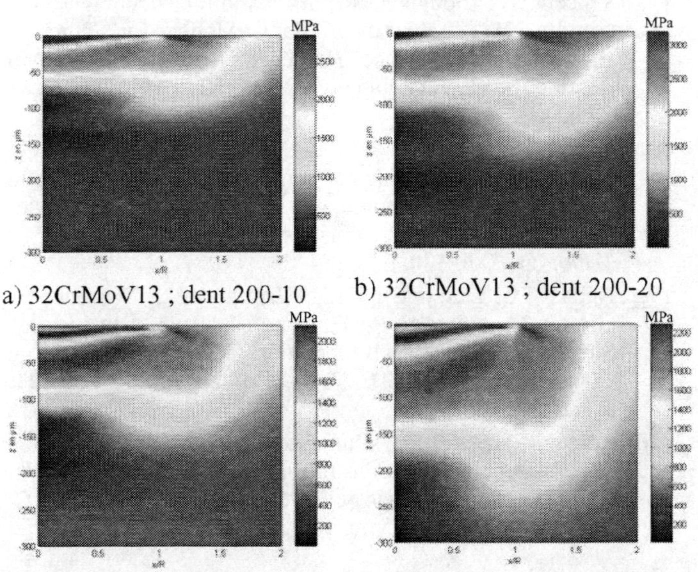

a) 32CrMoV13 ; dent 200-10 b) 32CrMoV13 ; dent 200-20

c) 32CrMoV13 ; dent 400-20 d) 32CrMoV13 ; dent 600-50

Figure 17. Von Mises residual stress introduced during indentation on nitrided 32CrMoV13 steel.

Residual stress. Figure 17 shows the residual stress field after unloading for the same dents as for Fig. 16. The first point is the very high magnitude of the Von Mises stress, always found higher than 2000 MPa and up to 3000 MPa for dent 200-20. The maximum stress is found just above abscissa x/R=1, very close to the surface. The second point is that the stressed zone is not limited to the plastically deformed one, but extends far away. For a surface point at abscissa x/R=2, the residual stress is found higher than 1000 MPa for all dents studied here.

For a better understanding of the role of residual stresses produced by the indentation each term of the stress tensor should be considered, as illustrated in Fig. 18 for dent 200-20. It can be concluded that the radial stress is essentially compressive within the first 150 μm. Conversely, terms $\sigma_{\theta\theta}$ and σ_{zz} are negative in the plastically deformed zone and positive elsewhere. Finally, the shear stress σ_{rz} is found maximum in the vicinity of the dent shoulder.

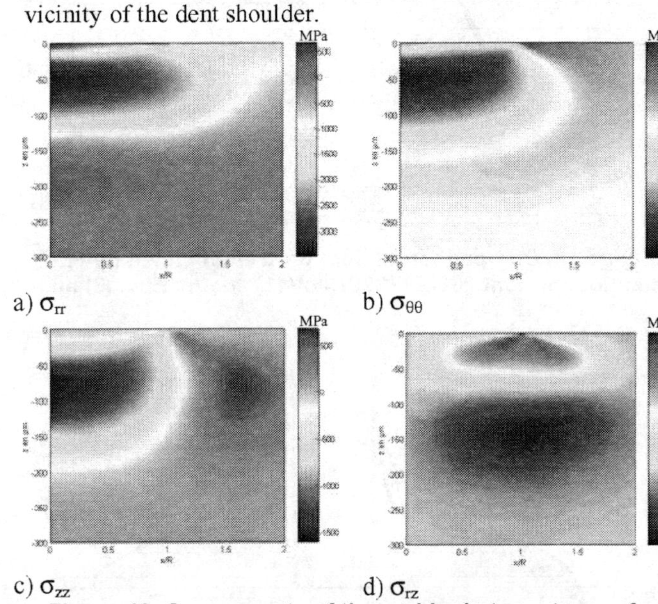

a) σ_{rr} b) $\sigma_{\theta\theta}$

c) σ_{zz} d) σ_{rz}

Figure 18. Components of the residual stress tensor for dent 200-20 on nitrided 32CrMoV13 steel.

Indent shape. Usually, the dent profile is approximated by the product of a cosine by a damping exponential function, Eq. (3). In this relation only one parameter, K also called the damping coefficient, is required to obtain the shoulder height.

$$h(r) = h_p.\cos(\frac{2\pi r}{4R})\exp(-K\frac{r^2}{(4R)^2}) \qquad (3)$$

Equation (3) has been plotted in Fig. 19 for the dent 200-20 and compared to those obtained by FE simulation for the nitrided material. It should be mentioned that the FEM simulation gives a dent profile that is globally close to the measured one. It is found that, while the dent depth and radius, and the height of its shoulder are well reproduced by Eq. (3), the shape of the dent is quite different from the FEM simulation (which is very close to the experimental profile).

For this reason a new function based on the previous one is proposed, Eq. (4). It requires 5 parameters, h_p, R, I_1, I_2 and I_3. Within the dent (r>R), Eq. (3) is kept. Parameters I_1, I_2 and I_3 allow to fit the slope of the dent around the shoulder top and its location. The continuity of the function at r=R is obtained by K which depends on these 5 parameters. It is observed in Fig. 19 a good agreement between this analytical function and the FEM simulated profile. Figure 20 presents the pressure distribution obtained using the analytical profile given by Eqs. (3) and (4).

It is found that Eq. (3) leads to a significant underestimation of the magnitude of the pressure peak obtained by using the result of FEM simulation or the equivalent geometry given by Eq. (4).

$r < R$

$$h(r) = h_p.\cos(\frac{2\pi r}{4R})\exp(-K\frac{r^2}{(4R)^2}) \quad \text{with } K = -16\ln\left(\frac{2RI_1(I_3 - I_2)}{\pi h_p}\right)$$

$r \geq R$

$$h(r) = I_1[\exp(-I_2(r-R)) - \exp(I_3(r-R))] \quad (4)$$

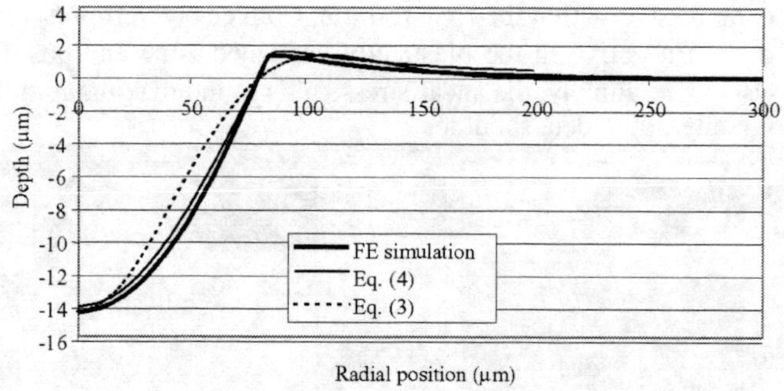

Figure 19. Comparison of dent profiles obtained by FEM simulation for dent 200-20 (32CrMoV13) and by Eqs. (3) and (4)

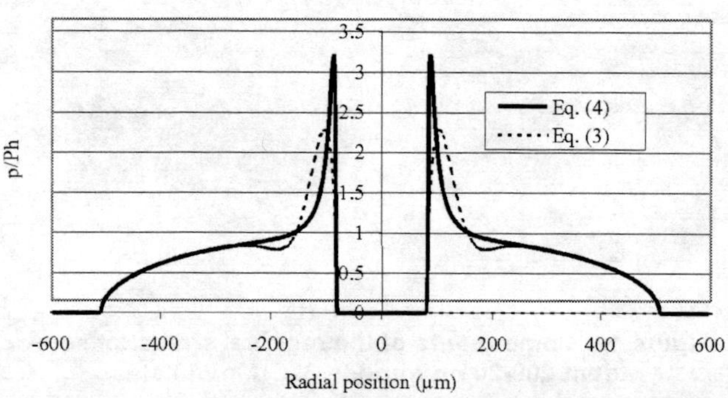

Figure 20. Pressure distribution which results from dent profiles given in Eqs. (3) and (4)

CONCLUSION

An original experimental procedure has been proposed to discriminate the RCF live performance of bearing steels. It consists of denting the contacting surfaces by indenters with spherical tips of radius 200, 400 and 600 µm, for loads ranging from 5 to 50 daN. Fatigue tests performed on a high-speed twin-disk machine have shown:

1) a better tolerance to surface dents for nitrided 32CrMoV13 compared to conventional M50 bearing steel.
2) A limited influence of the normal load applied during the fatigue test.
3) A slight effect of the hoop stress within the domain investigated (0 and 115 MPa).
4) A very detrimental effect of sliding.

Moreover, a FEM analysis has been performed to determine the plastic strain and residual stress fields due to indentation.

Finally a new equation is proposed to better describe the dent shape.

Additional numerical analyses are carried on in order to characterize the dent harmfulness from its geometry and material properties.

REFERENCES

[1] Tallian T.E., Failure Atlas for Hertz Contact Machine Elements, ASME Press, New York, 404p. (1992).

[2] Nélias D., Ville F., "Detrimental Effects of Dents on Rolling Contact Fatigue". *ASME Journal of Tribology*, Vol. **122**, n°1, 55-64 (2000).

[3] El Ghazal H., Lormand G., Hamel A., Girodin D., Vincent A.. "Microplasticity characteristics obtained through nano-indentation measurements: application to surface hardened steels," *Materials Science and Engineering A*, Vol. **303**, n°1, 110-119 (2001).

[4] D. Ghiglione, C. Leroux, C. Tournier. "Pratique des traitements thermochimiques – Nitruration, nitrocarburation et dérivés". Techniques de l'ingénieur : vol. MD, M 1222, pp. 1-44 (1996).

[5] Jacq C., Lormand G., Nélias D., Girodin D., Vincent A., "On the Influence of Residual Stresses in Determining the Micro-Yield Stress Profile in a Nitrided Steel by Nano-Indentation". *Materials Science and Engineering A*, Vol. **342**, 311-319 (2003).

[6] Flamand L., Berthe D., Godet M., "Simulation of Hertzian Contacts Found in Spur Gears with a High Performance Disk Machine," *ASME Journal of Mechanical Design*, Vol. **103**, n°1, 204-209 (1981).

[7] Vincent A., Fougères R., Lormand G., Dudragne G., Girodin D., "Endurance Limit Model for Through Hardened and Surface Hardened Bearing Steels," in Bearing Steel Technology, ASTM STP 1419, 459-473 (2002).

[8] Nélias D., Dumont M.-L., Couhier F., Dudragne G., Flamand L., "Experimental and Theoretical Investigation on Rolling Contact Fatigue of 52100 and M50 steels under EHL or micro-EHL conditions". *ASME Journal of Tribology*, Vol. **120**, n°2, 184-190 (1998).

[9] Nélias D., Dumont M.-L., Champiot F., Vincent A., Girodin D., Fougères R., Flamand L., "Role of Inclusions, Surface Roughness and Operating Conditions on Rolling Contact Fatigue". *ASME Journal of Tribology*, Vol. **121**, n°2, 240-251 (1999).

[10] Cercueil H., "Etude d'une nouvelle nuance d'acier à roulement pour conditions d'usage sévères et modélisation de son endommagement en présence d'une indentation". Ph.D. Thesis, INSA Lyon, 182 p. (1999).

Proceedings of
2003 STLE/ASME Joint International Tribology Conference
Ponte Vedra Beach, Florida USA, October 26-29, 2003

2003-TRIB-0281

THEORETICAL AND EXPERIMENTAL INVESTIGATIONS OF AN ELASTO-PLASTIC CIRCULAR CONTACT BOUNDED BY REAL ROUGH SURFACES

Dorel Prodan/University of Suceava, Romania

Emanuel Diaconescu/University of Suceava, Romania

ABSTRACT
To investigate the pressure distribution and the stress state in a real elasto-plastic circular contact, bearing balls of two different degrees of surface finishing were used. These were pressed against a polished flat surface made of softer steel. The surface microtopography data measured in each case by a laser profilometry was introduced in a numerical program developed to evaluate the pressure distribution, stress state and residual deformations for an elasto-plastic material with non-linear hardening. The dimension of calculated contact plastic deformations compare favorably with those found experimentally by aid of a laser profilometer.
Keywords: elasto-plastic contact, rough contact, pressure distribution, plastic deformation, real area.

INTRODUCTION
Many research workers dealt with the problem of asperity interaction on micro-contacts due to its crucial importance for reliability of rubbing surfaces.

Bowden and Tabor [1] pointed out that, because the real contact area is significantly smaller than the apparent area, high pressures are generated, which can induce plastic deformation. Archard [2] suggested that the contact between asperities may be plastic during the first loading cycles but afterwards it becomes predominantly elastic. To establish if a rough contact behaves elastically or plastically, Greenwood and Williamson [3] define a plasticity index ψ. In the range $0.6 < \psi < 1$ the elastic and plastic deformations are comparable and the contact behaves in an elasto-plastic fashion. Later on, Cheng, Etsion and Bogy [4] improved the Greenwood-Williamson model by considering that beyond the elastic limit the micro-asperity volume remains constant, in agreement to the principle of plastic incompressibility. Experimental evidence supplied by Evseev, Medvedev and Grigoriyan [5] do not confirm this theory probably the hypothesis of plastic incompressibility is applied in a domain where elastic and plastic deformation have the same order of magnitude. Halling and Nuri [6] defined an anisotropic plasticity index and considered ellipsoidal asperities. The contact of an asperity is assumed to obey Hertz theory up to a critical deformation. Once this deformation is exceeded, the micro-contact behavior becomes plastic.

All above models are based on the hypothesis that the deformation of a single asperity does not depend on neighbor microcontacts. To take into account the interaction between micro-contacts, Goryacheva and Dobychin [7] advanced a model made of an assembly of columns uniformly spread across contact zone and having a random distribution of heights. Their results reproduce the experimentally observed effect of saturation of real contact area at high loads.

Ike and Makinouki [8] extend the elasto-plastic behavior of a triangular asperity to the deformation of an assembly of identical micro-asperities.

Mayeur [9] offered a different approach to the problem. He started from an elastic rough contact and, by using a 2-D numerical modeling of an elasto-plastic contact, found a final model for a 3-D rough elasto-plastic contact. A supplementary material characteristic, namely the strain hardening modulus and small loadings, which do not generate plastic yields at Hertz depths and change insignificantly the surface pressure distribution, were used. As a result, the Hertz pressures are lower than the yield pressure and plastic deformations occur only in the surface layer having a thickness placed between a few and several tens of micrometers.

To evidence the effect of surface micro-topography ranging between isotropic and highly anisotropic cases, hardness and load upon deformation behavior of rough surfaces, Lee and Ren [10] used numerical methods and surface simulations. Recently, Bengisu and Akay [11] attempted to consider not only the elastic and plastic deformation modules, but also the chemical composition expressed by adhesion forces acting in local regions of the contact. Horng [12] advances a new model for the elliptic elasto-plastic micro-contact between asperities and finds values of separation, real contact area, plastically deformed area and plasticity index, which are compared with the results of other published models. Zhao and Chang [13] propose a micro-contact model to find the interaction between micro-asperities in an elasto-plastic rough contact. They underlined the effect of this interaction upon mean separation between surfaces, real contact area and the re-distribution of load across the contact.

Cretzu and Hatmanu [14] and especially Popescu [15], improved the model of Hartnett and Palazotto, and developed a numerical algorithm for 3-D smooth circular and line contacts under normal static loads and for rolling cylindrical contacts. They used both linear and nonlinear haredening.

The above review indicates that the elasto-plastic rough contact was investigated under simplifying assumptions concerning either the surface micro-topography or the constitutive laws of the material above the yield limit. Consequently, this paper aims to develop a numerical program that supplies the pressure distribution, stress state and the plastic displacement field in contacting bodies, by using a real surface geometry, determined by a laser profilometer and a model of nonlinear hardening for involved materials. Then, the calculated contact elements are compared with experimental results on specimens initially used to measure the roughness.

NOMENCLATURE

- E: Young modulus;
- $F(\sigma_{ij})$: function of plastic potential;
- W: potential energy;
- h: modulus of plasticity;
- f_{ijkl}: factor of proportionality between strain and stress infinitesimal increments;
- δ_{ij}: Kroneker delta;
- ε: total equivalent strain;
- λ: plastic multiplier;
- ν: Poisson's ratio;
- σ: Huber-Mises-Hencky (HMH) stress;
- σ_0: maximum pressure calculated assuming elastic behavior;
- σ_1, σ_2, σ_3: main (principal) normal stresses;
- σ_E: elastic limit of material;
- σ_{ij}: a component of stress tensor;
- $\sigma_{max} = 3\sigma_Y$: yield pressure;
- σ_Y; yield stress;
- τ_Y: shear yield stress;
- ψ: plasticity index;

PRINCIPLES OF NUMERICAL PROGRAM

The pressure distributed over contact surface influences directly the occurrence and development of plastic zones inside contacting bodies. In turn, the plastic deformations affect the surface deformations, and therefore the contact pressure, tending to reduce the pressure with respect to the elastic case. Therefore, of interest in a rough contact analysis is the determination of pressure distribution, stress state inside contacting bodies and the plastic deformations that occur at various load levels.

Mayeur [9] investigated the behavior of some specimens made of ball bearing steel in the elasto-plastic domain. By using an elasto-plastic model with linear hardening and Huber-Mises-Hencky (HMH) yield criterion, often called Von Mises, he evaluates the real contact pressure by a linear function of maximum elastic pressure:

$$\sigma_{real} = \sigma_{max} + K(\sigma_0 - \sigma_{max}), \qquad (1)$$

where K is a coefficient depending on the plasticity modulus of considered material and σ_{max} is the yield pressure.

For the ball bearing steels the value of plasticity modulus used in published investigations [9 and 15] is 188MPa, which implies a coefficient $K \approx 0.8$. On the other hand, for a through-hardened bearing steel the yield stress ranges between 1850MPa and 2200MPa, depending on tempering temperature. The surface yield surface is three times higher and, consequently, σ_{max}=5550…6600MPa.

A new program is developed inhere aiming to determine the pressure distribution, the stress state and the displacement field in an elasto-plastic rough contact by using a nonlinear hardening model. This is based on the method of a perfectly rigid half-space.

If the maximum contact pressure a material can support is given by Eq.1, there are two possibilities to find the contact pressure, namely modification of initial, un-deformed surfaces until the pressure does not exceed the yield pressure or imposing a maximum pressure equal to yield pressure and re-calculating the pressure distribution until the following contact conditions are satisfied:

$$p_{ij} \le \sigma_{max} \quad \text{on contact area;}$$
$$p_{ij} = 0 \quad \text{outside contact area.} \qquad (2)$$

The method proposed herein is a combination of these two possibilities because this way the pressure distribution and contact surface are obtained without other supplementary hypotheses concerning the surface geometry change due to plasticity phenomena.

The theories of Hertz and Boussinesq allow to build by digitization the following vector-matrix equation, valid for linear elastic contacts:

$$\{\mathbf{z_k}\} - [\mathbf{A}] \cdot \{\mathbf{p_k}\} + \delta_0 = \{\mathbf{h_k}\}, \qquad (3)$$

where z_k is the z ordinate of deformed body in the center of elementary rectangle k of contact area, p_k is the pressure on this element, h_k represents the initial ordinate in the point k, **A** is the matrix of influence coefficients, δ_0 is the rigid displacement of the elastic solid and k = 1, 2,…NT is the order number of one of the elementary digitization rectangles. In the points of contact area $z_k = 0$, whereas outside contact area $p_k = 0$. Consequently, the system 3 of equations possesses NT+1 unknowns.

By imposing δ_0, the solution of the system 3 of NT algebraic equations yields the elastic pressures p_k in the points of contact area and the ordinates of deformed surface outside this area. In the digitization points in which the pressure exceeds the yield pressure, the following equation that takes into account the nonlinear strain hardening is applied:

$$p_k = \sigma_{\mathbf{max}} + K(p_k - \sigma_{\mathbf{max}}). \qquad (4)$$

These values are then introduced into Eqs. 3 to find a new set of pressures. This procedure is repeated cyclically until contact conditions 2 are satisfied in all points of contact domain, chosen initially larger than Hertz area. The pressure distribution obtained this way generates a normal load smaller than that imposed initially, Q. Therefore, an incremental displacement is added to δ_0 until the force balance equation on load direction is satisfied:

$$\sum_{k=1}^{NT} p_k \Delta x \Delta y = Q. \qquad (5)$$

As suggested by Popescu [15], a Ramberg-Osgood type of equation is chosen to describe the correlation between stress and strain. Illustrated in Fig. 1, this has the following form:

$$\varepsilon = \frac{\sigma_E}{E} + \left(\frac{\sigma}{B}\right)^N, \qquad (6)$$

where B and N are typical constants of considered material.

With respect to Fig. 1, one can use the notations $\tan\alpha = E$ and $\tan\beta = E'$, E being Young modulus and E' apparent modulus.

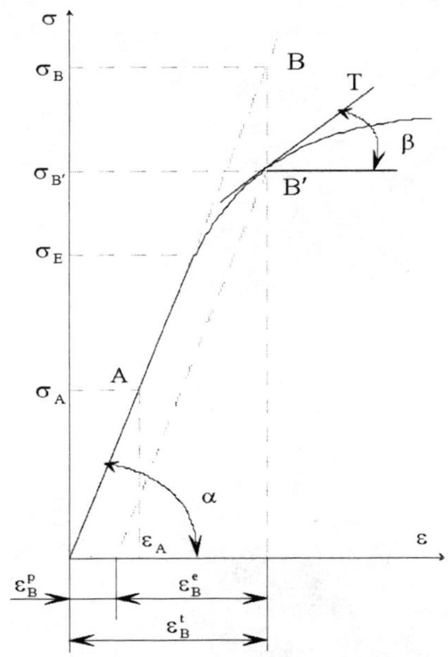

Fig 1. Evaluation of plastic deformation by compatibility theory

There are two methods to take into account the variable stiffness of materials stressed above the elastic limit, namely:
i. the use of an iterative solution starting from the initial elastic modulus to obtain finally the stresses and strains given by the real behavior of material in elasto-plastic domain;
ii. the use of incremental theory of plasticity, based on linear relation between stress increments and strain increments.

The second method allows the extension of linear elasticity in the plastic domain and therefore offers significant calculation simplicity. Consequently, this method is used for the purpose of this investigation. When this method is applied for the correlation stress-strain in contact problems, the following hypotheses are accepted:
- the infinitesimal increments of plastic deformation and of stresses are proportional:

$$d\varepsilon^p_{ij} = f_{ijkl} d\sigma_{kl}\,; \qquad (7)$$

- the contact is assumed frictionless and, consequently, the plastic displacements and deformations vanish in planes perpendicular on load direction;
- initial and final contact areas must be connected in the stress space, respecting the origin.

The total deformation increment is the sum of elastic and plastic increments:

$$d\varepsilon^t_{ij} = d\varepsilon^e_{ij} + d\varepsilon^p_{ij}, \qquad (8)$$

where the elastic increment is assumed to be given by the theory of linear elasticity by means of following operator:

$$d\varepsilon^e_{ij} = \frac{1+\nu}{E} d\sigma_{ij} - \frac{\nu}{E}\delta_{ij} d\sigma_{ii} \qquad (9)$$

and the plastic increment is obtained by Prandtl-Reuss equations, written for nonlinear hardening:

$$d\varepsilon^p_{ij} = \sigma'_{ij} d\lambda = \frac{\partial F(\sigma_{ij})}{\partial \sigma_{ij}} d\lambda\,. \qquad (10)$$

In Eq.10, the rate of plastic multiplier λ is expressed by:

$$\lambda = \frac{\dot{W}}{2\tau_Y^2}, \qquad (11)$$

where $\dot{W}$ is the energy rate.

The plastic stiffness, usually called equivalent plastic modulus is given, for a three-axial stress state, by following ratio of the increment oh HMH stress to that of plastic strain:

$$E'(\varepsilon^p) = \frac{d\sigma}{d\varepsilon^p}\,. \qquad (12)$$

By using the model of Hartnett and Palazotto, described and used by Cretu and Hatmanu [14] and improved by Popescu [15], and applying the same force-displacements method, a computing algorithm to find the plastic deformations in a 3-D circular contact between rough surfaces loaded normally by static forces is developed, as shown below.

The theory of compatibility assumes that the plastic deformations can be evaluated by means of elastic deformations. The increment of total deformation increment is expressed by Eq. 8 and, for an incrementally applied force, the elastic deformations are given by relation 9. According to Hill [16], if the HMH yield criterion is accepted, the plastic potential function is given by:

$$F = J'_2 - \tau_Y^2 = \frac{1}{3}\sigma^2 - \frac{1}{3}\sigma_Y^2, \qquad (13)$$

where J'_2 is the second invariant of stress deviator tensor. This tensor has the following components:

$$\sigma'_{ij} = \sigma_{ij} - \delta_{ij}\sigma_h, \qquad (14)$$

σ_h being the hydrostatic pressure:

$$\sigma_h = \frac{\sigma_{ii}}{3} = \frac{1}{3}(\sigma_1 + \sigma_2 + \sigma_3) = \frac{1}{3}(\sigma_x + \sigma_y + \sigma_z)\,. \qquad (15)$$

The HMH stress is given by the well-known relation:

$$\sigma^2 = \frac{3}{2}\sigma_{ij}\sigma_{ij} = \frac{1}{2}\left[(\sigma_1 - \sigma_2)^2 + (\sigma_2 - \sigma_3)^2 + (\sigma_3 - \sigma_1)^2\right]. \qquad (16)$$

If the expression 12 of modulus E' is used, the following expressions for global deformation increments is found:

$$d\varepsilon_{ij} = (1+\nu)\frac{d\sigma_{ij}}{E} - \nu\,\delta_{ij}\frac{d\sigma_{ii}}{E} + \frac{3\sigma'_{ij}}{2\sigma}\frac{d\sigma}{E'}\,. \qquad (17)$$

Equation 17 can be written in a matrix form in terms of stress components as follows:

$$[d\varepsilon_{ij}] = [A_{ij}]\cdot[d\sigma_{ij}], \qquad (18)$$

where the matrix A is given by the expression:

$$[A_{ij}] = \left[\frac{1}{E}\delta_{ij}\right] + \left[\frac{\nu}{E}(\delta_{ij} - 1)\right] + \left[\frac{a_{ij}}{4\sigma^2 E'}\right], \qquad (19)$$

in which:

$$a_{ij} = \frac{\partial\sigma^2}{\partial\sigma_i}\frac{\partial\sigma^2}{\partial\sigma_j}\,. \qquad (20)$$

The principle of compatibility theory is illustrated by aid of Fig. 1. When a first finite load increment is applied, as in the point A, the HMH stress σ_A is below the elastic limit, σ_E, and the deformation ε_A is purely elastic. Once an increment generates a stress $\sigma_B > \sigma_E$, the deformation is not anymore elastic. It possesses now two components ε^e_B and ε^p_B, respectively elastic and plastic, the corresponding point being

B′ on the characteristic of slope E′. In this situation, the real infinitesimal stress increment $d\sigma'_B$ can be determined by aid of Eq. 18 in which the deformation increment is equal to that corresponding to the elastic stress increment, $d\varepsilon_{B'} \equiv d\varepsilon_B$.

The solution of matrix equation 18 represents the real stress increment $d\sigma'_B$ which generates the real elastic deformation increment $d\varepsilon^e_{ijB}$ by means of Eq.9. Because the total deformation is known from the elastic solution, being denoted by ε^t_B in Fig.1, the plastic deformation increment is obtained from Eq. 8 as follows:

$$d\varepsilon^p_{ij} = d\varepsilon^t_{ij} - d\varepsilon^e_{ij}. \quad (21)$$

By using this procedure, the plastic deformations are found for all loading increments, up to a desired depth z_0. The total plastic deformation in a desired point is the sum of all incremental deformations. The plastic displacement of this point on z direction is given by following equation:

$$w^p_z = \sum_{z=0}^{z_0} \Delta\varepsilon^p_z \Delta z. \quad (22)$$

This mathematical model based on the theory of compatibility of deformations allowed these authors to develop a numerical program, reported in [17].

NUMERICAL ANALYSIS

In order to evaluate the pressure distribution, the stress state and the displacements in a rough frictionless contact, two sets of baring balls of 6.35mm diameter were selected having different degrees of finishing, namely ground and lapped. The measured micro-topography was introduced in the above-mentioned numerical program to find the pressure distribution, stress state and plastic deformations. Six full simulations were performed for each set, at calculated maximum Hertz pressures ranging between 3.5 and 6GPa, with a step of 0.5GPa.

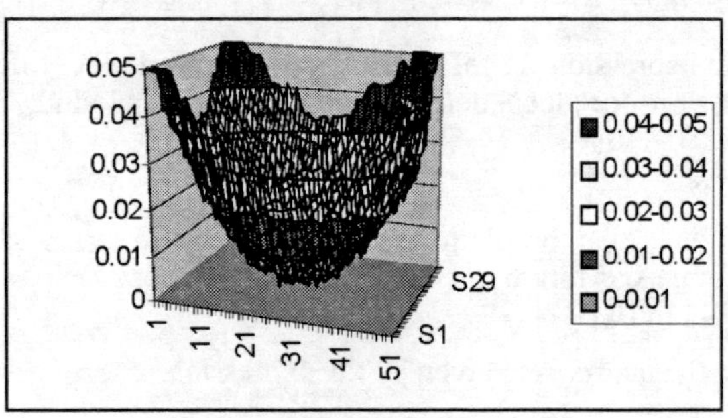

a: ground ball

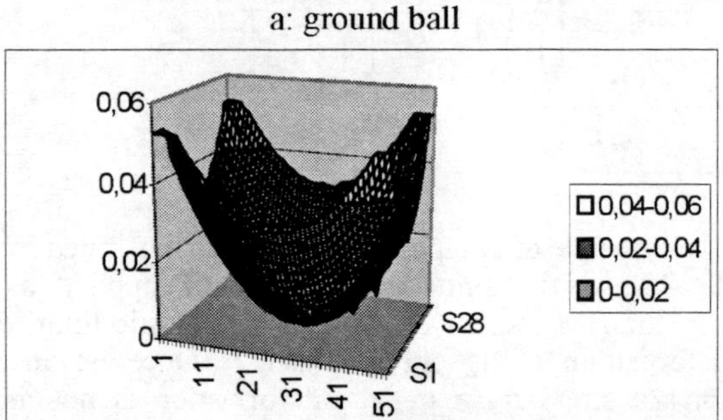

b: lapped ball

Fig. 2 Initial 3-D contact

Figures 2, a and b show the initial state of the contact between, respectively a ground and a lapped ball, and the rigid smooth half-space.

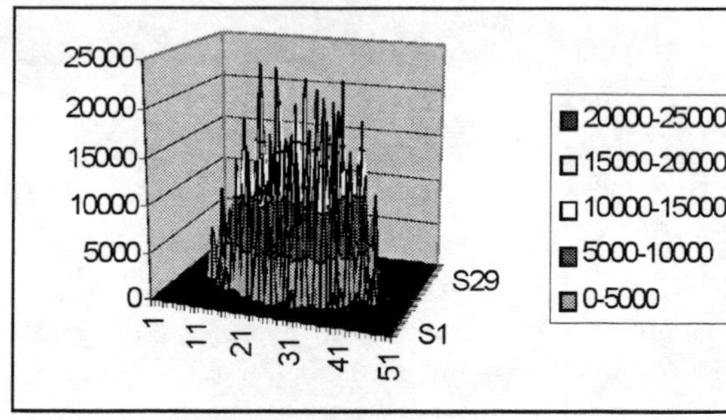

a: ground surface

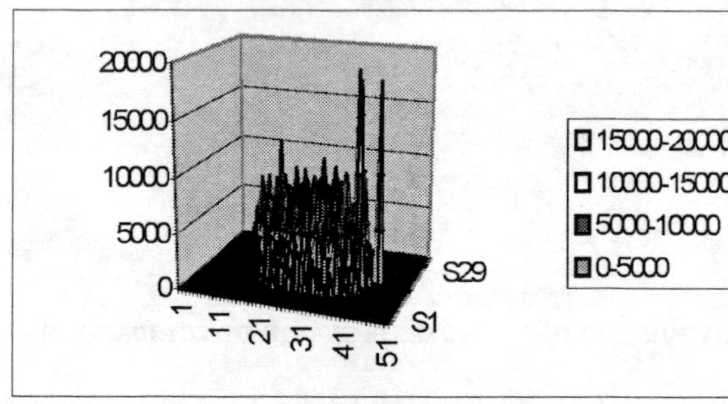

b: lapped surface

Fig. 3. Effect of surface finish upon pressure distribution at a theoretical maximum Hertz pressure of 6GPa

Color contour lines of normal stress σ_z are shown in Fig.4a and similar representation of orthogonal shear stress τ_{xz} can be seen in Fig.4b. Important stress risers occur in the immediate neighborhood of contact surface, whereas at large depths the stress field is practically similar to that occurring in smooth Hertz contacts, according to the well known Saint Vénant principle, frequently used in elasticity.

The effect of degree of surface finishing upon plastic displacements on z direction is evidenced in Fig.5, a and b for a calculated maximum Hertz pressure of 6 GPa.

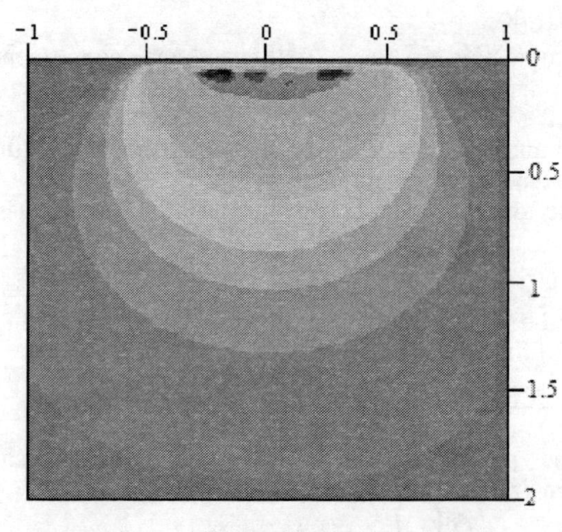

a: normal stress σ_z

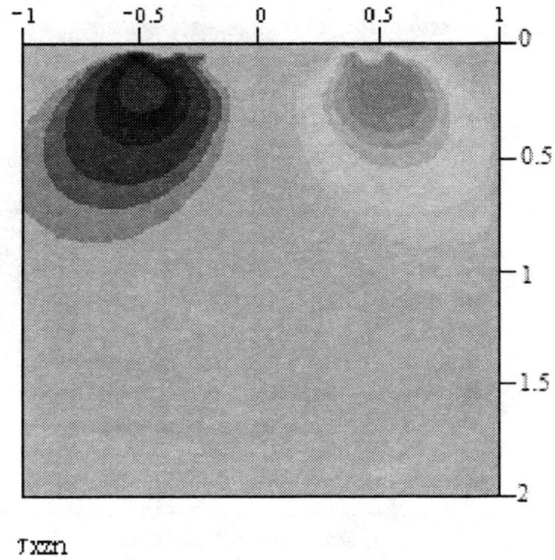

b: orthogonal shear stress τ_{xz}

Fig. 4. Subsurface shear stresses

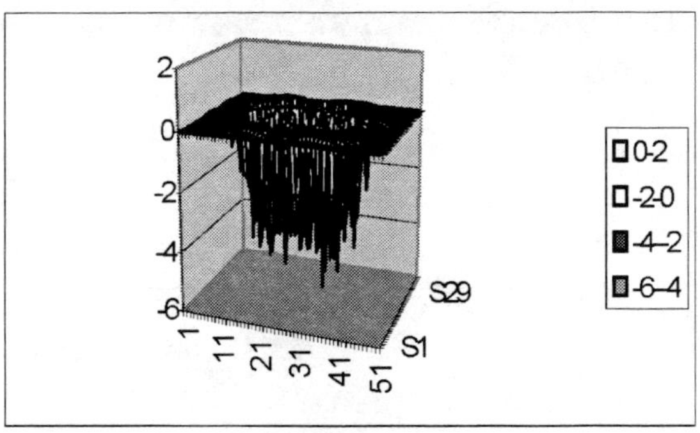

a: ground surface

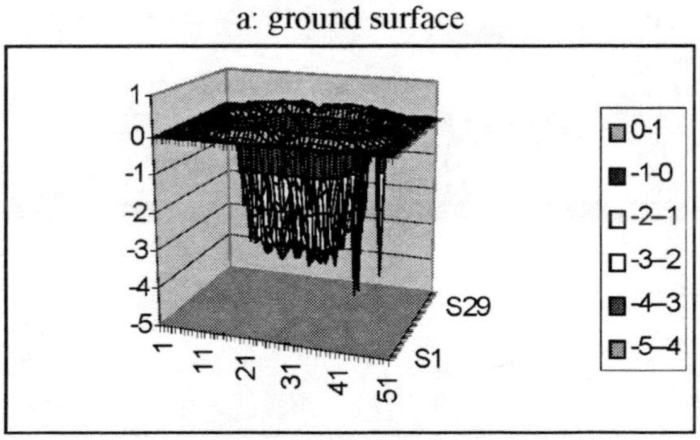

b: lapped surface

Fig. 5. Effect of surface finish upon surface plastic deformations

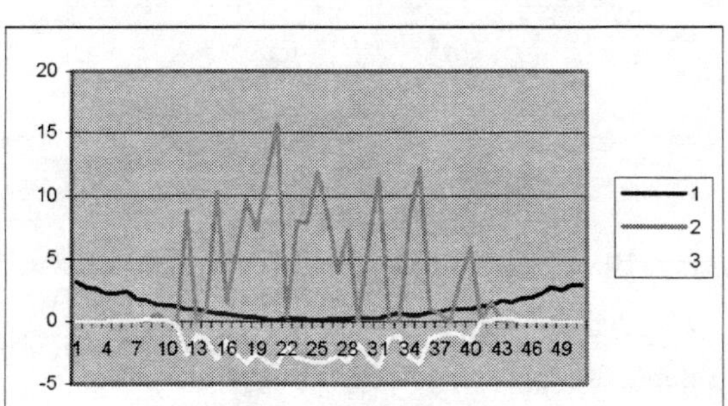

a: ground balls

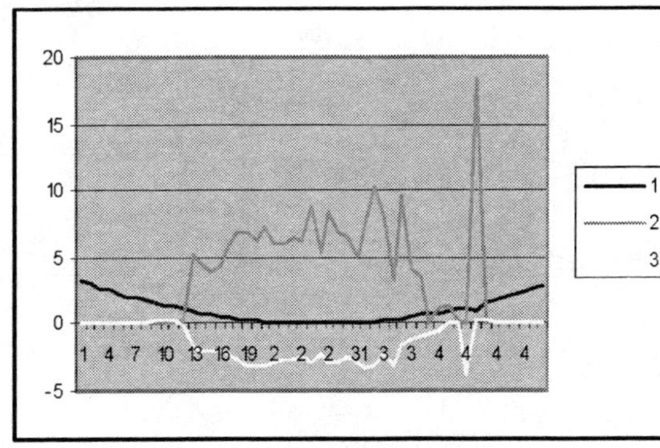

b: lapped balls

Fig. 6. Effect of surface finishing upon pressure distribution and surface plastic deformations

It is obvious that in the case of ground surface the pressure drops frequently to zero, whereas for lapped surface such drops are very rare.

Figure 7 shows plots of maximum pressure for the six load levels, evaluated numerically for ground and lapped surfaces as well as for a smooth ball surface. It can be seen that, although the lapping is superior to grinding, qualitatively, both finishes are considerably below an ideally smooth surface.

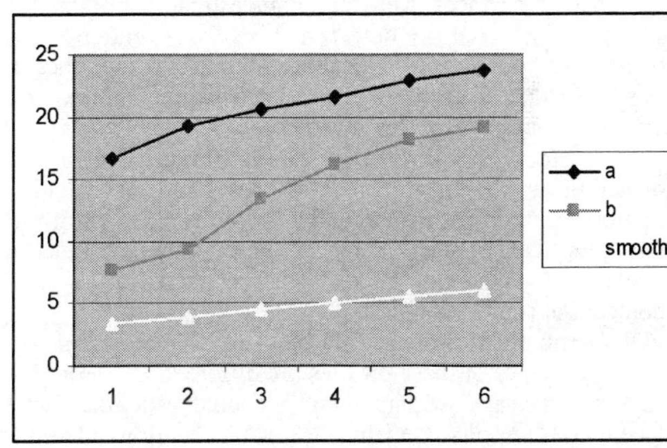

Fig. 7. Effect of surface finishing upon maximum pressure

Similar plots are shown in Fig.8, a and b for, respectively, central and maximum depth of plastic deformation expressed in micrometers, in comparison to the smooth case.

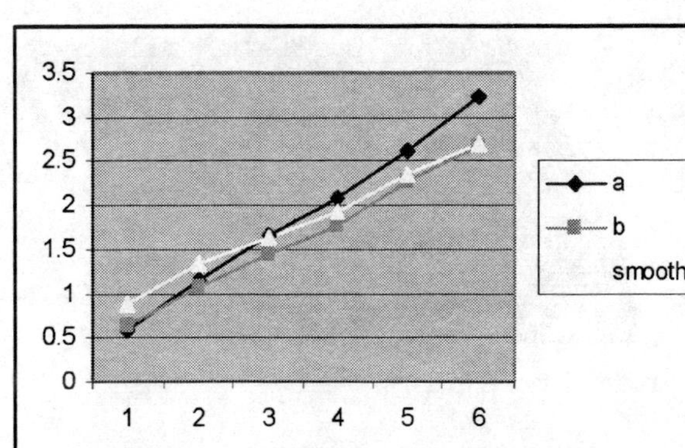

a: central plastic deformation

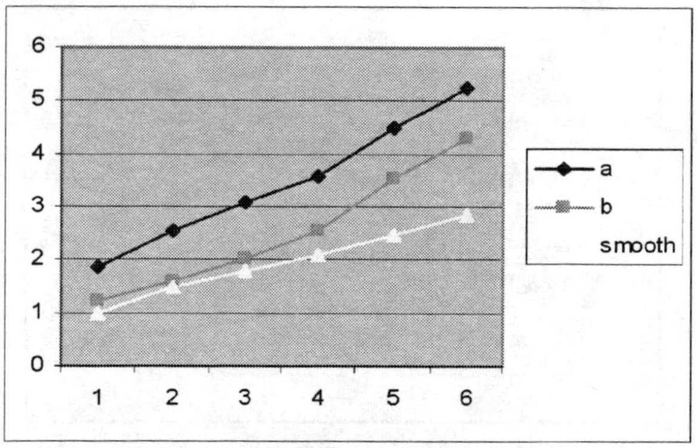

b: maximum plastic deformation

Fig. 8. Effect of surface finishing upon the depth of plastic deformation

EXPERIMENTAL RESULTS

Experimental data concerning the micro-topography of real surfaces was obtained by aid of a laser profilometer UBM 14. This apparatus stores the data in its own characteristic format. In order to use the information, these files were converted in ASCII format which can be recognized by FORTRAN 100 in which the actual numerical program was developed.

After each contact loading into the elasto-plastic domain, the contacting surfaces were again investigated on the laser profilometer in order to detect and record eventual plastic deformations. These measurements allowed to evidence the effects of surface finishing upon contact parameters and some of the results obtained are reviewed below.

The effect of surface finishing upon 3-D plastic deformation is shown in Fig.9, a and b for a calculated maximum Hertz pressure of 6GPa. In the case of ground surface the real contact area is significantly smaller than apparent area, whereas lapped surface leads to a closer agreement between these areas.

Diametric profiles of the 3-D plots are shown in Fig. 10, a and b. The shapes of these profiles are different: very rough for ground surface and considerably smoother for the lapped surface. These results confirm the conclusions drawn from the 3-D representations.

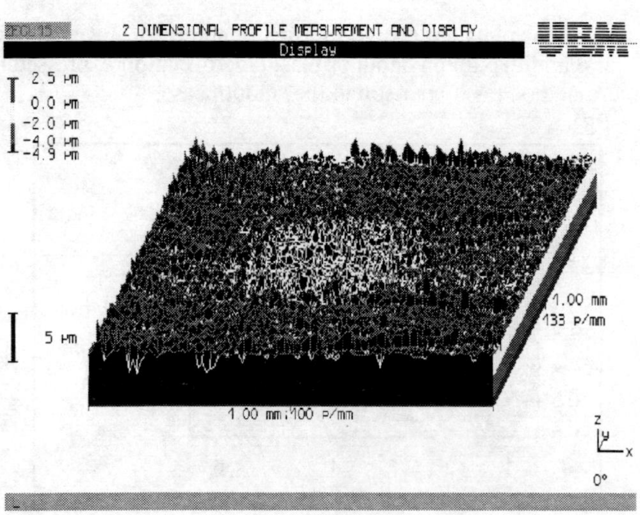

a: ground surface

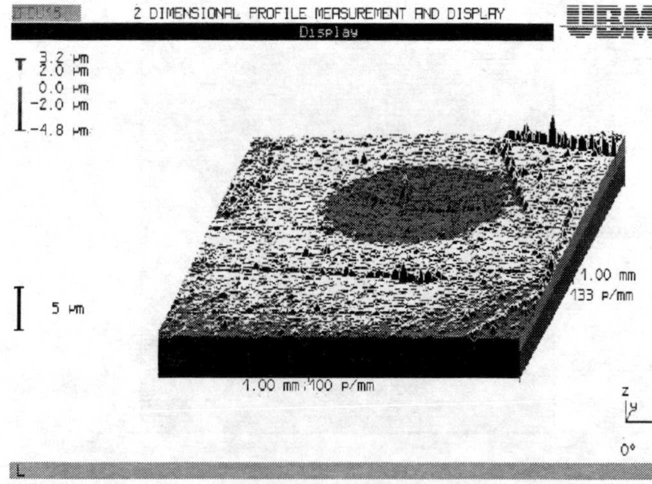

b: lapped surface

Fig. 9. Effect of surface finish upon 3-D plastic deformation

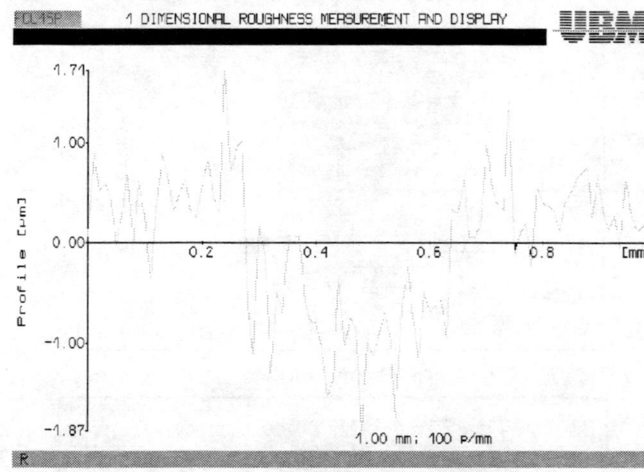

a: ground surface

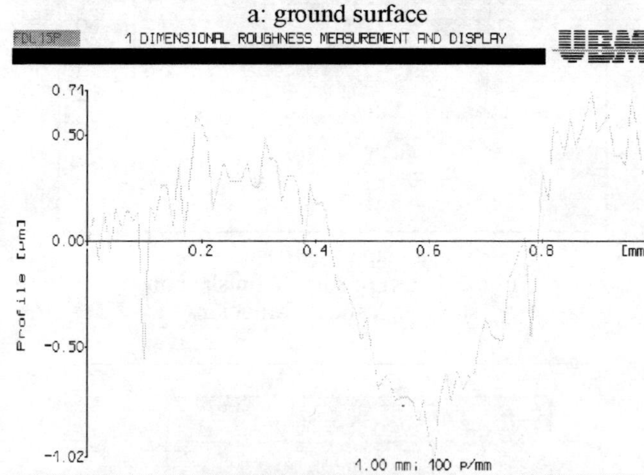

b: lapped surface

Fig. 10. Profiles of plastically deformed surfaces

A direct comparison between theoretical and experimental central depths of plastic deformations is shown in Fig. 11 for the two surface finishes, having the applied normal load as a parameter.

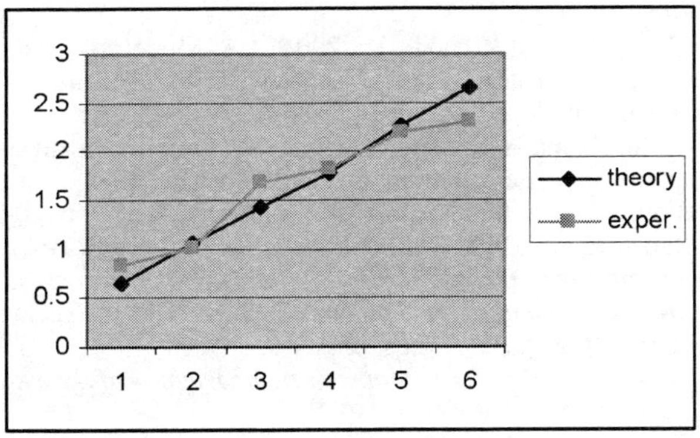

a: ground surface

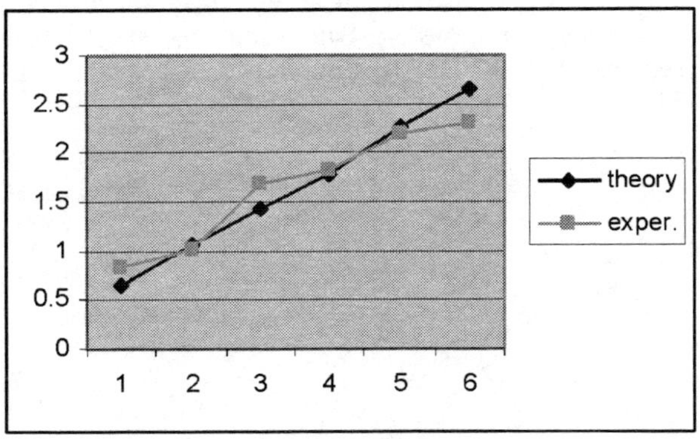

b: lapped surface

Fig, 11. Theoretical and experimental central depths of plastic deformations

A quite good agreement can be seen between theoretical predictions and experimental measurements. Surprisingly, at lower loads, the agreement is better for ground surface, whereas at higher loads the lapped surface yields better theoretical results. Overall, the agreement between theory and experiment is better in the case of lapped surface.

During plastic deformation the material yields sub-superficially and the shape and volume of asperities changes. Therefore, the geometric parameters of surface roughness suffer modifications during loading into the elasto-plastic domain, obviously tending to diminish. To evidence these changes, roughness measurement was performed before and after loading for both surface finishes. The results obtained for arithmetic average value of roughness, Ra and for total roughness depth, R_t, are plotted graphically in Fig. 12, a and b.

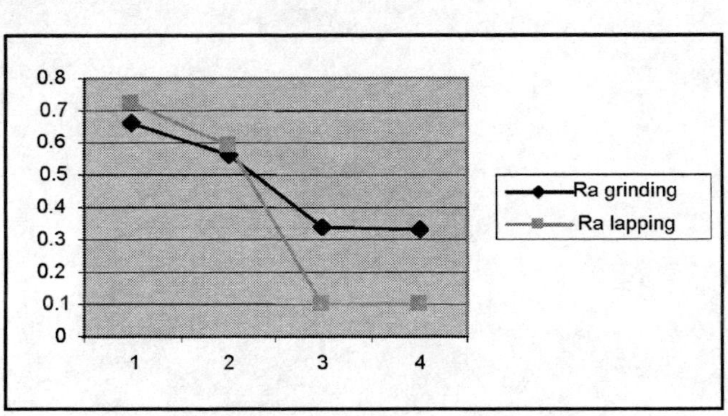

a: arithmetic average roughness

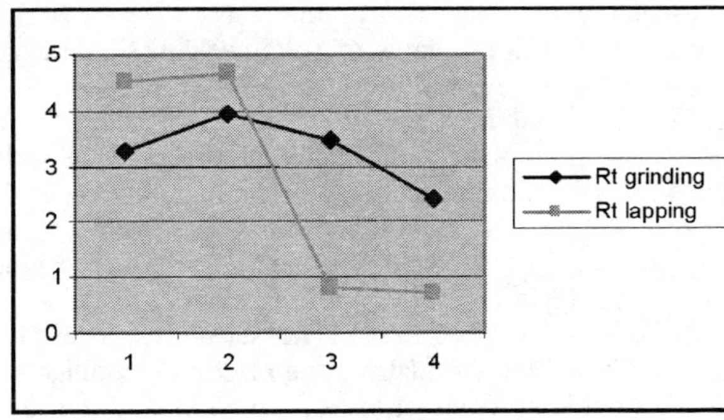

b: total depth of roughness

Fig. 12. Effect of load upon roughness parameters

CONCLUSIONS

The following conclusions can be drawn from the work herein reported.

- A numerical program written in FORTRAN 100 was developed to determine the pressure distribution, the stress field and the plastic deformations occurring in a rough contact.
- The microtopography of two sets of bearing balls, ground and lapped was recorded by aid of a laser profilometer and was introduced in the numerical program as input data.
- The numerical results show that maximum pressure calculated from the real surface microtopography diminishes as the degree of surface finishing improves. The central plastic deformation for a smooth ball is larger than for rough balls at pressures bellow 5GPa, but at higher pressures the situation reverses, probably due to an existing valley in this point. The maximum plastic deformation is usually reached outside contact centre, diminishes as the surface roughness increases and is considerably larger then for smooth surfaces.
- The surface regions subjected to contact were investigated with the laser profilometer, measuring the 3-D plastically deformed surface and its profile. The experimental results are in a good agreement with theoretical predictions.
- Both, theoretical and experimental results show that as the geometric surface quality improves, the plastically deformed surface is smoother and the roughness parameters decrease.

REFERENCES

1. Bowden, F. and Tabor, D., 1950, *Friction and lubrication of solids*, Oxford: Clarendon Press, Vol. 1.

2. Archard, J.F., Hunt, R.T. and Onions, R.A., 1975, Stylus profilometry and the analysis of the contact of rough surfaces, in *The mechanics of the contact between deformable bodies*, Ed. J.J. Kalker, Delft: Delft University Press, pp. 282-303.

3. Greenwood, J.A. and Williamson, J.P.B., 1966, "Contact of nominally flat surfaces", Proceedings of Royal Society, Series A, pp. 300-319.

4. Cheng, W.R., Etsion, I. and Bogy, D.B., 1987, "An elastic-plastic model for the contact of rough surfaces" ASME J. of Tribology, **119**, 2, pp. 257-263.

5. Evseev, D.G., Medvedev, B.M. and Grigoryan, G.G., 1991, "Modification of the elastic-plastic model for the contact of rough surfaces", Wear, **150**, Nr. 1-2, pp. 79-88.

6. Halling, J. and Nuri, K., 1991, "Elastic-plastic contact of surfaces considering ellipsoidal asperities of work-hardening multi-phase materials", Tribology International, **24**, Nr. 5, pp. 311-319.

7. Goryacheva, I.G. and Dobychin, M.N., 1991, "Multiple contact model in the problems of tribomechanics", Tribology International, **24**, Nr. 1, pp. 29-35.

8. Ike, H. and Makinouki, A., 1990, "Effect of lateral tension and compression on plane strain flattening processes of surface asperities lying over a plastically deformable bulk", Wear, **147**, Nr. 1, pp. 17-38.

9. Mayeur, C., 1995, *Modelisation du contact rugueux élastoplastique*, PhD, INSA Lyon, p. 179.

10. Lee, S.C. and Ren, N., 1994, "The Subsurface Stress Field Created by Three-Dimensionally Rough Bodies in Contact with Traction", Tribology Transactions of the STLE, **37**, Nr. 3, pp. 615-621.

11. Bengisu, M.T. and Akay, A., 1997, "Relation of Dry-Friction to Surface Roughness", Journal of Tribology, Transaction of the ASME, **119**, pp. 18-25.

12. Horng, J.H., 1998, "An Elliptic Elastic-Plastic Asperity Microcontact Model for Rough Surfaces", ASME Journal of Tribology, **120**, pp. 82-88.

13. Zhao, Y. and Chang, 1., 2001, "A Model of Asperity Interactions in Elastic-Plastic Contact of Rough Surfaces", , Journal of Tribology, **123**, pp. 857-864.

14. Crețu, S. and Hatmanu, V., 1985, "A numerical analysis of the permanent deformation in elasto-plastic line contact", Buletinul I.P.Iaşi , tom XXXI (XXXV) fasc. 1-4, IV, pp. 20-25.

15. Popescu, G., 1998. *Theoretical and experimental research concerning upon behavior of frictionless Hertz contacts in the elasto-plastic domain*, in Romanian, Ph.D. Thesis, Technical University "Gh. Asachi", Iaşi, pp. 84-95.

16. Hill, R., 1955, *The Mathematical Theory of Plasticity*, Clarendon Press, Oxford.

17. Prodan, D. and Diaconescu, E., - Numerical calculation of stress state and deformations in an elyptical contact between wavy surfaces in elasto-plastic domain, in Romanian, Proceedings of VAREHD 10 Conference, Suceava, 2000, pp. 321-330.

Proceedings of
2003 STLE/ASME Joint International Tribology Conference
Ponte Vedra Beach, Florida USA, October 26-29, 2003

2003-TRIB-0282

NANO-SCRATCHING AS A MEANS TO STUDY WEAR DEBRIS GENERATION IN POLYMERS

Rosli Bin Mohamed Sani, Sujeet K. Sinha*
Department of Mechanical Engineering, National University of Singapore, Singapore 117576

Joyce Pei Ying Tan and Kai Yang Zeng
Institute of Materials Research and Engineering, Singapore

* corresponding author (mpesks@nus.edu.sg)

ABSTRACT

Scratches in the range of 1000 nm deep were produced on the surfaces of two polymers using diamond tip indenter to study the material deformation and wear debris generation characteristics. Our results show that the process of wear debris generation depends upon the manner in which scratches are produced on the surface. Single track scratches do not produce wear debris despite several passes of the indenter tip whereas intersecting scratches accelerate the wear debris generation process. The results indicate that the material near the four corners of intersecting scratches experience extremely high strain leading to extrusion and eventual breakage from the bulk of the polymer forming a debris particle.

KEYWORDS: Nano-scratching, polymethylmethacrylate, polytetrafluoroethylene, wear

INTRODUCTION

Wear is defined as a process of material removal when two contacting surfaces experience relative motion. The process by which material removal takes place has been described by mechanisms such as abrasive wear, adhesive wear, delamination wear, fatigue wear and so on [1]. These mechanisms of wear have been studied extensively on macro and micro scale. However, nano scale study of wear process of materials has been very few [2,3]. Nano-scale studies are important as by using these investigations we can determine the rate controlling mechanism among several mechanisms that are characteristics of any single wear process. Currently there are several applications such as data storage and hip/knee joint replacement where nano to micron scale wear is a major technical challenge.

The current paper will describe the nano-scale wear mechanism studied on two polymers using scratching technique. The scratch technique has been used by several researchers to model abrasive wear process for metals and polymers [4]. However, in a large majority of cases, the scratching has been carried out on a single track with one or multiple passes. In a previous study involving ultra-high molecular weight polyethylene (UHMWPE) we have shown that the process of material deformation can be very severe if scratches are produced in an intersecting manner [5]. This paper will extend these findings to other polymer systems and present several evidences that intersecting scratches can increase the wear rate by several orders when compared with single track scratching even if all other parameters are kept constant.

EXPERIMENTAL DETAILS

Sample preparation

Teflon (polytetrafluoroethylene) and PMMA (polymethylmethacrylate) of commercial grade was used for the tests. Several polymer coupons with 1.5 x 1 cm surface area and varying thicknesses were cut using a micro lathe. The samples were mechanically polished to an average surface roughness (R_a) of 0.3 μm with a variation of ± 0.2 μm. The samples were handled with latex gloves in order to prevent contamination. A total of five samples were prepared in this manner. The samples were washed with distilled water and dried using a chemical air-duster after polishing. Each sample was then fixed to a sample holder using a cyanocrylate adhesive for the scratching experiments. Upon completion of the scratch tests, the samples were carefully removed from their holders using an acetone solvent. For examination under AFM, the samples were fixed onto mica disks attached to the AFM sample holder.

Scratch experiments

Scratch experiments were carried out on a MTS Nano Indenter XP with a Lateral Force Measurement option that is capable of making multiple indentation or scratch tests at micron to sub-micron levels. Conical diamond scratch tips with 90° cone angle and 1μm radius of curvature were used for the experiments conducted during this project. Three different types of experiments were carried out.

Unidirectional – In this test, scratching was carried out on one track making several cycles of passes.

Orthogonal multi-pass – In this type of test the scratches were produced on orthogonal tracks intersecting each other. First a fixed number of cycles of passes were given in one direction followed by same number of passes in orthogonal direction.

Orthogonal omni-pass – Similar to the orthogonal mulit-pass, in this type of test the scratches were produced on orthogonally intersecting tracks, however, direction of scratching was changed after every single pass by 90°.

A typical single scratch test involved pre-scratch profiling and actual scratching followed by post-profiling and finally cross-sectional profiling at selected points. The tests were carried out at normal loads of 5 mN and 10 mN.

RESULTS

Unidirectional scratching

Figure 1 shows a scratch made on the surface of PMMA at a normal load of 10 mN. The material shows plastic deformation forming pile-ups on the sides of the scratch. However, no wear particle formation is observed. This shows that the volume of the material that is removed by the tip is merely displaced to the sides forming pileups.

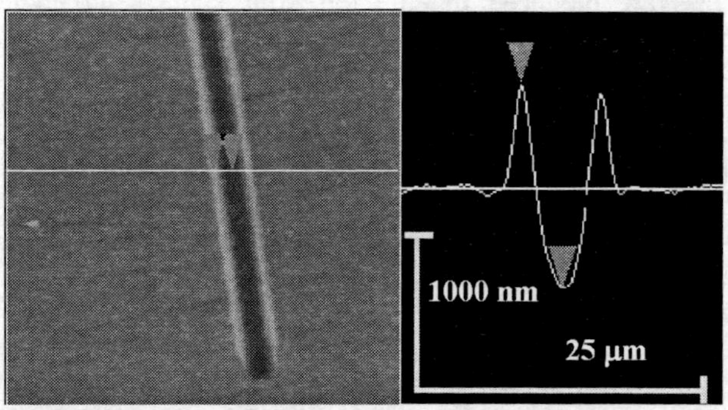

Figure 1: AFM image of a unidirectional scratch on PMMA surface. The picture on right hand side shows a cross sectional profile of the scratch.

Figure 2: The optical microscope image of the diamond indenter tip taken after unidirectional scratching. This image is same as that for a clean indenter indicating now wear debris attached to the surface of the tip.

Figure 2 shows an optical micrograph of the tip immediately after scratching. The surface of the tip resembles that of freshly cleaned one. There is no evidence of any wear debris that might have become attached to the indenter tip.

Orthogonal multi-pass scratching

Figure 3 shows AFM images of the orthogonal scratches with 5 pass for each direction indicating 5 times scratching in one direction followed by 5 times scratching in the orthogonal direction.

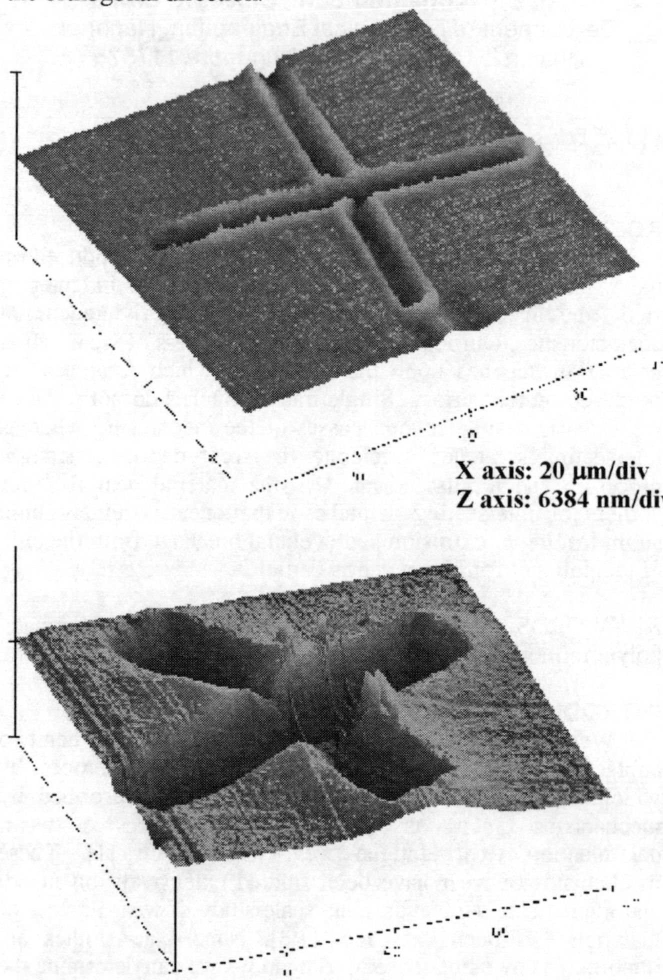

Figure 3: AFM scan images of orthogonal multi-pass scratches on PMMA (upper image) and PTFE (lower image). For both cases there are large localized strain regions at the corners of the scratches with some extrusion of the material in the vertical upward direction.

Comparing the cross sectional scratch profiles for PMMA and PTFE we can see that for PMMA the pileup of material is much sharper and takes place within narrow region from the scratch. Also, PMMA being harder than PTFE the scratch depth for PMMA is small compared to that for PTFE. Figure 4 shows the cross sectional profile on PTFE after 5 passes. The noticeable feature of orthogonal scratching compared to unidirectional scratching is the formation of "wall" when the direction of scratching changes. The "wall" is formed when the indenter tip meets the pileup formed by the previous scratch at 90° and drags the material in the direction of scratching. This process leads to extensive plastic deformation of the material present in the vicinity of the intersection. Such extensive plastic deformation leads to material extrusion in the upward direction at the four corners of the scratch intersection. This was observed by taking a line scan on one of the pileup along the scratch length passing through the intersection. Figure 5 shows an example of this scan on PMMA pileup. The extrusion of material at the intersections is much larger for PMMA than for PTFE as measured by the profile of the pileup near the scratch intersections. The formation of "wall" and the material extrusion for orthogonal scratching gives rise to generation of debris material. This is because the material can detach from the bulk of the polymer if local strains are extremely high. High strain also leads to strain hardening and thus increases the tendency of the materials to fail in brittle manner by crazing and fracture.

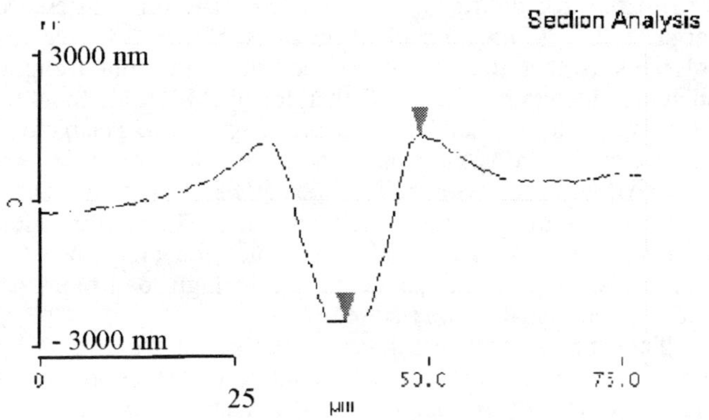

Figure 4: Cross sectional profile of the scratch made on PTFE after five passes.

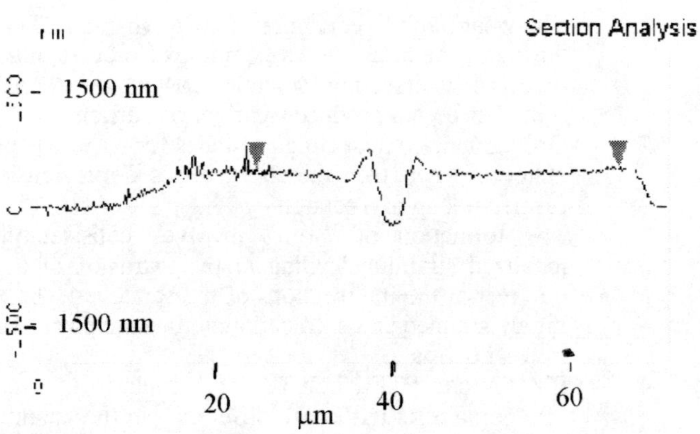

Figure 5: AFM line scan on the pileup of PMMA scratch passing through the scratch intersection area. The upward peaks near the intersection indicate that the material extensively deform and extrudes in the upward direction at the four corners of the scratch intersection.

Hence it is observed that the wear and the generation of debris particles is because of the orthogonal scratching, a process which causes severe material deformation at the four corners. However, the amount of wear will depend upon the number of times that the scratching direction is changed, in other words, how many times the "wall" is made and re-made during the orthogonal scratching.

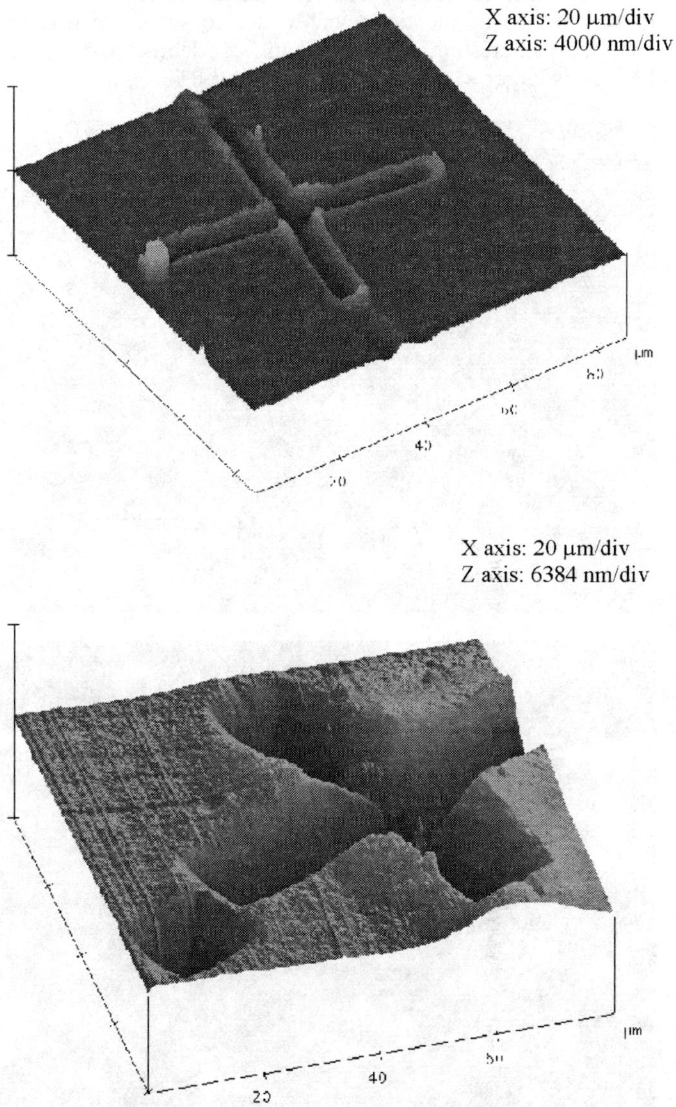

Figure 6: AFM scanned images of orthogonal omni-pass scratched PMMA (upper) and PTFE (lower) surfaces. These images indicate that there can be extensive deformation of materials at the intersection due to several orthogonal passes of the indenter tip. After several orthogonal passes, the "wall" can virtually disappear leaving two intersecting scratches with highly deformed and piled-up material at the intersections.

Orthogonal omni-pass scratching

Figure 6 shows scratches of the orthogonal omni-pass type for PMMA and PTFE. It is observed that although the depth of the scratch does not change much with the successive pass of the tip on a scratch, the omini-pass scratching produces "wall" and extrusion of the material for every change of the direction of scratching. Hence, the total accumulated strain of the localized volume of material near the intersection far exceeds those of the orthogonal multi-pass scratching. The increase in the wear rate for omni-pass scratching can be observed from the optical images of the indenter tip after scratching. Figure 7 gives these optical images for PMMA and PTFE. Hence, even though there is no presence of wear debris near the scratches, we can clearly see loose particles at the tip which are attached to the surface of the tip probably due to some electrostatic charging. Interestingly, the amount of debris particle for PMMA is larger when compared to that for PTFE.

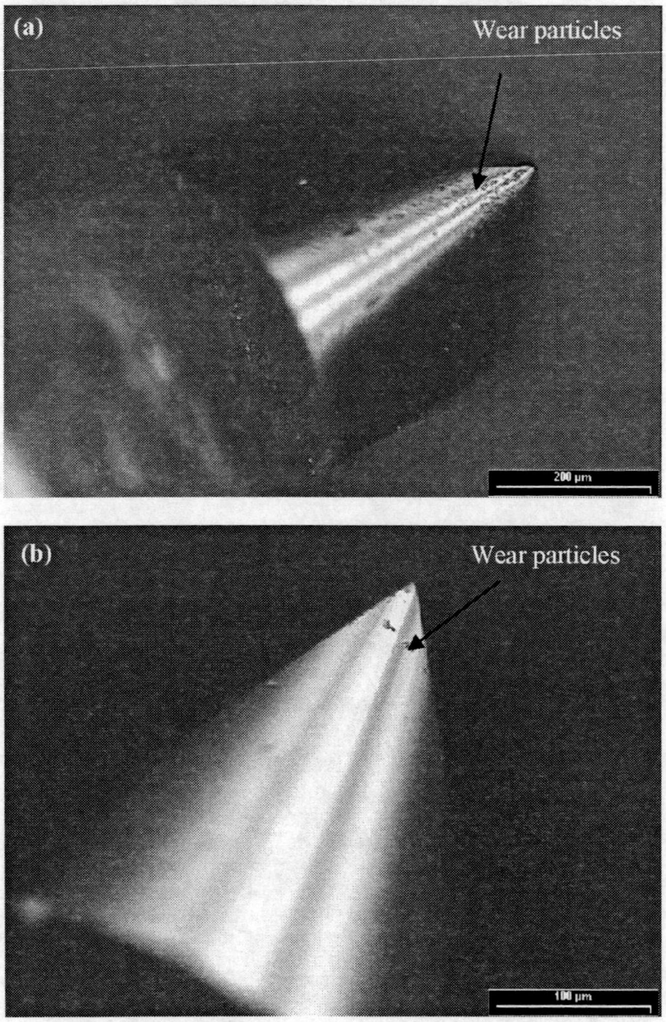

Figure 7: Optical micrographs of the diamond tip after scratching for (a) PMMA and (b) PTFE for orthogonal omni-pass scratching.

DISCUSSION

In this study, we have shown that the wear mechanism for two body wear does not only depend upon the presence of asperities but how they produce scratches or grooves on the surface of the softer material. Several passes on a single scratch by the tip does not increase the scratch by any appreciable amount showing no wear particle generation. The orthogonal omni-pass scratching introduces much larger amount of strains at very localized regions than compared to uni-directional scratching. Therefore, we see the formation of wear particles only in the case of omni-pass scratching because of the higher number of change of direction than for orthogonal multi-pass which gave the change of scratching direction only once. Hence, it is obvious that the wear rate is dependent upon the number of times the scratching direction is changes for intersecting scratching. Wear rate does not depend upon the number of passes made on the same scratch in unidirectional type.

The quantification of wear rate for the nano-scratching experiment is very challenging as it is not possible, in the present case, to collect these micron size wear particles from the tip of the indenter. However, a relative difference between the wear rate for PMMA and PTFE can be observed from the number of the wear debris particles attached to the tip. In the present case, PMMA shows much higher wear rate that PTFE even though PTFE has lower scratch hardness as measured from the profiles of the scratches.

The scratch profiles as observed from the AFM scanned pictures do not show any fracture but only ductile or plastic deformation for both PMMA and PTFE. The only difference between the deformation characteristics of the two selected polymers is that the volume of deformed material forming pileup is much larger for PTFE than for PMMA. This could be related either to the fact that the depth of scratching is larger for PTFE than for PMMA for the same normal load or to the fact that PMMA is amorphous while PTFE exists in semi-crystalline form at room temperature. The deformation behaviour of materials including polymers depends upon the depth of scratching. Larger depth due to high load promotes crack initiation and brittle fracture.

The present work has revealed the actual mechanism by which material is removed when a single hard asperity passes over the surface of two polymers, PMMA and PTFE.

CONCLUSIONS

Following conclusions can be drawn form the experimental work described above.

1. The wear particle generation for PMMA and PTFE during scratching is a complex process and involves intersecting scratches as unidirectional scratches do not produce wear debris particles.
2. Orthogonally intersecting scratches form "walls" of material when the scratch traverses a previously formed pileup of a scratch.
3. The formation of "wall" involves considerable localized straining leading to the extrusion of the polymer at the intersections of the scratches. These highly strained and extruded materials are precursor to wear debris.
4. PMMA and PTFE behave very similarly in nano scratching tests with only difference in the volume over which the pileups of the scratches are distributed. For PMMA the pileup is much closer to the scratch while for PTFE the pileup takes much larger volume probably due to the larger depth of scratching.

REFERENCES

[1] I.M. Hutchings, Tribology: Friction and Wear of Engineering Materials, Edward Arnold, UK (1992), (co-published by CRC Press, Boca Raton, USA).

[2] R. Kaneko and E. Hamada, Wear 162-164 (1993), 370-377.

[3] Michael J. Adams, Alexander Allan, Brian J. Briscoe, Peter J. Doyle, David M. Gorman, Simon A. Johnson, Wear 251(2001), 1579-1583.

[4] S. P. Ho, R. W. Carpick, T. Boland and M LaBerge, Wear 9213(2001), 1-11.

[3] Brian K. P. Wong, Sujeet K. Sinha, Joyce P. Y. Tan and Kai Yang Zeng, "Nano-wear mechanism for ultra-high molecular weight polyethylene sliding against a model hard asperity", submitted to Tribology Letters.

Proceedings of
2003 STLE/ASME Joint International Tribology Conference
Ponte Vedra Beach, Florida USA, October 26-29, 2003

2003-TRIB-0283

A COMPUTATIONAL STUDY OF NANO-SCALE CONTACT AND FRICTION

Jeffrey L. Streator
G. W. Woodruff School of Mechanical Engineering
Georgia Institute of Technology
Atlanta, GA 30332

Ilya I. Kudish
Kettering University
1700 West Third Avenue
Flint, MI 48504-4898

ABSTRACT

A simulation is undertaken to consider the effects of the scale of surface roughness on elastic contact and friction. In the simulation, a cylindrical rigid indenter of a prescribed profile is brought into contact with an elastic half-space. Interaction between the indenter and the half-space occurs through an assumed surface potential derived from the 6-12 Lennard-Jones intermolecular potential. The hypothesis under investigation is that by assuming the frictional stress is proportional to the magnitude of the normal stress, irrespective of sign, one can account for the enhanced frictional stress at the nano-scale vs. the micro-scale that has been observed experimentally. Results of the simulations indicate qualitative agreement with recent experimental findings.

INTRODUCTION

It is well known that the topography of contacting surfaces is a significant factor in determining failure, durability, and reliability of various joints. The standard requirements on surface roughness for bearings and gears in auto, air, and other similar applications are on the micrometer scale. Surface roughness is especially important in such applications as MEMS, NEMS, and in various magnetic recording devices. In the latter cases the surface roughness must remain in the nanometer scale. Recently, it was experimentally established [1-4] that in dry single nano- and micro-asperity contacts of the same elastic materials the values of the friction coefficient differ by more than an order of magnitude. The first attempt to explain this phenomenon based on single/multiple dislocation slip, is made in recent papers by Hurtado and Kim [5,6].

Let us consider the basics of the dry friction phenomenon in a contact of two elastic solids. First we consider the ideal case of perfectly smooth elastic solids sliding one with respect to another in a contact subjected to a normal force P. The friction force T created in such a contact may be related to the normal load P through a coefficient of friction λ, which is determined by the material properties at the interface. In most cases the relationship between the tangential stress τ and pressure p satisfies the Amontons-Coulomb law

$$\tau = \lambda p \quad \text{and} \quad T = \lambda P \tag{1}$$

The friction coefficient λ characterizes an average resistance of molecules to relative motion tangential to the interface. In the case considered, the contact pressure p is nonnegative practically everywhere in the contact. It may assume a small negative value in very narrow zones adjacent to the contact boundaries where the contact surfaces separate. However, for perfectly smooth surfaces, the effect of these zones with negative pressure is negligibly small. To emphasize that λ is the coefficient of friction for purely smooth surfaces (that are in a contact with predominantly positive pressure) we may view it as the nominal coefficient of friction as opposed to actual coefficient of friction μ that is realized in contacts of real rough surfaces.

A similar situation takes place for elastic solids with rough surfaces that are subjected to a sufficiently high normal load P. In this case the presence of surface roughness does not contribute to the sliding friction force T besides shaping the distribution of the positive pressure and the real and nominal friction coefficients are equal, i.e., $\mu=\lambda$. This is a correct representation of the dry friction phenomenon until the normal load P applied to the contact drops so low that the molecular attraction between the two solids can no longer be ignored. For such loads P, the topography of the contact surfaces becomes increasingly significant as negative values of pressure may be realized in regions that are comparable in size with the regions of positive pressure. In such contacts, the Amontons-Coulomb friction law as expressed in (1) is no longer valid. That is, the factor relating shear stress to pressure is not the same as that relating friction force to normal load. The difference between the local and global factors in the presence of negative pressure stems from the fact that the direction of the frictional stress is independent of the sign of contact pressure p. In other words, regions under tension resist sliding just as do regions under compression.

Therefore, the Amontons-Coulomb friction law as expressed in (1) should be replaced by another one that is valid even when there is significant negative pressure in the interface. The simplest generalization of the Amontons-Coulomb friction law (1) that leaves intact the proportionality between the frictional stress τ and pressure p is as follows:

$$\tau = \lambda|p| \quad \text{and} \quad T = \mu P \tag{2}$$

It can be expected that always

$$\mu \geq \lambda \tag{3}$$

Obviously, when positive pressure is dominant in the contact Eq. (2) coincides with the Amontons-Coulomb friction law (1) and $\mu=\lambda$. However, when the pressure changes its sign, Eq. $(2)_1$ is nonlinear in p and $\mu>\lambda$. This situation suggests different physical effects. In particular, in nano-scale contacts subjected to a sufficiently low normal load P, the contact surfaces are located in such proximity that molecular attraction is essential, leading to the presence of zones with significant positive and negative pressure.

In the present paper we are proposing a model that, in principle, is capable of predicting the value of the actual friction coefficient μ in elastic contacts ranging from the nano- to the micro-scale provided that the value of the nominal (local) friction coefficient λ is known. This model is based on elastic continuum mechanics coupled with molecular interactions according to the Lennard-Jones potential. The purpose of this paper is to investigate the role of roughness scale in the development of friction in the context of the hypothesis $(2)_1$. In particular, we compare the ratio μ/λ at the nano- and micro-scales.

NOMENCLATURE

A	=	asperity amplitude
a	=	radius of punch = radius of contact zone
b	=	local equilibrium spacing
c_1	=	dilatational wave speed
h	=	local separation
L	=	asperity wavelength
N	=	total number of periods
n	=	period number
P	=	normal load
P_n	=	normal load due to nth period
p	=	pressure
p_{avg}	=	average pressure
$\|p\|_{avg}$	=	average pressure magnitude
r	=	radial position
T	=	friction force
u	=	dummy variable of integration
v	=	dummy variable of integration
z	=	vertical position
$\Delta\gamma$	=	work of adhesion
δ	=	interference
λ	=	nominal (local) coefficient of friction
μ	=	actual (global) coefficient of friction

INTERFACE MODEL

Geometry, Properties and Forces

To model the effect of length scale on the apparent friction coefficient, we consider the contact between a rigid cylindrical punch and an isotropic elastic half-space, the surface of the punch being nominally flat, but rough. The imposed roughness has an idealized sinusoidal geometry and therefore is characterized by a given amplitude, wavelength, and phase. The phase of the roughness is taken so that the surface has a peak value along the axis of the punch. The properties of the half-space are characterized by its modulus of elasticity, E, and Poisson's ratio, ν. For all of the simulations, the Poison's ratio was taken as 1/3.

The force interaction between the rigid punch and the elastic half-space arises from a surface potential based on the classic Lennard-Jones 6-12 intermolecular potential (see, for example, [7]). This force law can be obtained by triple integration of the Lennard-Jones potential. The result is given by [8] as

$$p(h) = \frac{8}{3}\frac{\Delta\gamma}{b}\left[\left(\frac{b}{h}\right)^9 - \left(\frac{b}{h}\right)^3\right] \tag{4}$$

where p is the pressure exerted by one surface on the other, h is the local separation between the two surfaces, b is the local equilibrium separation, and $\Delta\gamma$ is the work of adhesion, or surface free energy. In this study, b is a normalizing factor whose value need not be specified, but a value of 0.3 to 0.5 nm is physically realistic. The value of $\Delta\gamma/Eb$ is set—somewhat arbitrarily—in all cases to 0.00244, which, according to (4), corresponds to a maximum possible tensile stress given by $-p_{min}/E = 0.0025$. Thus the magnitude of the tensile stress can remain under a typical yield stress, thinking in terms of metals. Similarly, the load ranges considered are selected to keep the stresses in the elastic region—again, thinking in terms of typical metals.

The normal force P that the rigid body exerts on the half-space arises solely from the above interaction potential, and is given by:

$$P = \int_0^a p(r)2\pi r dr = \int_0^a \frac{16\pi}{3}\frac{\Delta\gamma}{b}\left[\left(\frac{b}{h(r)}\right)^9 - \left(\frac{b}{h(r)}\right)^3\right] r dr \tag{5}$$

where r is the radial coordinate, a is the radius of the contact zone, and $p(r)$ represents the pressure distribution.

The hypothesis of this paper is that there exists a local friction coefficient λ relating the local shear stress to the magnitude of the local pressure as given by Eq. $(2)_1$. (Presumably this local coefficient could be determined by an appropriate measurement.) Thus the friction force, or traction, T, can be determined as follows:

$$T = \int_0^a \lambda|p(r)|2\pi r dr = \int_0^a \lambda\frac{16\pi}{3}\frac{\Delta\gamma}{b}\left|\left[\left(\frac{b}{h(r)}\right)^9 - \left(\frac{b}{h(r)}\right)^3\right]\right| r dr \tag{6}$$

Invoking the concept of the average value of a function in (5) and (6) we obtain

$$P = \int_0^a p(r) 2\pi r dr = p_{avg} \pi a^2 \quad (7a)$$

$$T = \int_0^a \lambda |p(r)| 2\pi r dr = \lambda |p|_{avg} \pi a^2 \quad (7b)$$

So that the friction coefficient μ can be written as,

$$\mu = \frac{T}{P} = \lambda \frac{|p|_{avg}}{p_{avg}} \quad (8)$$

In light of the above equation, we introduce a "friction ratio," which is given by

$$\frac{\mu}{\lambda} = \frac{|p|_{avg}}{p_{avg}} \quad (9)$$

The above ratio represents the degree to which the actual (global) friction coefficient μ exceeds the nominal (local) friction coefficient λ. By inspection of the right-hand side of Eq. (9) we see that the friction ratio is necessarily greater than or equal to unity. It is of interest in this paper to investigate how surface topography—in particular asperity amplitude and wavelength —affect the friction ratio.

Computational Algorithm

While this paper is concerned with static equilibrium configurations, the solutions computed here are based upon an elastodynamic model developed earlier [9] that was used to simulate the interaction between a rigid sphere and an elastic half-space. Complete details of the elastodynamic model can be found in the foregoing reference, but here we provide a brief summary of that model. The half-space was discretized in radial and vertical coordinates and the resulting equations of motion for the corresponding volume elements were integrated using the classical 4th order Runge-Kutta scheme (e.g., [10]). To prevent any influence of the boundary of the finite calculation domain, the simulations were stopped before any waves reflected from the boundaries entered the domain of interest (i.e., the contact zone). In this manner the model simulated a true half-space, the boundaries of which are infinitely remote.

In the current analysis, a modification was made to damp the vibrations so that the static equilibrium configuration could be found relatively quickly. While there exist purely static iterative methods to solve for contact configurations [11], it was more convenient to modify our dynamic simulation as it was readily available. To accomplish the damping, we artificially attenuated the nodal velocities each time step by 1%. That is, once the numerical integration algorithm was applied at a given time step to predict the velocities at the succeeding time step, we multiplied each of the velocities by a factor of 0.99. This procedure was found to yield convergence to the static equilibrium configuration in a reasonable time—typically around 1000 units of simulation time, where one time unit is defined as the time required for a dilatation wave to traverse a distance b, the local equilibrium spacing (see Eq. (4)).

The dimensionless time step of the integration routine was set to 0.1, meaning that there were 10 integration time steps per one unit of simulation time. Owing to the damping algorithm describe above, the simulations could be run arbitrarily long without there being a problem caused by waves reflecting from the boundaries of the calculation domain. Depending on the particular simulation, the number of grid points as well as the interval size was varied. In any case the total number of nodes did not exceed 1681.

Simulation Procedure

There were two general categories for simulation runs: (i) short asperity wavelength and (ii) long asperity wavelength. For the short wavelength cases, the asperity wavelength was set at 500b, and the asperity amplitude was varied. In addition, the radius of the punch surface was set at precisely 3 wavelengths, or a distance of 1500b. For the long wavelength case, the asperity wavelength was set at 5000b and its amplitude was set at 10b. In order to maintain sufficient grid resolution and in light of the limitations on the number of nodal points, the long wavelength cases were computed using just one half of a wavelength and the results were extrapolated to yield results corresponding to a large number of wavelengths in contact. (The extrapolation procedure is described below.) For each simulation, the rigid surface is started at a distance 3b from its final position. The rigid surface is then moved toward the half-space at a speed of $0.01c_l$ for 300 time units, where c_l is the dilatational wave speed. At this point the motion of the punch is stopped and the simulation is continued for 900 more time units, which allows time for (near) convergence to the equilibrium configuration. At the end of the simulation, the various quantities of interest are computed. Starting the punch at some distance away and "easing" it toward the half-space, as opposed to starting with the punch at its final position, decreases the magnitude of the vibrations at the beginning of the simulation, thereby speeding up convergence.

It was observed in preliminary simulations that the equilibrium configurations tended to produce highly periodic pressure profiles that, within a period, were nearly reflection symmetric, as illustrated in Fig. 1.

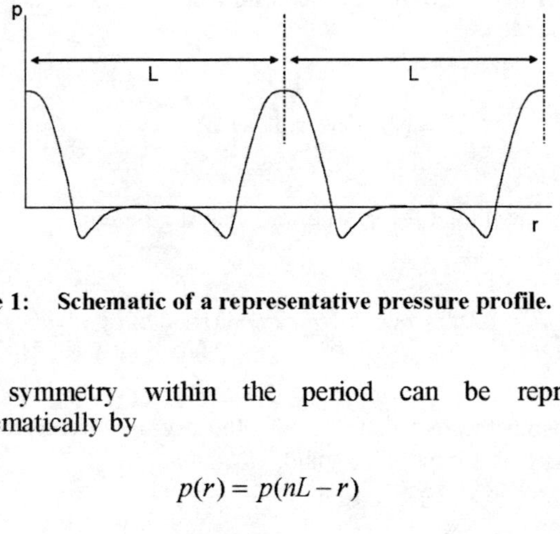

Figure 1: Schematic of a representative pressure profile.

This symmetry within the period can be represented mathematically by

$$p(r) = p(nL - r) \quad (10)$$

where n is the period number. So the load P_n contributed by the nth period can be represented as

$$P_n = \int_{(n-1)L}^{nL} p(r)2\pi r dr = \int_{(n-1)L}^{(n-\frac{1}{2})L} p(r)2\pi r dr + \int_{(n-\frac{1}{2})L}^{nL} p(r)2\pi r dr \quad (11)$$

But by making the change of variable, $r = nL - u$, in the last integral, we obtain

$$P_n = \int_{(n-1)L}^{(n-\frac{1}{2})L} p(r)2\pi r dr + \int_0^{\frac{L}{2}} p(nL-u)2\pi(nL-u)du \quad (12)$$

Now using the symmetry relation (10) in the second integral of (12) and expanding, we get

$$P_n = \int_{(n-1)L}^{(n-\frac{1}{2})L} p(r)2\pi r dr + \int_0^{\frac{L}{2}} 2\pi nLp(u)du - \int_0^{\frac{L}{2}} p(u)2\pi u du \quad (13)$$

Then making a change of variable, $r = v + (n\text{-}1)L$, the first integral in (13) becomes

$$\int_{(n-1)L}^{(n-\frac{1}{2})L} p(r)2\pi r dr = \int_0^{\frac{L}{2}} p(v+(n-1)L)2\pi v dv + \int_0^{\frac{L}{2}} 2\pi(n-1)Lp(v+(n-1)L)dv \quad (14)$$

Since the pressure is assumed periodic with period L, the expression for P_n can be represented in the form

$$P_n = \int_0^{\frac{L}{2}} p(v)2\pi v dv + \int_0^{\frac{L}{2}} 2\pi(n-1)Lp(v)dv + \int_0^{\frac{L}{2}} 2\pi nLp(u)du - \int_0^{\frac{L}{2}} p(u)2\pi u du \quad (15)$$

Finally, recognizing that the parameters u and v are dummy variables of integration that can be replaced by r, (15) readily simplifies to

$$P_n = (2n-1)2\pi L \int_0^{\frac{L}{2}} p(r)dr \quad (16)$$

Now the total load P for N periods is just given by

$$P = \sum_{n=1}^{N} P_n = \left(2\pi L \int_0^{\frac{L}{2}} p(r)dr \right) \sum_{n=1}^{N} (2n-1) = \left(2\pi L \int_0^{\frac{L}{2}} p(r)dr \right) N^2 \quad (17)$$

Then using the fact that $a = NL$, the average pressure, $P/\pi a^2$, can easily be computed to yield,

$$p_{avg} = \frac{2}{L} \int_0^{\frac{L}{2}} p(r)dr \quad (18)$$

Note that the above expression is independent of N, the number of periods. Therefore, under the assumed symmetry conditions, it is sufficient to compute the pressure profile for a half wavelength to obtain the average pressure for an arbitrary number of wavelengths. The same reasoning holds for computing the average magnitude of stress.

RESULTS AND DISCUSSION

Let us consider some simulation results. Figure 2 shows the damping characteristics of the simulation. The hashed curved represents the position of the origin of the surface of the half-space, while the other curve shows the average pressure in the interface during the simulation. This case corresponds to the short wavelength case (L=50b) with an amplitude A equal to $2b$ and an interference δ of $0.4b$, where the interference is defined as the amount by which the peaks of the rigid punch get closer to the original half-space plane than one equilibrium spacing b. As shown, the half-space origin deflects upward under attractive forces as the rigid punch approaches. Once the punch gets sufficiently close, the forces become repulsive and the half-space deflects downward. Eventually, both the position of the half-space origin and the average interfacial pressure reach nearly steady values.

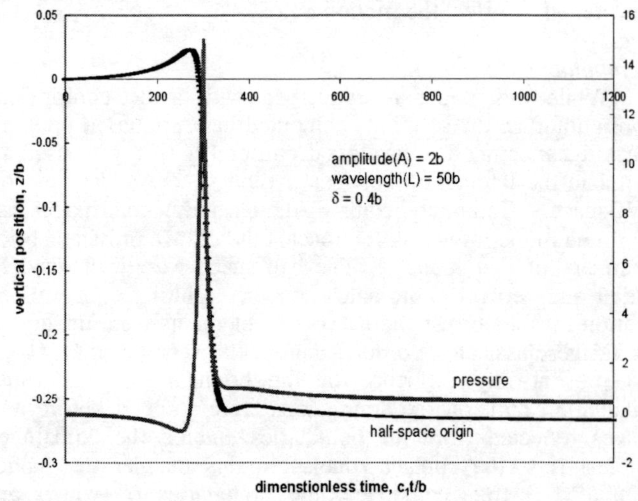

Figure 2: Typical simulation result showing damping characteristics.

Figure 3 depicts representative surface and pressure profiles. The case considered here is the same as in the previous figure. As observed, the half-space surface profile and pressure profiles are periodic away from the origin and the outer radius. The higher pressure seen at the origin is likely due to the fact that the final equilibrium position of the half-space origin has not yet been reached by the end of the simulation (see Fig. 2).

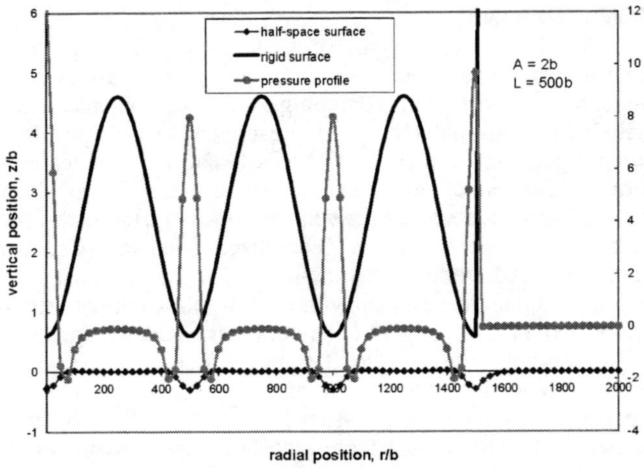

Figure 3: Representative half-space surface deformation and interfacial pressure profiles for the short-wavelength case.

At the outer radius, the sharp edge is responsible for the higher finite pressure there, as it follows from results for a punch with a sharp edge and a coating that possesses a local response to applied pressure (e.g., [12,13]). We note that the symbols depict the grid resolution used in the computations.

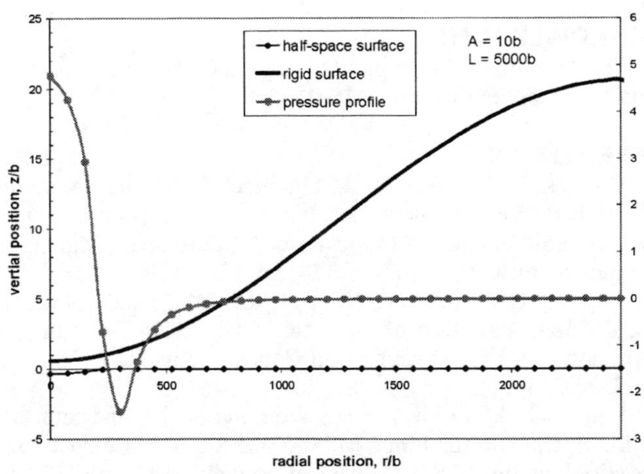

Figure 4: Representative half-space surface deformation and interfacial pressure profiles for the long-wavelength case.

Figure 4 depicts a typical result for the long wavelength case. For this particular simulation, the amplitude is 10b and the wavelength is 5000b. At this large wavelength, it can be seen that the deformation fields due the next and subsequent asperities would be independent of the single asperity shown here. This fact justifies the analysis leading to Eq. (18). Both Figures 3 and 4 show sizable contact regions occupied by significant negative pressure, a fact that illustrates the importance intermolecular attraction in the contact.

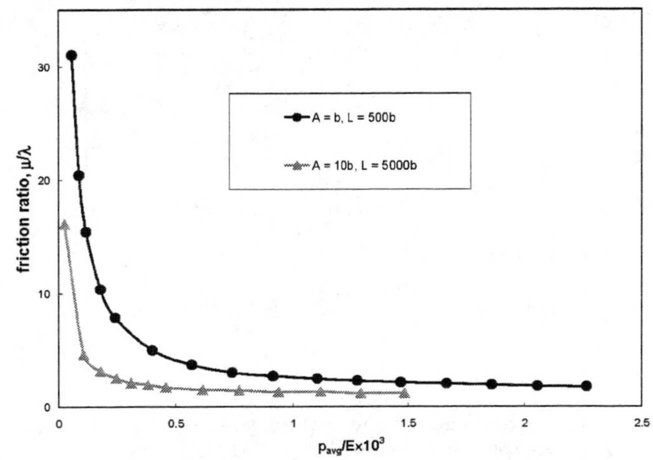

Figure 5: Behavior of the ratio of the actual (μ) and nominal (λ) friction coefficients for the short- and long-wavelength cases.

Figure 5 shows the dependence of the friction ratio μ/λ on the load in the interface, which is represented by the average pressure. It is observed that that as p_{avg} increases, the ratio of the actual (μ) and nominal (λ) coefficients of friction approaches unity for either combination of asperity amplitude A and wavelength L. The values of μ and λ for large p_{avg} represent the value of the actual friction coefficient for perfectly smooth surfaces in contact with no significant regions of negative pressure. For smaller values of p_{avg} the departure of the friction ratio μ/λ from unity reflects the influence of surface geometry. In other words, the actual (global) friction coefficient μ is equal to the nominal (local) one times a factor due to the surface geometry, for a given applied normal load. For the asperity amplitude $A=b$, the ratio μ/λ is higher than for $A=10b$. That is due to the fact that for $A=b$ the surfaces of the elastic half-space and punch are, on average, closer to each other than in case of $A=10b$. In both cases of $A=b$ and $A=10b$ both positive and significant negative pressures are developed. However, for $A=b$ the negative pressure occupies proportionally greater contact area because of the proximity of the contact surfaces. For $A=10b$ the half-space surface is unable to closely follow the surface of the punch due to its relative remoteness. The latter case leads to a much smaller influence of the negative pressures.

The above considerations also explain the monotonically decreasing behavior of the friction ratio μ/λ for increasing asperity amplitude A depicted in Figure 6. The greater is the amplitude, the smaller the extent of the regions with large negative pressure. Note, that according to (4), the pressure becomes most negative at a spacing (h) equal to $1.20b$, with rapidly decreasing magnitude with increasing spacing. Thus, the higher amplitudes put much of the rigid surface "out of reach" of the molecular attraction.

Figure 7 represents the graphs of p_{avg} and $|p|_{avg}$ versus the interference. The trends of p_{avg} and $|p|_{avg}$ are also caused by the presence of significant regions of negative pressure. For small interference δ and for the asperity amplitude $A=b$, the

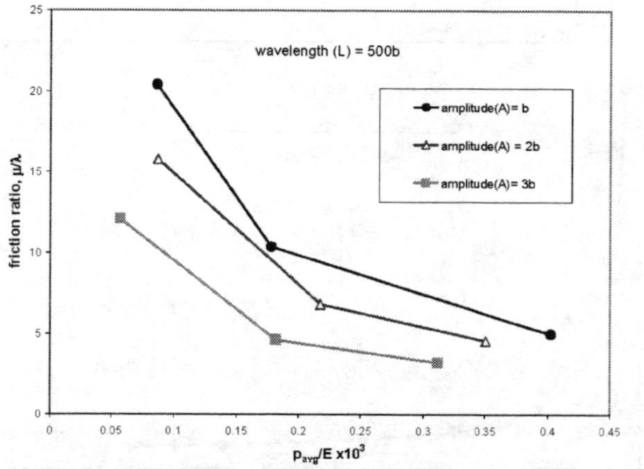

Figure 6: Dependence of the friction ratio μ/λ on the asperity amplitude A for the short wavelength case.

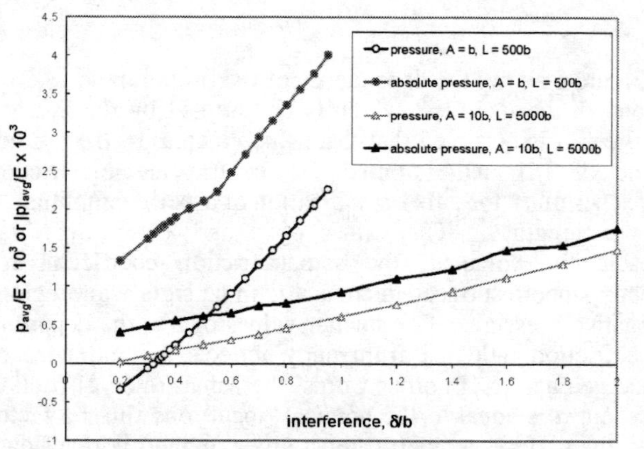

Figure 7: Behavior of the average interfacial pressure p_{avg} and the average of the absolute value of the contact pressure $|p|_{avg}$ as a function of interference δ for the short-and long-wavelength cases.

average pressure p_{avg} is smaller than for the case with A=10b, while the average of the absolute value of pressure $|p|_{avg}$ for A=b is higher than for A=10b. This is a clear indication of the presence of significant negative pressures. Moreover, p_{avg} and $|p|_{avg}$ asymptotically approach the same curve as the interference δ increases. For higher interference δ, the results are expected to be in close agreement with Hertzian contact analysis (e.g., [14]) which predicts that, for the same interference, the average interfacial pressure will decrease with increasing radius of curvature. For a sinusoidal profiles the asperity radius of curvature is readily shown to be $L^2/(4\pi^2 A)$ so that the long wavelength case has a radius of curvature 10 times that of the short wavelength case. In fact Hertzian analysis predicts that the average pressure for the short wavelength case should be greater by a factor of $10^{1/2}$ = 3.16, which is approximately borne out. While Hertzian analysis is not expected to hold precisely for these cases primarily because of the existence of negative pressures, the higher the interference, the more closely the analysis holds, provided the asperity interaction zones remain mutually independent.

CONCLUSIONS

The effects of the scale of surface roughness on elastic contact and friction are considered using an elastodynamic simulation model with damping. In the simulation, a cylindrical rigid indenter of a prescribed surface profile is brought into contact with an elastic half-space. The interaction between the punch and the half-space occurs through an assumed surface potential derived from the 6-12 Lennard-Jones intermolecular potential. A generalized Amontons-Coulomb friction law is introduced that assumes that the frictional stress is proportional to the magnitude of the normal stress, irrespective of its sign. The behavior of the interfacial pressure, surface deformation, and coefficient of friction coefficient as functions of various parameters, including surface interference, asperity amplitude and asperity wavelength has been investigated. The simulation results show that by assuming the generalized Amontons-Coulomb friction law one may be able to account for the enhanced frictional stress at the nano-scale versus the micro-scale that has been observed experimentally. For the cases studied here, the degree of frictional enhancement at the nano-scale was significant, but less than an order magnitude and therefore not as high as that observed experimentally [1-4]. Nevertheless, this simple study may provide insight as to the effect of roughness scale on contact and friction.

ACKNOWLEDGMENTS

This work was supported by the National Science Foundation under Grant # CMS-0200681.

REFERENCES

[1] Carpick, R. W., Agrait, N., Ogletree, D. F., and Salmeron, M., 1996a, "Measurement of Interfacial Shear (Friction) with an Ultrahigh Vacuum Atomic Force Microscope," Journal of Vacuum Science Technology, **B14**, pp. 1289-1295.

[2] Carpick, R. W., Agrait, N., Ogletree, D. F., and Salmeron, M., 1996b, "Variation of the Interfacial Shear Strength and Adhesion of a Nano-Meter Sized Contact," Langmuir, Vol. 12, pp. 3334-3340.

[3] Lantz, M. A., O'Shea, S. J., Well and M. E., and Johnson, K. L., 1997, "Atomic Force Microscope Study of Contact Area and Friction on $NbSe_2$," Physical Review, **B55**, pp. 10776-10785.

[4] Homola, A. M., Israelashvili, J. N., McGuiggan, P.M., and Hellgeth, J.W., 1990, "Fundamental Studies in Tribology," *Wear*, Vol. 136, pp. 65-84.

[5] Hurtado, J. A. and Kim, K.-S., 1999a, "Scale Effects in Friction of Single-Asperity Contacts. I. From Concurrent Slip to Single-Dislocation-Assisted Slip," Proceedings of the Royal Sosiety of London, **A455**, pp. 3363-3384.

[6] Hurtado, J. A. and Kim, K.-S., 1999b, "Scale Effects in Friction of Single-Asperity Contacts. II. Multiple Dislocation Cooperated Slip," Proceedings of the Royal Society of London, **A455**, pp. 3385-3400.

[7] Atkins, P. W., 1990, *Physical Chemistry*, W. H. Freeman & Co., New York.

[8] Muller, V. M., Derjaguin, B. V., and Toporov, Y. P., 1983, "On Two Methods of Calculation of the Force of Sticking of an

Elastic Sphere to A Rigid Plane," Colloids and Surfaces, **7**, pp. 251-259.
[9] Streator, J. L, 2003, "Dynamic Contact of a Rigid Sphere with an Elastic Half-Space: A Numerical Simulation, Journal of Tribology, **125**, pp. 25-32.
[10] Gear, C. W., 1971, *Numerical Initial Value Problems in Ordinary Differential Equations*, Prentice--Hall, New Jersey.
[11] Greenwood, D., 1997, "Adhesion of Elastic Spheres," Proceedings of the Royal Society, London, **A453**, pp. 1277-1297.
[12] Aleksandrov, V. M. and Kudish, I. I., 1979, "Asymptotic Analysis of Plane and Axisymmetric Contact Problems Taking into Account Surface Structure of Contacting Bodies," Mechanics of Solids, **14**, no.1, pp. 58-70.
[13] Kudish, I. I., 2000, "Contact Interactions for an Elastic Half-Space with Rough Surfaces," Dynamic Systems and Applications, **1**, no.1, pp. 229-246.
[14] Timoshenko, S., and Goodier, S. N., 1951, *Theory of Elasticity*, 3rd edition, McGraw-Hill, New York.

Proceedings of
2003 STLE/ASME Joint International Tribology Conference
Ponte Vedra Beach, Florida USA, October 26-29, 2003

2003-TRIB-0284

AN ASSESSMENT OF ARCHARD AND PERSSON'S MODELS FOR THE ELASTIC CONTACT OF ROUGH SURFACES

M. Ciavarella **G. Demelio**

CEMEC-PoliBA - Centre of Excellence in Computational Mechanics
V.le Japigia 182, Politecnico di Bari, 70125 Bari - Italy
http://cemec.poliba.it/

ABSTRACT

The Archard and the Persson models for elastic contact of rough surfaces are critically assessed for a Weierstrass series profile, finding that they both do not take into account of redistribution of load and interaction effects fully, unless scales are separated enough.

INTRODUCTION

The main deterministic and statistical elastic contact theories for rough surfaces are reviewed. In particular, in the first class are the classical Archard model (1957) and its modern interpretation using the Weierstrass series (Ciavarella et al 2000, Ciavarella and Demelio, 2001). Archard (1957) was the first to explain with the effect of multiscale roughness why elastic contacts seem generally to still obey Amonton's law for friction. Till then, the classical explanation by Bowden and Tabor (1950) required full yielding, whereas Hertz' theory would show a 2/3 power law with load. Archard suggested a fractal model *ante literam*, where for every sphere at a given resolution, there are a large number of smaller spheres of much smaller radius. Because of the discrete nature of the multiscale construction, it was assumed that spheres at scales below were not affecting the pressure distribution at the larger scale, and load could be redistributed simply by applying Hertz theory recursively for every local level of pressure, neglecting all interaction effects even within each single scale. This process leads to a linear relation real contact area to total load, but also, as noticed by Ciavarella & Demelio (2001), and probably unnoticed by Archard himself, to zero contact area and therefore infinite mean pressure.

In the second class, are the classical Greenwood-Williamson theory (1966), and the more recent Persson theory (2001). The former (GW) firstly showed that if "independent" identical asperities having an arbitrary local area-load behaviour have a Gaussian distribution of heights, the ratio between total normal and tangential forces is nearly constant. The GW model, or subsequent more sophisticated statistical theories using random process theories had success because its features are more naturally recognized from measurements of real surfaces or profiles, but it contains strong assumptions such as the absence of interaction effects. Viceversa, the original Archard model needs identification of "scales" which seem at first quite arbitrary, and probably the concept of a resolution-dependent contact area decreasing without limit if more and more scales were added, was not easy to accept. The GW-type models have often been questioned because profiles showing a multiscale nature lead to strong scale-dependence of "asperity" features (see for example Greenwood & Wu, 2000), and since the radius of the asperity for a fractal seems to decrease without limit, the GW model also predicts zero contact area in the limit, as does the Archard model. In fact, these conclusions, paradoxically, show that the original Bowden & Tabor model (1950) was not, after all, necessarily more remote from the truth than these elastic models predicting infinite pressure!

A recent statistical theory (Persson, 2001) proposes an attractive approximate solution for a Gaussian random profile with continuous spectrum, which gives a resolution dependent contact area as a function of the variance of the slopes of the profile up to that resolution. Persson's model, under a set of initial assumptions, seems to develop remarkably simple results, but *contains account of some but not all interaction effects,* since the contact area at a given resolution is assumed to depend only on features larger than the scale defined by the resolution level. A full numerical analysis shows that the Persson method leads to similar results of the old Archard model, particularly for a deterministic Weierstrass series, and both models may lead to very poor predictions of the real contact area. It is argued that the success of present engineering calculations where roughness is not taken into account suggests that the inclusion of a gradual number of increasingly smaller features, i.e. the original Archard's idea of describing a given surface as having shape, waviness and roughness, may be still appealing today.

PERSSON'S MODEL APPLIED TO A WEIERSTRASS PROFILE

Persson's results, is that the proportion of area A_0 in contact at stage Y is

$$A(Y)/A_0 = Erf\left(1/2\sqrt{G(Y)}\right) \cong \left(\pi G(Y)\right)^{-1/2} \quad (1)$$

where the second approximate expression is valid for small mean contact pressure σ_0 *or* large magnification Y (defined as L/?, where L is some characteristic length and ? is the wavelength of the scale of interest). Also, the frequencies q in the rough surface range from q_0 to $q_0 Y$, and

$$G(Y) = \tfrac{1}{2}\left\langle \sigma_{fc}^2(x,Y)\right\rangle / \sigma_0^2 \quad (2)$$

where $\left\langle \sigma_{fc}^2(x,Y)\right\rangle$ is the mean square pressure required to achieve full contact over surfaces with this roughness. In this form, the expressions are equally applicable to 2D and 3D cases. The $\left\langle \sigma_{fc}^2(x,Y)\right\rangle$ can be expressed in terms of the spectrum of the surface, but for discrete deterministic processes, it is easier to obtain it directly.

The Weierstrass profile is given by

$$z(x) = \sum_{n=0}^{\infty} g_n \cos(q_n x) = g_0 \sum_{n=0}^{\infty} \gamma^{(D-2)n} \cos(q_0 \gamma^n x) \quad (3)$$

A sinusoidal profile of amplitude g and circular frequency q is flattened by a sinusoidal pressure of amplitude $E^*qg/2$ (Johnson, 1985), hence

$$\sigma_{fc}(x) = (E^* q_0 g_0/2) \sum_{n=0}^{\infty} \gamma^{(D-2)n} \gamma^n \cos(q_0 \gamma^n x) \quad (4)$$

Squaring this, and integrating over a whole number of wavelengths, gives a mean square value to which only the $\cos^2(.)$ terms contribute, and they average half their squared amplitude, hence,

$$\left\langle \sigma_{fc}^2(x)\right\rangle = \frac{1}{2}\left(\frac{E^* q_0 g_0}{2}\right)^2 \sum_{n=0}^{\infty} \gamma^{2(D-1)n} \quad (5)$$

We are interested in terms up to $\gamma_n = Y$, and summing the series up to this limit gives,

$$\left\langle \sigma_{fc}^2(x,Y)\right\rangle = \frac{1}{2}\left(\frac{E^* q_0 g_0}{2}\right)^2 \frac{Y^{2(D-1)}\gamma^{2(D-1)} - 1}{\gamma^{2(D-1)} - 1} \quad (6)$$

and for large γ, and D>1, the last term dominates, giving approximately,

$$A(Y)/A_0 = \frac{2}{\sqrt{\pi}}\left(\frac{2\sigma_0}{E^* q_0 g_0}\right) Y^{-(D-1)} \quad (7)$$

The term in brackets is the same non-dimensionalised mean pressure as $\widetilde{p}$ in Ciavarella et al. (2000), and given $q_0 = 2\pi/\lambda_0$, can be expressed as

$$\widetilde{p} = \frac{2\sigma_0}{E^* q_0 g_0} = \frac{\sigma_0 \lambda_0}{\pi E^* g_0} \quad (8)$$

The Ciavarella et al (2000) Archard-like solution is

$$A(Y)/A_0 = K_\infty \widetilde{p} \gamma^{-(D-1)(n-1)} = K_\infty \widetilde{p} \gamma^{D-1} Y^{-(D-1)} \quad (9)$$

Hence, the solutions (7),(9) are of the safe form, but here we have a factor of $2/\sqrt{\pi}$ = 1.1284 instead of 0.39139, i.e. the Persson solution exceeds the Ciavarella et al one by a factor of 2.88, but because of the difference between n and n-1, the discrepancy will actually depend on γ, and remains a constant factor as Y (and n) tend to infinity.

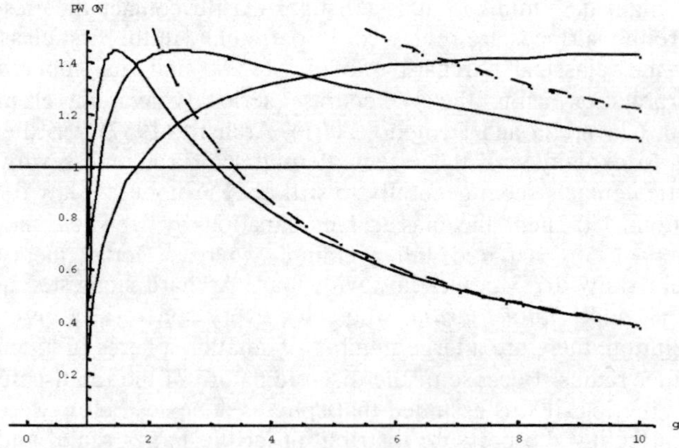

Fig. 1 The ratio of the Persson and the Archard solutions for the total contact area for a Weierstrass series contact at the limit resolutions. Solid line is the correct ratio (6,9), whereas dotted line is the limit for large γ, using (7,9).

Plotted in Fig.1 is the ratio of the Persson and the Archard predictions of the contact area for Y (and n) tend to infinity, as a function of the parameter γ. It could be noticed that the difference can be quite large and in both directions. The most part of the contribution to this difference comes from the treatment of the discrete sinusoidal terms in Persson's theory. In fact, it is immediate to show that Persson predicts for a single sinusoid the contact area

$$A(Y)/A_0 = Erf(\tilde{p}) \qquad (10)$$

and when this is divided by the correct Westergaard's solution

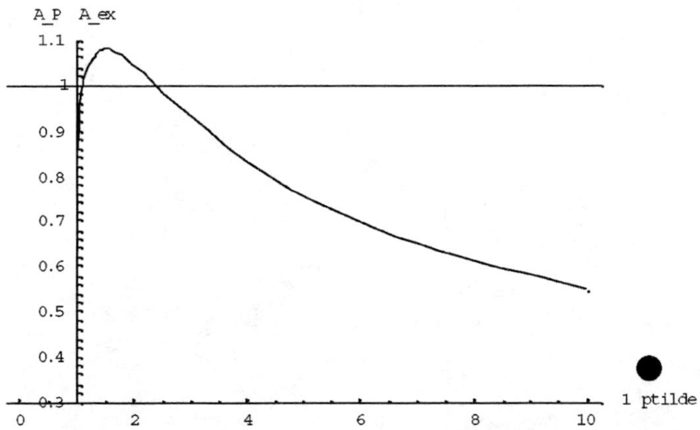

Fig. 2 The ratio of the contact areas for Westergaard (exact) and Persson's solutions for a single sinusoid.

(see Johnson, 1985), we obtain the ratio plotted in Fig.2, as a function of the inverse of the pressure. It is clear that Persson's solution is satisfactory, even for this limit case, for large pressures up to the full contact case (although the error is not negligible) but becomes very poor for light pressures and separated contact. It is for exactly this reason that the full Weierstrass case, which necessarily implies more and more separated contact when we increase the resolution, is likely to show a much larger contact area than what predicted by either Archard of Persson's theories, for light loads (which corresponds to the large x-values in Figg.2 and 3). Fig.3 shows an example case, and the result look very close to the case in Fig.2 of a single sinusoid, simply because for that value of ?=5, the Weierstrass series essentially reproduces the features of a single sinusoid in an amplified manner.

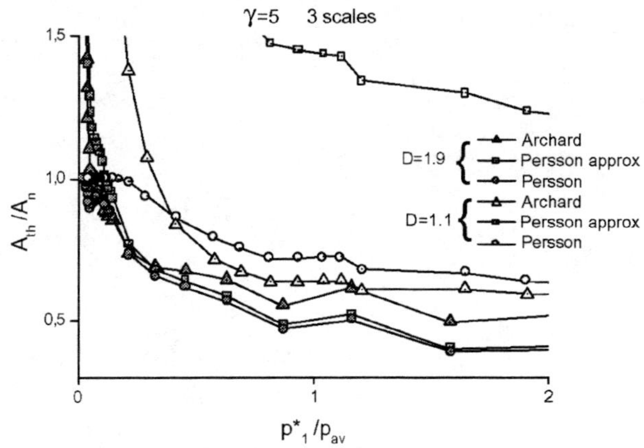

Fig. 3 The ratio of the contact areas according to the Archard of Persson's theories (in both the forms of eqt.1, the correct one involving erf and the approximate one) divided by the true contact area obtained with a full numerical method. For an example case of a Weierstrass series with 3 terms only, fractal dimension D=1.1 or 1.9, and ?=5.

CONCLUSIONS

When applied to a discrete fractal function such as Weierstrass, Persson's theory looks very similar to the old Archard theory, and they both do not predict interaction effects well enough. It was remarkable to find that Archard theory works very poorly even for sufficiently widely spaced scales (?=5), whereas it is in principle expected to work correctly for much larger ?=5. It may be argued that Persson's theory applies to a gaussian random surface rather than the contrived fractal case of the discrete sinusoidal waves in Weierstrass, but it remains important to notice how large the error may be in applying such an apparently sophisticated theory to such as simple representative case.

REFERENCES

Archard, J.F. (1957), "Elastic deformation and the laws of Friction," *Proc. R. Soc. Lond.* A243, pp.190—205.

Bowden, F. P., and Tabor, D., 1950, Friction and Lubrication of Solids, Clarendon Press, Oxford, UK.

M.Ciavarella, G.Demelio, J.R.Barber and Yong Hoon Jang, Linear elastic contact of the Weierstrass profile, Proc.Roy.Soc. (London), Vol. A 456 (2000), pp. 387-405.

M.Ciavarella, G.Demelio, 2001 Elastic Multiscale Contact of Rough Surfaces: Archard's Model Revisited and Comparisons With Modern Fractal Models, J. of Appl. Mech. Transactions of the ASME} Vol. 68, 496-498

Greenwood J.A. and Williamson J.B.P., (1966), "The contact of nominally flat surfaces," *Proc. Roy. Soc.* (London) A295, pp.300-319.

Greenwood JA, Wu JJ (2001) Surface roughness and contact: An apology MECCANICA 36 (6): 617-630

K.L.Johnson, Contact Mechanics, Cambridge Univ Press, Cambridge (UK), 1985

B.N.J. Persson, Elastoplastic Contact between Randomly Rough Surfaces, Phys. Rev. Lett. 87 , 1161 (2001); J. Chem. Phys. 115 , 3840 (2001).

Proceedings of
2003 STLE/ASME Joint International Tribology Conference
Ponte Vedra Beach, Florida USA, October 26-29, 2003

2003-TRIB-0285

ON RATCHETTING-BASED MODELS OF WEAR AND ROLLING CONTACT FATIGUE (RCF)

M. Ciavarella, **L.Afferrante**

CEMEC-PoliBA - Centre of Excellence in Computational Mechanics
V.le Japigia 182, Politecnico di Bari, 70125 Bari - Italy
http://cemec.poliba.it/

ABSTRACT

Recent efforts to develop simple unified models of both wear and RCF (Kapoor & Franklin, 2000, Franklin et al., 2001) are discussed, in view of previous theoretical and experimental results on ratchetting in rolling contact. At sufficiently high contact pressures, surfaces deform plastically with unidirectional cumulation of "ratchetting" strains (Johnson, 1985, Ch.9). However, the modelling of ratchetting strains as a function of plastic material properties has turned out more complicated than what originally suggested by the first attempts (Merwin & Johnson, 1963), as recently discussed by Ponter *et al.* (2003). Wear due to surface ratchetting occurs for sufficiently high friction, whereas RCF is mainly due to ratchetting subsurface. It appears that experimental data on ratchetting strains in the literature unfortunately do not show a clear and unique trend, and various proposed fitting equations differ significantly in quantitative and qualitative terms, particularly at large number of cycles. It is shown that ratchetting in rolling contact is a combination of "structural ratchetting" (that modelled with the perfect plasticity model) and "material ratchetting", and the latter is very sensitive to the hardening behaviour of the material. Also, the surface and subsurface flow regimes are very different: in pure rolling, a simplified model of the stress cycle condition is a fully reversed cycle of shear superposed to an out-of-phase pulsating compression in a extended region below the surface (neglecting other two components also of pulsating compression); increasing the friction coefficient, a mean shear stress is induced as well as a tensile component in the direct stress, and for friction $f>0.3$ the maximum moves at the surface, but the highly stressed zone becomes a thin surface layer which suffers uniquely of "material ratchetting". In the limit of very high friction, we have the critical condition on the surface which obviously gives a pulsating shear stress cycle in phase with a pulsating compression, but in addition we have a nearly fully reversed cycle of tension-compression (although the tensile peak is very localized also in the longitudinal direction). Such multiaxial stress fields and their largely different features introduced cause a response of the material which has not been studied enough, perhaps both in terms of ratchetting rates and in terms of the failure condition. In particular, the ductility for ratchetting surface flow as used in wear models seems apparently much higher than that for RCF ratchetting models. Also, RCF at large number of cycles in the C&S experiments (Clayton & Su, 1996, Su & Clayton, 1997) seems not well correlated with shakedown theory, and accordingly, simple ratchetting equations based on excess of shakedown such as that of Tyfoor et al (1996), do not seem well suited a Wohler SN life curve. However, these conclusions are only very qualitative as the materials in the two tests are different, and at present empirical separate models for wear and RCF based on hardness of materials and a posteriori data fitting seem the only quantitative way forward for engineering purposes.

INTRODUCTION

At sufficiently high contact pressures, surfaces deform plastically with unidirectional cumulation of "ratchetting" strains (Johnson, 1985, Ch.9). However, the modelling of ratchetting strains as a function of plastic material properties has turned out more complicated than what originally suggested by the first attempts (Merwin, 1962, Merwin & Johnson, 1963), as recently discussed by Ponter *et al.* (2002, 2003). Moreover, wear is such a complex phenomenon that a truly predictive model based on plasticity constants is not foreseeable: the standard approach is to recur to maps identifying various possible regimes and in each of them use empirical models (Lim&Ashby, 1987). Archard's wear law, for example, suggests wear to be proportional to normal load and inversely proportional to material hardness, and is used as a fitting equation to be calibrated with experimental data.

With this background, it is not surprising that present engineering design practise involving either wear or RCF generally neglects detailed plastic deformation analysis, and the key single relevant material constants is identified to be the material hardness. For example, in gears design standards, fatigue limit is defined according to Buckingham's equation and the entire fatigue curves specific to RCF (pitting strength) are given for various materials in terms of their surface hardness (and their surface treatments). It has been found recently that Buckingham's equation provides a reasonable accuracy also for RCF in railways materials, so that a similar philosophy could possibly be applied (Afferrante et al, 2003).

Recently, there have been interesting attempts to correlate the plastic ratchetting processes to both dry wear (Kapoor & Johnson, 1994, 1995, Kapoor et al., 1996, 2000, 2002) and RCF (Franklin et al., 2001). In particular, Kapoor and Franklin (2000) discretized the near-surface material into "tribological layers", each subjected to a cycle of high shear stress as predicted by the elastic solution essentially independent from the other layers. The amount of ratchetting is assumed to be a *linear function of the local "excess of shakedown"* which in turn depends on the local yield stress. This "linearity" seems to result experimentally, although just based on 5 data points from Tyfoor *et al.* (1996) with BS11 rail steel. Linearity does not imply constant ratchet rates, because the material in the layers strain hardens, and eventually fails if it cumulates a critical amount of strain. Results in Tyfoor *et al.* (1996) are a remarkable example of careful experiments and interpretation, but remain limited to a particular material (BS11 pearlitic steel) and to a special loading condition (one value of pressure, 1500 MPa). Clearly, linearity from a shakedown condition is necessarily a good approximation very near shakedown itself (which needs to be defined appropriately), but it is unclear if linearity would hold for a much wider set of conditions.

Attention needs to be paid to the value of friction coefficient. It is believed that RCF occurs only under lubricated conditions either because dry conditions induce high wear removing all initiated cracks, or because lubrication induces crack pressurization, as suggested already by Way (1935). However, this may not be always the case, and this is a strong reason to try to model the interaction of wear and RCF more carefully. In modelling wear, friction coefficient is typically high so that the plastic flow is mainly on surface, and similarly the hardening is faster on surface. In Kapoor and Franklin (2000), when a layer is worn out, the entire structure is "shifted" upwards, and the process continues. A step towards modelling both RCF fatigue and wear is (Franklin et al., 2001) which essentially makes use of the usual Tyfoor *et al.* (1996) ratchetting equation but instead of a layered structure, the model is built of a number of bricks so that the ambiguity between RCF failure and wear failure is apparently solved because failure is defined not just by a single brick reaching the critical strain, but by an ensemble of bricks failing in some way or another. This model is expected to work also when friction is low, so that RCF is prevalent over wear, the condition which was prevalent in the classical experiments in the literature on rolling contact ratchetting on similar materials. In particular, those of Merwin (1962), Clayton & Su (1996) and Su & Clayton (1997), which show unexpectedly significant discrepancies with each other. It would seem that the progress of ratchetting modelling is not particularly fast: in fact, it took more than 20 years to notice that Merwin's calculation (Merwin, 1962) based on perfect plasticity were largely wrong, first with FEM analysis (Bhargava *et al.*, 1985) and then with distributed dislocations (Hearle & Johnson, 1987). Then, the use of much more refined non-linear kinematic hardening models (Bower & Johnson, 1989) still found difficulties in predicting the long term ratchetting behaviour from the plasticity constants identified in biaxial tests, mainly because specimen generally fail after only few hundreds of cycles at the strain levels needed for ratchetting. Although very sophisticated non-linear hardening models exist (Jiang et al., 1999, 2002), the problem of fitting the appropriate large number of constants still remains. Hence, it is not clear yet if the approach remains competitive with respect to more empirical rules, perhaps based on a single material constant (hardness) like in gears design standards.

Ponter et al (2003) in analizing Merwin's results for subsurface flow (negligible friction), propose a distinction between the initial stages of "structural ratchetting" (SR) and the later stages of "material ratchetting" (MR). The former plastic flow (SR) is well described by the perfectly plastic prediction, where the yield limit is increased according to the level of deformation in an essentially isotropic component, and leads to a quick saturation. The following deformation is attributed to the steady-state material ratchetting response (MR). It is immediate to apply this calculation to the case of high friction, as saturation occurs within the first cycle (the perfectly plastic solution in this case leads to a thin layer of material being sheared in one direction of an unconstrained amount). Therefore, the response would be purely MR. It is concluded that while wear is strongly dependent on MR, friction and the various surface conditions (including roughness), the features of subsurface ratchetting should be less sensible to these factors, and has a component of SR.

ROLLING CONTACT TESTS

RCF experimental results do not show a perfect agreement between fatigue limit and shakedown, unless the yield stress value is "adjusted" to fit the fatigue limit. For example, tests in Clayton & Su (1996) and Su & Clayton (1997), don't show a distinct fatigue limit, and at 2 millions cycles are still around $p_0 / k_0 = 2.5$ (p_0 being the Hertzian peak pressure and k_0 the nominal yield limit in shear). This is in contrast to classical theory, which predict shakedown and ratchet limits at $p_0 / k_0 = 4$ for plane contact and pure rolling, and although Clayton & Su's experimental results are obtained for a 10% creep ratio, with a friction coefficient of $\mu = 0.2 \div 0.25$, this is unlikely to justify the discrepancy, particularly as the effect of hardening has not been taken into account in the figure. In fact, the value $p_0 / k_0 = 2.5$ is more likely to correspond to a much lower value when the actual value of yield stress in shear after the proper number of cycles is taken into account. According to plasticity theory, the yield point and the plastic collapse mechanism occur at the surface for sufficiently large friction coefficient ($\mu > 0.25$), and otherwise subsurface.

Many tests were run over the years by K.L Johnson's students in Cambridge, and in particular Merwin (1962), in the 1960's for a range of materials (copper aluminium and mild steel), and Bower in the 1980's (Bower & Johnson, 1989) with copper and rail steel BS11. These tests were studying subsurface flow under lubricated conditions. Merwin (1962) found a constant ratchetting strain in the layer of thickness of approximately $h = 0.75a$, where a is contact semi-width, suggesting the resulting forward displacement δ per cycle to be

$$\frac{G\delta}{p_0 a} = 4.2 * 10^{-5}\left(\frac{p_0}{k}\right)^{4.58} \quad (1)$$

where G is shear modulus. Notice that this equation works for the yield strength chosen by Merwin, which, as remarked by Ponter et al. (2003), was particularly high for copper with respect to the monotonic curve available (corresponding to nearly 25% of permanent deformation).

Turning to Tyfoor *et al.* (1996)'s tests, they were run at $p_0 =$ 1500 or 1200 MPa, and a rail disc creepage of -1 %, with an initial dry phase after which water lubricant was applied. Dry cycling caused an increase of a coefficient of friction to around 0.12, where it stabilised for a short period before it started to rise again. It rapidly reached a maximum value of around 0.48 and gradually dropped again. An effective ratchetting loads P_{eff} was defined as the "distance" from the shakedown limit, and the accumulated unidirectional plastic strain Rd found as a linear function of net ratchetting load P_r and number of dry cycles N_{eff} when the coefficient of friction is greater than 0.25, by fitting to 5 experimental data measured very near the surface, giving

$$R_d = 0.00237(P_r)N_{eff} = 0.00237(P_{eff} - Ps)N_{eff} \quad (2)$$

where P_s is the plastic shakedown limit corresponding to μ_{eff} (the average coefficient of friction during N_{eff}).

In recent papers (Franklin et al., 2001) and (Kapoor and Franklin, 2000) this was translated into

$$\Delta\gamma = 0.00237\left(\frac{\tau_{max}}{k_{eff}} - 1\right) \quad (3)$$

where k_{eff} is the effective hardened local yield strength, which was measured subsurface. The derivation of (3) from (2) is probably motivated by the consideration that under surface flow conditions, i.e. for high friction, $\tau_{max} = \mu\, p_0$ and so $\tau_{max} / k = \mu\, p_0 / k$ and since the shakedown limit for 2D contact and large friction is $p_0 / k = 1/\mu$, (see Johnson, 1985) then $\tau_{shake} / k = 1$. On the other hand, if subsurface maxima conditions are considered, then from shakedown limit $p_0 / k_0 = 4$, since $\tau_{max} = 0.25\, p_0$ (see Johnson, 1985), we get again $\tau_{shake} / k = 1$. Obviously there is a small approximation neglecting the transition between subsurface to surface conditions, but from shakedown maps this is clearly very small (see Johnson, 1985).

The remarkable feature of the equation (3) of identifying the condition of shakedown of what are really two very different multiaxial condition, motivated the use of (4) as a starting point for more sophisticated models (Franklin et al., 2001) and (Kapoor and Franklin, 2000) where it is used in a point-wise fashion.

Moving to Clayton and Su's data (Clayton & Su, 1996, Su & Clayton, 1997), they are somewhere intermediate between Merwin's and Tyfoor's, in that they are obtained under water lubrication and a creep ratio of 10%, and friction is found around $\mu = 0.2 \div 0.25$. They suggested and measured with many data points a power-law relation

$$\gamma = \gamma_0 N^b \quad (4)$$

where $b = 0.5 \div 0.6$ *and more precisely* was found a decreasing function of the pressure itself as

$$b = 1.1023 - 0.093\frac{p_0}{k_0} \quad (5)$$

where by removing the dependence on friction coefficient as obtained in (Afferrante et al., 2003), i.e. supposing that for lower friction coefficient the equation would continue to give a reasonable accuracy of the ratcheting strain (for high friction, things may change significantly)

$$\gamma_0 = 8.3275\times10^{-8}\, Exp\left(1.812\frac{p_0}{k_0}\right) \quad (6)$$

EMPIRICAL EQUATIONS FOR RATCHETTING

In this paragraph, we attempt to compare the various measurements of ratchetting in rolling contact and their corresponding proposed fitting equations. Despite this comparison may be made on different materials and loading conditions, we shall consider that largely different results or predictions indicate a sensitivity of the posed problem and a corresponding difficulty in deriving a simple but robust model, rather than a weakness of the comparison itself.

Fig. 1 shows a comparison of C&S (dashed) vs. Tyfoor's results (solid line) for a given $\mu = 0.235$ (C&S did not look at dependence on μ, so we use the value corresponding to their actual tests) for increasingly higher $p_0 / k = 5, 8, 10$ and it is clear that the two equations not only differ quantitatively, but also qualitatively, as the power laws are linear and almost square root with N_f.

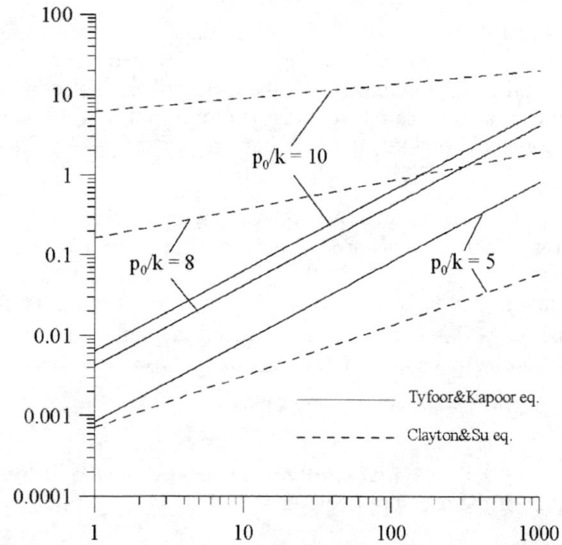

Fig. 1 Comparison of C&S (dashed) vs. Tyfoor's results (solid line) for $\mu = 0.235$ and for increasingly higher $p_0/k = 5, 8, 10$.

This difference is even clearer in Fig. 2 where the comparison of Clayton&Su (dashed) vs. Tyfoor's results (solid line) is for a given $N_f = 10, 100, 1000$ and still $\mu = 0.25$ but as a function of p_0/k and it seems clear the dependence is very different.

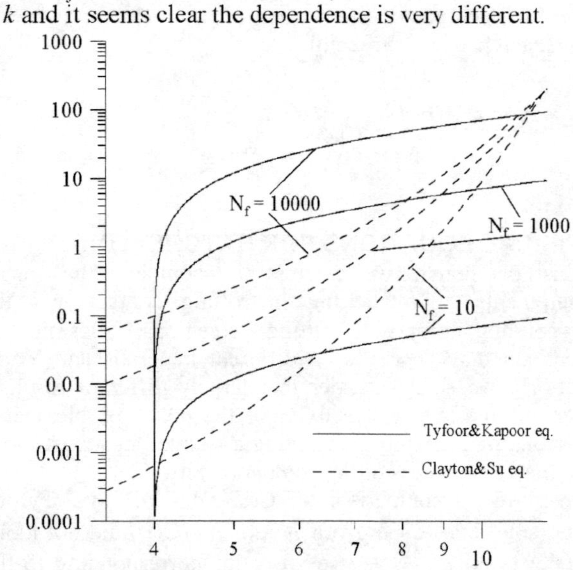

Fig. 2 Comparison of Clayton&Su (dashed) vs. Tyfoor's results (solid line) is for a given $N_f = 10, 100, 1000$ and $\mu = 0.235$ as a function of p_0/k.

A striking difference appears also in the "critical" ratchet strain. C&S seem to suggest a "critical" ratchet strain of the order of 1.5, which is consistent with other data on biaxial tests of steel (Kapoor, 1994) whereas Tyfoor et al. (1996) in the wear tests seem to suggest a "critical" ratchet strain nearly of order of magnitude larger (11.5). A possible explanation of this difference is that there is a strong difference in the gradient in the stress field, so that the stressed area is much larger for subsurface ratcheting condition than at surface. Also, it may be that since at the surface the material is "constrained" only from below, it can reach higher values of strain before failing. Finally, that the wear detachment "failure" is strongly different from the fatigue "crack initiation" failure. However, these conclusions are only very qualitative as the materials in the two tests are different, and this difference is not clear.

These remarkable differences over the prediction of ratcheting strains and over the critical conditions reflect into the life prediction capabilities for RCF. In fact, if we take the life SN curve results in C&S (Fig. 3) on STD steel tests (from Clayton & Su, 1996) and compare with the critical strain hypothesis using Tyfoor Beynon & Kapoor (TBK) equation (solid line) obtained with a $k_0 = 311$ MPa and the Clayton&Su (C&S) equation (dashed line), clearly different results are obtained. Both the C&S and the TBK curves are plotted for two values of critical strain $\gamma_c = 1.5 \div 11.5$ and a coefficient of friction $\mu = 0.235$. It can be noticed that the TBK curves tend to an apparent fatigue limit for the shakedown limit, at around $p_0/k = 3.1$ and shows a very poor estimate of the experimental data, not just in quantitative but also in qualitative terms. It is to be noticed that the effect of hardening in TBK has not been considered, and clearly it would tend to shift the curves upward. However, this would certainly make the predictions at long lives even poorer. For example, if the final hardened yield stress is twice the original, the shakedown limit would move in this plot to the value $p_0/k = 6.2$ where the actual data show a life of 20-30 thousand cycles, and certainly not infinite!

The C&S equation seems to fit the trend much better, and this is not *per se* particularly surprising, given the fatigue life and ratcheting measurements are made on the same tests. However, this is true even outside the range for which the data were obtained ($p_0/k = 5 \div 6.7$ and $N = 10^3 \div 10^5$) and is probably expected to work well in the very low number of cycles range by extrapolating. However, at the long lives, the decrease of the RF curve is too rapid.

In the pure "best fit" sense, a power law curve in the Wöhler diagram seems to work well in an extended region from 10000 to the long lives end of the data as plotted in Fig. 3. However, notice that at the low number of cycles, it is the best fit Wöhler curve to give wrong extrapolation, as static failure would be predicted at an unrealistically high value of $p_0/k = 42$! Clearly, a Wöhler curve approach would need to change slope at the very low number of cycles regimes. Probably, the ratcheting mechanism would require to be included in an extended Coffin-Manson equation.

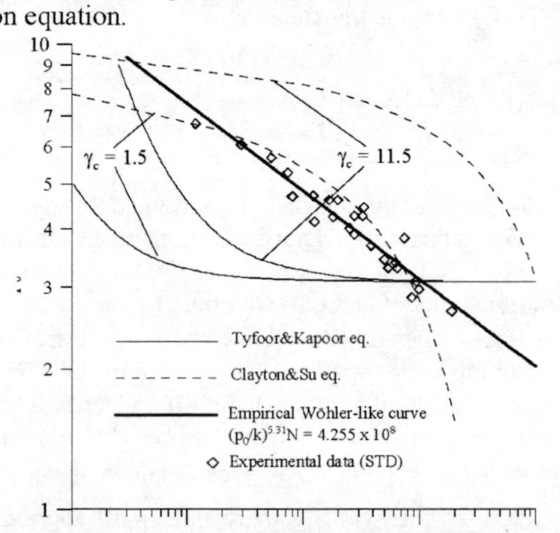

Fig.3 Comparison of Clayton&Su equation, Tyfoor's equation and power law curve in the Wöhler diagram.

DISCUSSION

The reasons for the discrepancy in the TBK equation prediction seem to be linked to the oversimplification in assuming linearity of the ratchet strain rate with excess of shakedown, which in turn is simply assumed from the maximum of the in-plane shear stress only. Finally, in terms of the stress field induced, the Hertz problem and the rolling contact are well-known, if anything from the Johnson's seminal book (1985). However, it may be useful, having to deal with the fatigue aspect to remark some details which are less obvious than just the maximum values of the stress components.

In Fig.4 we plot the maximum of the τ_{xz}, σ_{xx} and σ_{zz} stress components as well at their minimum value (computed at the depth for which we obtained the maximum τ_{xz}), as a function of the friction coefficient (assuming full sliding). It is clear that the surface and subsurface flow regimes are very different: for pure rolling the limit cycle of fully reversed in-plane shear stress is superposed to an out-of-phase pulsating compression in a extended region below the surface (neglecting other two components also of pulsating compression); increasing the friction coefficient (and for full sliding conditions), a mean shear stress is induced in the shear stress cycle as well as a tensile component in the direct stress, and for high friction the highly stressed zone becomes a thin surface layer which suffers uniquely of "material ratchetting". Indeed, for very high friction the minimum would be zero and the cycle becomes pulsating shear stress at the surface in phase with pulsating compression, but in addition we have a nearly fully reversed cycle of tension-compression. Notice however that the tensile peak very localized both in depth and in the longitudinal direction, and therefore the differences between the two limit cases of pure rolling and of high friction sliding are quantitative and qualitative in both the magnitude, the components of stress, their respective phases, and finally the extension and the gradients of the stresses. These reasons may be sufficient to make it very hard to find unified models for wear and RCF.

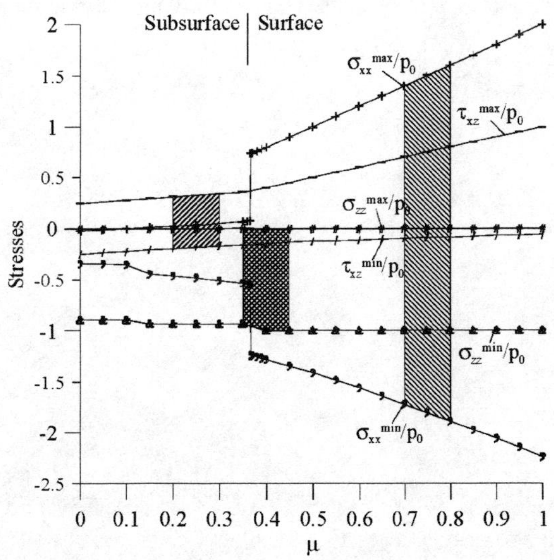

Fig. 4 Maximum of the τ_{xz}, σ_{xx} and σ_{zz} stress components as well at their minimum value (computed at the depth for which we obtained the maximum τ_{xz}), as a function of the friction coefficient (assuming full sliding).

Plotted in the Fig. 5 are the maximum and minimum value of the stress components τ_{xz} , σ_{xx} and σ_{zz}, computed at the depth for which we obtained the maximum of σ_{xx} and σ_{zz}. Notice that in this case the shear stress cycle is always pulsating, whilst the compression stress cycle becomes alternating for high coefficient of friction.

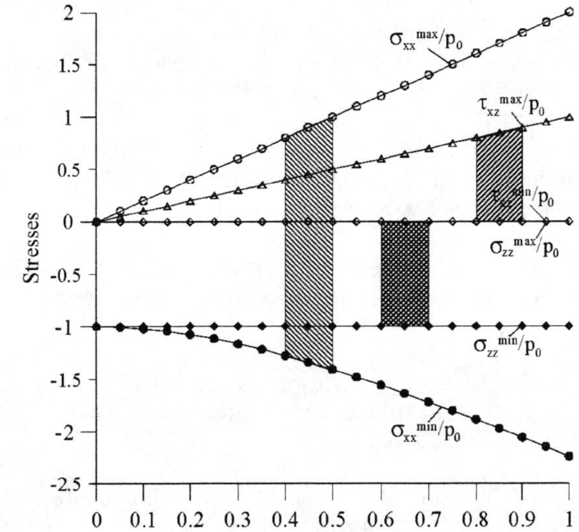

Fig. 5 Mmaximum and minimum value of the stress components τ_{xz} , σ_{xx} and σ_{zz}, computed at the depth for which we obtained the maximum of σ_{xx} and σ_{zz}.

Therefore, with such a peculiar range of possible multiaxial stress fields different in extension, magnitude and gradients, the response of the material can differ greatly perhaps both in terms of ratchetting rates and in terms of the failure condition.

It is certain, in the meantime, that the present simplified equations for ratchetting as linear extrapolation from excess of shakedown can only be applied in qualitative terms, and show surprising inconsistencies when applied to the RCF case. It seems still difficult to derive a robust and consistent quantitative model valid for both wear and RCF in a general condition, but the only way to make progress would be to derive more experimental data.

REFERENCES

L. Afferrante, M. Ciavarella and G. Demelio, (2003), A Re-examination of Rolling Contact Fatigue Experiments by Clayton & Su with Suggestions for Surface Durability Calculations, accepted on *Wear*

ANSI/AGMA 2001-B88, (1988) Fundamental rating factors and calculation methods for involute spur and helical gears.

BS ISO 6336-2 (1996) Calculation of load capacity of spur and helical gears (Part 2 Calculation of surface durability (pitting); Part 5: Strength and quality of materials).

Bower, AF, Johnson, KL, (1989) The influence of strain-hardening on cumulative plastic-deformation in rolling and sliding line contact, *J MECH PHYS SOLIDS* 37 (4): 471

Ciavarella, M., (2003) "Comments on "A. Ekberg, E. Kabo and H. Andersson, (2002) An engineering model for prediction of rolling contact fatigue of railway wheels", submitted to *Fatigue Fract Eng Mat Struct*.

Clayton, P. and Su, X. (1996) Surface initiated fatigue of pearlitic and bainitic steels under water lubricated rolling/sliding contact *Wear* **200**, 63-73.

Franklin, F.J., Widiyarta, I., Kapoor, A., (2001) Computer Simulation of Wear and Rolling Contact Fatigue, *Wear* **251**, 1-12, 949-955

Hamilton, G. M. (1963). Plastic flow of rollers loaded above their yield point. *Proc. Inst. Mech. Engrs.*, **177**, p. 667.

Hearle, A. D. and Johnson, K. L., (1987). Cumulative plastic flow in rolling and sliding line contact. *J. Mech. Phys. Solids*, **33**, p. 61.

Jiang YY, Sehitoglu H A model for rolling contact failure WEAR 224 (1): 38-49 JAN 1999

Jiang YY, Xu BQ, Sehitoglu H Three-dimensional elastic-plastic stress analysis of rolling contact J TRIBOL-T ASME 124 (4): 699-708 OCT 2002

Johnson, K. L. (1985). Contact Mechanics. Cambridge University Press, Cambridge.

Kapoor, A. (1994) A Re-evaluation of the Life to Rupture of Ductile Metals by Cyclic plastic Strain, *Fatigue Fract. Engng. Mater. Struct.*, **17**, 2, pp. 201-219.

Kapoor A, Franklin FJ Tribological layers and the wear of ductile materials *Wear* **245** (1-2): 204-215 Sp. Iss. SI OCT 2000

Kapoor, A., Franklin, F.J., Wong, S.K., Ishida, M., (2002) Surface Roughness and Plastic Flow in Rail-Wheel Contact *Wear* **253**, 1-2, 257-264

Kapoor A (1997) Wear by plastic ratchetting *Wear* **212** (1): 119-130

Kapoor A, Johnson KL, (1995) Plastic ratchetting as a mechanism of erosive wear, *Wear* **186** (1): 86-91

Kapoor A, Johnson, KL (1994) Plastic ratchetting as a metallic wear *P ROY SOC LOND A MAT* **445** (1924): 367-381

Kapoor A, Johnson KL, Williams JA (1996) A model for the mild ratchetting wear of metals *Wear* **200** (1-2): 38-44

Lim S.C., and Ashby, M.F., (1987), Wear-Mechanism maps, Overview n.55, *Acta Metall.*, **35** (1), 1-24.

Merwin, J. E. and Johnson, K.L., (1963). An analysis of plastic deformation in rolling contact. *Proc. Inst. Mech. Engrs.,* **177**, p. 676.

Ponter, A.R.S., Afferrante, L. and Ciavarella, M. (2003) A note on Merwin's analysis of forward flow in rolling contact, accepted on *Wear*

Su, X. & Clayton, P. (1997) Ratchetting strain experiments with a pearlitic steel under rolling/sliding contact *Wear* **205**, 137-143.

Tyfoor W.R., Beynon, J.H., Kapoor, A., (1996), Deterioration of Rolling Contact Fatigue Life of Pearlitic Rail Steel due to Dry-Wet Rolling-Sliding Line Contact *Wear* **197**, 1-2, 255-265

AUTHOR INDEX

Contact Mechanics — Friction: Modeling and Experiment

Book Number: I00675